KT-477-325

THE MODERN THEORY OF THE TOYOTA PRODUCTION SYSTEM

A SYSTEMS INQUIRY OF THE WORLD'S MOST EMULATED AND PROFITABLE MANAGEMENT SYSTEM

THE MODERN THEORY OF THE TOYOTA PRODUCTION SYSTEM

A SYSTEMS INQUIRY OF THE WORLD'S MOST EMULATED AND PROFITABLE MANAGEMENT SYSTEM

Phillip Marksberry, PhD, PE

CRC Press is an imprint of the
Taylor & Francis Group, an **informa** business

CRC Press
Taylor & Francis Group
6000 Broken Sound Parkway NW, Suite 300
Boca Raton, FL 33487-2742

Printed in the United States of America on acid-free paper
Version Date: 20120605

International Standard Book Number: 978-1-4665-5674-4 (Hardback)

Library of Congress Cataloging-in-Publication Data

Marksberry, Phillip.
The modern theory of the Toyota production system : a systems inquiry of the world's most emulated and profitable management system / Phillip Marksberry.
p. cm.
Includes bibliographical references and index.
ISBN 978-1-4665-5674-4
1. Industrial management. 2. Industrial productivity. 3. Toyota Shatai Kabushiki Kaisha. I. Title.

HD30.3.M36647 2013
658.001--dc23 2012020738

Visit the Taylor & Francis Web site at
http://www.taylorandfrancis.com

and the CRC Press Web site at
http://www.crcpress.com

Contents

Foreword

The Toyota production system (TPS) is often recognized for revolutionizing manufacturing, in much the same way that Henry Ford's assembly line revolutionized manufacturing a half century before. Many companies have adopted and adapted aspects of TPS. For some of these companies, the goal has been to lead their industry towards efficiency and gain a competitive edge, and for some of these companies, elements of TPS are adopted in hopes of keeping up with their industry competition. Companies have brought TPS to their organizations with mixed success, as there is often a misunderstanding of how and why TPS works. What was often lacking was an understanding of the Toyota Production System as not just the implementation of tools and techniques, but rather as a system that drives problem solving and eventually impacts the entire organization's operation and culture.

During the early 1990s, Toyota approached the University of Kentucky (UK) with an invitation to study TPS. Since that time, Toyota has provided us with an opportunity to work closely with the plant and with Toyota personnel toward the goal of better understanding the Toyota system, and to then communicating this understanding to organizations beyond Toyota. Dr. Marksberry joined the UK team with a background from industry and academia to help companies transform to Lean. Over several years he undertook a deep study of TPS, developing an understanding of the operations of TPS and how TPS works. The result is this book, in which Dr. Marksberry looks at TPS in the context of organizational behavior, social sciences, management science, and engineering sciences. He presents new insights into how the elements of TPS work together as a system across an organization. Both scholars and practitioners of TPS and Lean manufacturing will find this book to be thought provoking and revealing about how successful TPS operates.

Larry Holloway
Professor and Chair of Electrical and Computer Engineering
University of Kentucky

Preface

The science of industrial engineering is one of the most multidisciplinary fields of all time. Spanning many technical, social, and political domains, industrial engineering is often used as the primary method to improve how organizations perform. Over the years, many names have been used to describe the field of industrial engineering in an effort to revitalize how companies should improve. Organizational development (OD) is one field that has been influenced the most by industrial engineering. OD evaluates the inner workings of the business by looking at the largest picture possible. Management science is a field dedicated to helping managers and leaders to improve the way organizations perform by drawing on the various disciplines of organizational development and industrial engineering.

I have written this book because most managers and leaders are not introduced to the classical ideas of industrial engineering and organizational development. The field of contemporary industrial engineering is leading managers away from basic skills that make improvement simple. Contemporary industrial engineering calls for optimization, numerical analysis, and a complete overhaul of an organization's information network. Classical industrial engineering employs techniques that can be used by anyone in the company—skills that can be taught and applied by the executive, all the way down to the team member on the shop floor. Organizational development is a field that tries to make the complex workings of a company simple and easy to understand. OD uses abstraction and generalization to help managers and leaders make sense of the world around them. The best managers can only hold in their mind six or seven variables at one time, which makes the need to generalize an important skill in decision making. One of the most recent business strategies that emulate industrial engineering and OD is the Toyota Production System (TPS).

I have also written this book to explain that Toyota's success is not because of their production system but their management system. The technical aspects of TPS are not responsible for making the company profitable. While Toyota does provide some interesting alternatives technically, it is how the mandatory management system of the business is run that makes it different.

I have realized that learning a management system is not something that can be easily done without understanding systems theory. The theory of systems, such as general systems theory (GST) and living systems theory (LST), represent the best way to learn and study interdependent structures. TPS does not work when only one part of the system changes, TPS works when the system, and all its parts, is working together.

The need to understand how to design and implement systems is overdue. There have been many exciting and interesting perspectives from researchers, authors, consultants, and employees from Toyota, but there has not been an attempt to show how the pieces work together. From the outside, there have been many accounts of Toyota's behaviors. From the inside, there have been many accounts of Toyota's activities. These accounts are educational and helpful for learning new ways of thinking, but they do not help leaders at managing the variables to be successful. Systems theory, and the properties of systems, explain to leaders how to break down the parts of the TPS and how they work together.

I have written this book to help make the study of TPS more comprehensive and advanced. There has been a real gap bridging implementation from theory. Lean activists are frustrated by copying Toyota without understanding the factors that make it successful. There have been too many failures implementing Lean or TPS-like strategies by replicating the peripheral and nonessential aspects of Toyota. Most managers and leaders are not aware that Toyota's management system is similar to many of the present-day management science theories. While the combination of these theories and balance are not the same for each company, leaders and managers should be aware that they exist. This book tries to make some of these classical and elementary theories known so that Lean enthusiasts can study Toyota's techniques with a starting point and a destination point.

Acknowledgments

This book could not have been completed without the support of my family. The ongoing encouragement from my wife and three boys has made the task of writing a book possible. Their backing and sometimes "Are you done yet?" kept me going. Most of all, I am fortunate to have their patience and understanding. I dedicate this book to them for their patience and support.

This book could also not have been completed without the support from the University of Kentucky and Toyota Motor Manufacturing Kentucky (TMMK). I have been fortunate to have worked with the best Lean manufacturing minds of all time and the best students. I would like to thank Toyota for its ongoing support and the University of Kentucky's Lean Manufacturing Group for allowing me to be part of a great team.

Introduction

The Toyota Production System (TPS), more formally known as Lean manufacturing, has been one of the most widely researched and emulated business strategies in modern times. Companies today are in a frenzy to try to be like Toyota in hopes of increased profits and improved competitiveness. Toyota has been open about its production system and has identified that the secret to TPS is that it is a management system that impacts all areas of the business. Possibly a more suitable and less-confusing name would have been the Toyota Management System (TMS).

Like most management systems, how the pieces are combined is more important than any single aspect or piece of the system itself. This has been problematic for Lean practitioners because most of the emphasis in Lean implementation has been the identification of a part or piece of TPS. Most companies find it less appealing to hear that the solution has to be an entire one, instead of a partial one. Holism, rather than reductionism, is fundamental to understanding TPS.

The best way to study the pieces of TPS and how they work together is through systems theory, also referred to as general systems theory (GST) or living systems theory (LST). Systems theory is an interdisciplinary science used to explain, describe, and predict complex situations by comparing trade-offs within a system to achieve the best overall capability. Lean practitioners can use system theory to improve their decision making when applying Lean or Toyota-like management systems.

Systems theory is helpful because most managers tend to believe that their problems are independent, simple, and without influential forces and can be managed by understanding how the parts operate by themselves. Managers who have been entrenched by older systems, or systems that have not changed over the years, believe that new and emerging properties cannot occur. The thinking is that because the system is old problems can be solved with a narrow view just by analyzing single causes.

Systems theory provides the ultimate criteria for deciding the performance or output of a whole set of resources assembled for a given purpose (Gigch, 1978). Systems theory can solve problems that relate to stability and complexity, problems that are unpredictable, organic, or nonlinear. Systems theory is also suited for aiding decision making when it is difficult to know when to optimize or suboptimize, centralize (integrate), or decentralize (isolate). Systems theory is widely known for its use in organizational development (OD), specifically challenges that make it difficult to solve technical and social issues simultaneously. Last, systems theory can be used to solve problems that relate to innovation and continuous improvement.

Interestingly, most attempts to understand TPS fail because the glue that holds the pieces together is not viewed as important compared to all the individual pieces. This is mainly due to how TPS was described in the early years and the way consultants entered into the business. The term *Lean* was first used in a book, *The Machine That Changed the World*, by Womack, Jones, and Roos in 1990. This book described a phenomenon that seemed to be occurring in Japanese automotive companies where both quality and production improved simultaneously. Typically, when production rises, quality declines, and when quality rises, production declines. Six years later, Womack and Jones wrote another book, *Lean Thinking*, with the intention of extending the idea of Lean beyond the automotive industry by establishing five Lean principles. These new concepts provided insight into explaining the phenomenon; however, most companies treated these ideas as a philosophy similar to total quality management (TQM). Other companies used these ideas as initiatives to boost short-term profits. Many companies believed that if they could supply a short burst of profits, they would be considered on the Lean journey.

Lean initiatives have become widespread over the years mainly because they produce something that is tangible. Lean initiatives stimulate improvement in organizations without a lot of planning or consensus from everyone. They are often a knee-jerk response to solving issues that could not be solved by the entire organization, which makes them popular for assigning a group to work the problem alone. As long as the initiative shows profits, there is really no concern how the improvement is done. At the organizational level, most employees do not have to participate when there are initiatives. Leadership does not have to teach, and managers do not have to agree on the improvement approach. Initiatives are short term and voluntary and are replaced when leadership changes. *Simply, initiatives are an attempt to delegate what the existing management system cannot do day to day.*

Lean is not the same as TPS. Lean is a continuous improvement journey; TPS is a mandatory management system that dictates how to achieve continuous improvement by all members. Specifically, getting on the journey to continuous improvement is not nearly as important as first defining how improvement will be achieved by the organization. Not all improvement is good improvement. Improvement at the local level does not necessarily produce a positive result at the organizational level. This is why so many attempts to emulate TPS have failed. The good news is that the management system that Toyota employs can be installed relatively quickly in organizations if they understand systems theory. The problem is that most authors and consultants are not trained in systems theory as they are in performing continuous improvement.

The goal of this book is to share an in-depth study of how Toyota has designed its management system from an organizational system theory perspective. Current theoretical models are used to explain and justify many of Toyota's business practices and organizational preferences. This is significant in current research because many of Toyota's practices can already be explained using classical organizational theory. It is argued that the characteristics that make TPS successful at Toyota are present in most companies. The only difference is the degree of balance and emphasis placed on those elements. This is crucial because most businesses believe that Toyota has something that they do not. While the consultancy model has convinced most organizations they need something unique, it is shown in this book that they do not. Simply, the balance of the elements is more important than the presence of an individual element.

Systems theory is the modern theory of TPS. Systems theory is the best framework that can be used today to simplify how TPS works because any other model would be as complex as the real thing. The issue is the degree of simplification that is sensible in understanding TPS using GST or LST. As people, we all have different worldviews that affect what we see and how we describe our experiences. This means that multiple models are needed to describe a single view.

The use of existing theoretical models (which have already shown creditability in the OD community) is a practical starting point in understanding Toyota's management system. Existing theories have been used over many decades to illustrate, compare, and simplify the driving objectives of organizational systems.

The modern theory of TPS is essential for understanding the basic properties of Toyota's management system. According to living systems theory, an organism cannot survive unless it contains the basic biological properties. Organizations are examples of living systems because they can interact with their environment and regulate their own purposeful behavior.

It is believed that systems theory can best describe Toyota's management system more accurately than previous attempts because the properties of TPS have not yet been defined. To now, researchers could not answer the following questions about TPS:

What is the identity and purposeful behavior of TPS?
What regulates TPS?
Which aspects of TPS have to be differentiated or cannot be substituted?
Are there aspects of TPS that have to be as complicated as the company's surroundings?
What gives TPS order and disorder?
What are the basic transformation processes in TPS?
How much interdependence is needed for TPS to function?
Are there equally effective ways in applying TPS?

The modern theory of TPS, systems theory, is a framework that can explain how all systems must survive and grow. It is believed that a systems approach for studying Toyota's management system will provide new insight and understanding of TPS. It is hoped that researchers and practitioners will not stop with the ideas presented in this text but continue with the systems approach to better understand and design effective management systems.

REFERENCES

Gigch, J. (1978) *Applied General Systems Theory*, Harper & Row, New York.
Womack, J., Jones, D., and Roos, D., (1990) *The Machine That Changed the World*, Harper Perennial, New York.
Womack, J., and Jones, D., (1996) *Lean Thinking*, Free Press, New York.

About the Author

Dr. Phillip Marksberry is the vice president of engineering and quality for American LaFrance, a custom manufacturer of emergency vehicles and modern fire engines. He received his BS in mechanical engineering at the University of Kentucky in 1997 and became board certified as a professional engineer in 2001. In 2000, he received his master's degree in business management at Brescia University and in 2004 his PhD in mechanical engineering at the University of Kentucky. Dr. Marksberry has more than 20 years experience working in the automotive industry and holds several patents in metal cutting and product design. He has also held roles in production engineering, product design, industrial engineering, and operations. Prior to American LaFrance, he was a faculty member of the College of Engineering, with a joint appointment in the Department of Mechanical Engineering and Center for Manufacturing. Dr. Marksberry has published over 30 technical journal articles and has been a member of various technical societies. Dr. Phillip Marksberry can be contacted via e-mail at phillip.marksberry@yahoo.com.

1

Systems Theory and the Relation to TPS

1.1 WHY SYSTEMS THEORY FOR THE TOYOTA PRODUCTION SYSTEM?

For many years improvement activities, initiatives, and programs have been implemented throughout organizations in an effort to achieve companywide goals. Traditionally, managers focus on one part of the organization and, once a task is completed, shift their attention to another area.

Initially, this strategy works in isolated areas of the company, but on a larger scale, the organization suffers. The reason is that organizations are systems. Changing one aspect of the organization without knowing how it can impact the whole is not only risky but also dangerous.

Unfortunately, the Toyota Production System (TPS) is treated the same way. When standardized work is implemented in manufacturing without ergonomic standards, TPS can lead to worker intensification. Even the simplest adjustments in an organization can drastically harm other areas if not properly understood. TPS is a system that has organized components and interacting elements that form an integrated whole. Without understanding how the pieces fit together, TPS is more likely to weaken a system rather than strengthen it.

Another analogy that can be used to describe interdependency in organizations is gardening. Plants cannot be successfully transplanted unless the soil and environment are the same. Unsuccessful transplanting occurs when a gardener makes decisions based on a plant's outward appearance rather than decisions on what keeps a plant alive. A successful gardener has to know how the entire system works if a plant is to be replanted and live (Figure 1.1).

Unfortunately, many companies make the same mistake when implementing Lean. Organizations are mesmerized with Japanese words and

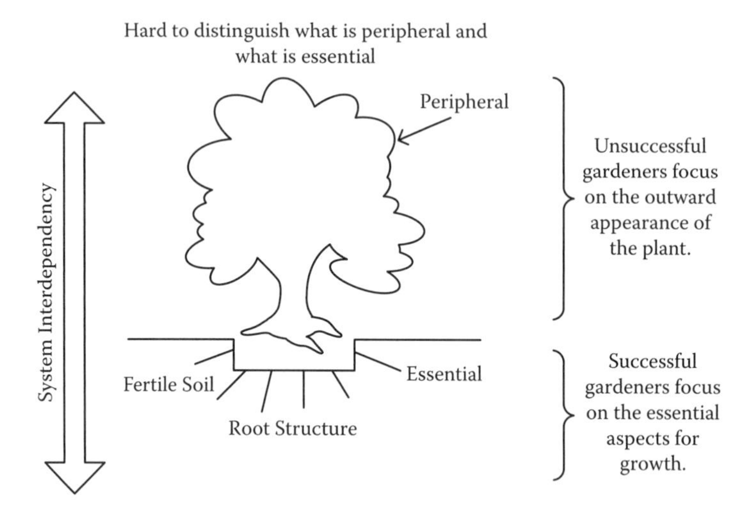

FIGURE 1.1
Systems thinking and gardening.

new tools, almost everything that can be seen on the surface. When organizations implement the peripheral aspects of Lean, they are left wondering why they did not get the same benefits as Toyota. Systems thinking prevents users from making decisions on the peripheral aspects of Lean and focusing on the true aspects that keep the system alive and healthy. Without systems thinking, it is hard to distinguish what is peripheral and what is essential (Figure 1.2).

The same situation can be explained using abstract algebra. When two objects show a relationship and are structurally identical, they are called *isomorphic*. The isomorphic property is used in mathematics to explain how some relationships might work in uncharted territories when they have been proven to work in others. This one-to-one mapping function suggests that what works at Toyota can work anywhere (Figure 1.3). Or, if one aspect of TPS can work, then all other components and aspects can work as well. Of course, this one-to-one mapping concept for Lean implementation only works if both organizations are completely identical. This is why copying Toyota does not work.

A less-complicated view of TPS is the implementation of a Lean technique or tool. Often, organizations choose not to treat TPS as an organizational strategy or business practice. In this case, industrial engineering tools such as 5S (i.e., five Japanese works starting with the letter "S" that describe workplace organization: *seiri*, *seiton*, *seiso*, *seiketsu*, and *shitsuke*), hoshin kanri, and standardized work are used to implement Lean.

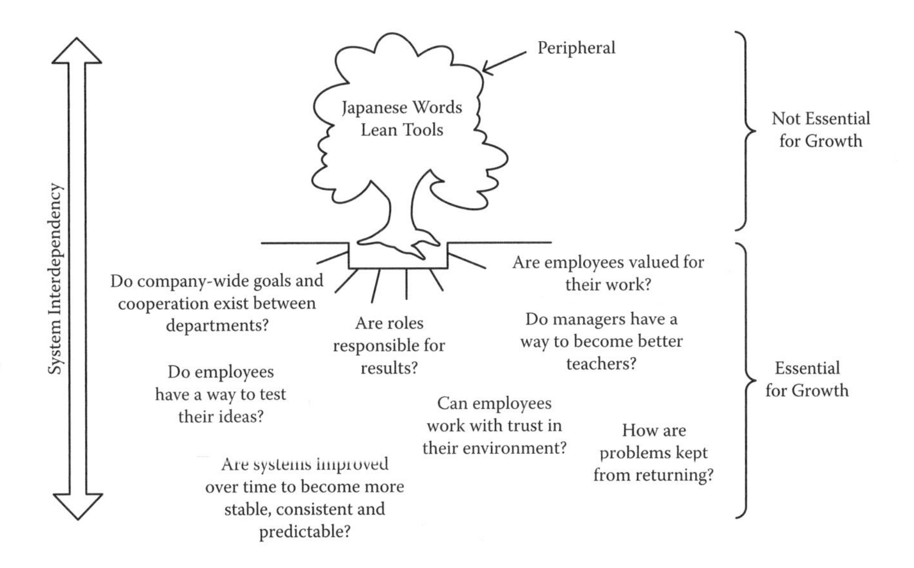

FIGURE 1.2
Systems thinking and TPS.

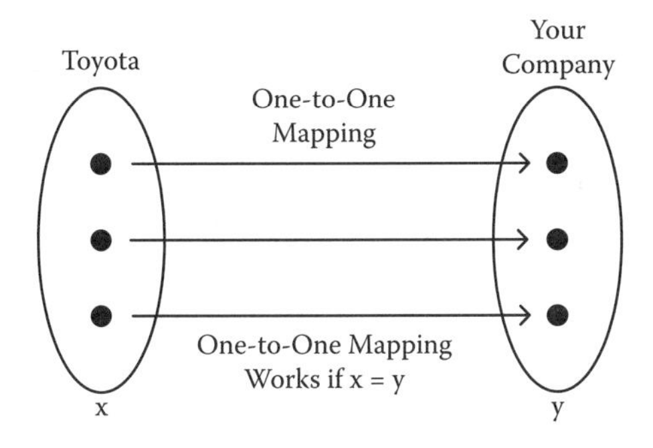

FIGURE 1.3
One-to-one mapping function for Lean implementation.

The view is that if the organization adopts the same tools as Toyota, then it is Lean. In algebraic terms, this is known as many-to-one mapping (Figure 1.4); the use of any tool is considered Lean in an organization trying to emulate Toyota. Unfortunately, adopting one or many of Toyota's tools does not provide an organization a competitive advantage.

A more appropriate way to understand how TPS works is by examining the features that make it universal. By treating TPS as a system, practitioners can share a broader perspective for how TPS functions and operates. Systems

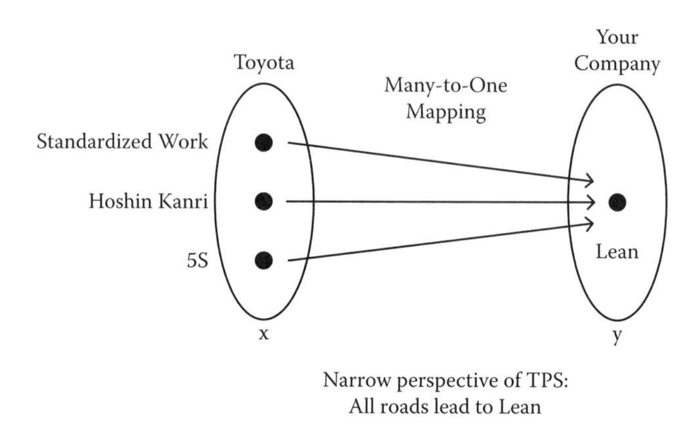

FIGURE 1.4
Many-to-one mapping function for Lean implementation.

thinking can help bring new insights and thinking for understanding TPS and provide a universal framework for organizational change. Systems thinking can also help in avoiding unintended consequences that can weaken the organization when one part of a system negatively impacts another. Most important, systems thinking can provide a big picture in solving complex problems by seeing the patterns instead of events for interpreting success.

The most universal features of TPS are its principles. In the 1950s, Taiichi Ohno, the creator of TPS, illustrated the company's principles using a house, known today as the Toyota House. The house is constructed using the company's four most basic principles: standardization, just in time, jidoka, and kaizen. The image of the house was used to teach suppliers about TPS and simply to illustrate that no part of the house can be built without the other, and a house can only be built from the ground up. Taiichi's house is illustrated in Figure 1.5.

The universal nature of TPS can be illustrated by mapping Toyota's principles using the one-to-many mapping function (Figure 1.6). In this context, the principles of TPS are universal to any organization, business function, or industry. Principle basis thinking represents the rules the organization chooses when making decisions. If every person shares the same rules, decisions will be made consistently in alignment with the company's goals and objectives. Importantly, Ohno was more concerned about teaching the system of TPS (the house) rather than the features (such as the tools) that are specific in many ways to Toyota.

Finally, systems thinking offers a way to identify the same characteristics shared by Toyota and other companies. It is argued that TPS does not

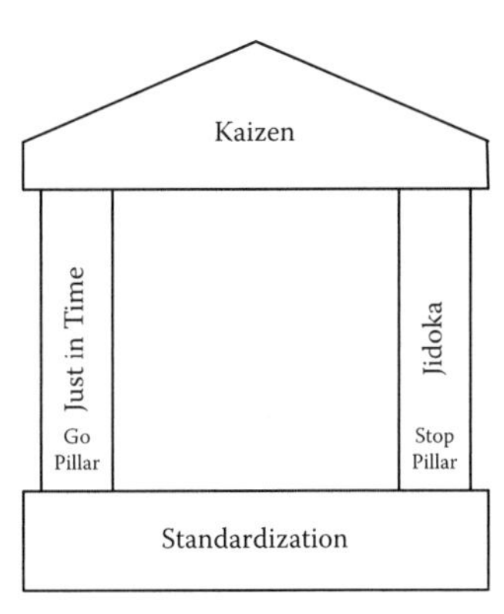

Kaizen is continuous improvement. Also refers to the series of activities whereby instances of waste are eliminated one by one at minimal cost, by workers pooling their wisdom and increasing efficiency in a timely manner.

Jidoka refers to the ability of production lines to be stopped in the event of problems such as equipment malfunctions, quality problems or work being late either by machines which have the ability to sense abnormalities or by workers who push a line-stop button. Jidoka helps defects from being passed on which allows quality to be built into the production process.

Just-In-Time refers to the manufacturing and conveyance of only what is needed, in the amount needed. This enhances efficiency and enables quick responses to change.

Standardization is the most basic management tool organizations can use to minimize the use of resources. Standardization is the process of establishing standards, processes and/or work methods that have proven to yield success.

FIGURE 1.5
Ohno's representation of TPS: Toyota House (House Thinking = Systems Thinking).

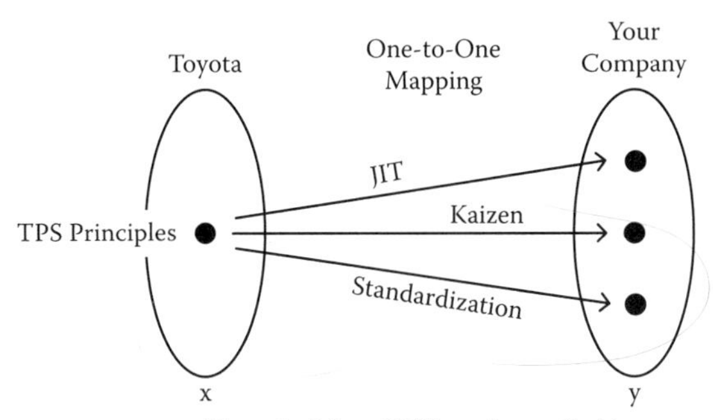

The principles of TPS can be applied to any organization

FIGURE 1.6
One-to-many mapping function for Lean implementation. JIT, just in time.

contain any elements or aspects that are not already present in organizations today; rather, it is the balance of how those elements interact that makes TPS unique. This can be illustrated using the many-to-many mapping function (Figure 1.7). The one aspect that is completely the same in all organizations today and at Toyota is people. Organizations have people doing work. Toyota has people doing work. The common denominator is people doing work. Surprisingly, many organizations emulate Toyota's technical systems rather than their people systems. Systems thinking can

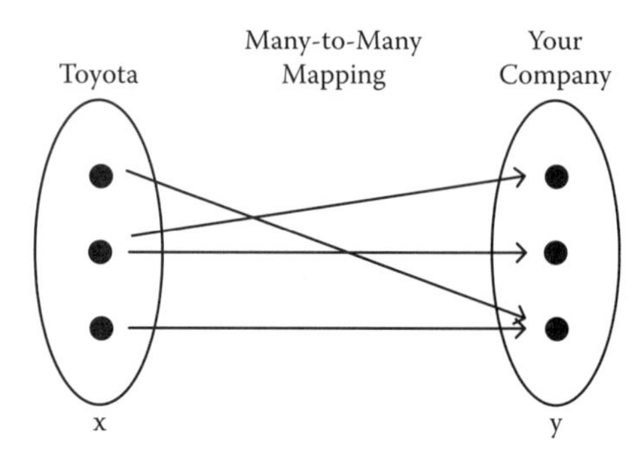

FIGURE 1.7
Many-to-many mapping function for Lean implementation.

help practitioners find the common denominators of TPS that are more easily applicable to their organization.

1.2 WHY SYSTEMS THINKING IS NOT POPULAR WHEN IMPLEMENTING TPS

Systems thinking typically provides a long-term perspective when studying complex problems. Most organizations want quick gains that do not require extensive effort over the long term. It is not uncommon for organizations to want to implement Lean overnight and receive the same success as Toyota when applying TPS-like systems. Using systems theory to characterize dramatic gains goes against the concept of systems thinking. The systems approach views systems as stable over time, which means that innovations are often characterized as system anomalies rather than a normal function of the system.

It is also difficult to think in terms of systems. It is human nature to think one dimensionally, and the task of managing several variables is cognitively challenging. Trying to study the interdependence of variables across the organization is too much for one person to understand completely themselves. One systems thinker in the organization is not going to drive organizational systems. System thinking has to be shared among leadership and management if they are going to be successful as a group understanding system interdependencies. Last, organizations are a special type of system, which makes them more difficult to analyze as systems. Organizations are open systems, which means they are dynamic, always moving and changing. Inputs that change in open systems do not mean that the outputs will change, and predications cannot be made about the outputs by studying the inputs. The study of organization development is complex. Most professionals will agree that you cannot learn anything about the organization until you try to change it.

1.3 SYSTEMS THEORY BACKGROUND

Various definitions and frameworks have been proposed in the study of general systems theory. Each body of theory emphasizes different viewpoints, disciplines, and outcomes. Ackoff (1981) said that a system composed of two or more elements should satisfy the following three conditions:

1. The behavior of each element has an effect on the behavior of the whole.
2. The behavior of the elements and their effects on the whole are interdependent.
3. When subgroups are formed, these groups (acting together) have an effect on the whole, but none of the groups has an independent effect on the whole.

Hegel (1929) formulated that systems work as:

a. The whole is more than the sum of the parts.
b. The whole defines the nature of the parts.
c. The parts cannot be understood by studying the whole.
d. The parts are dynamically interrelated or interdependent.

Boulding (1964) added to general system theory by postulating that systems are orderly, regular and non-random acting. Bowler (1981) described systems as having hierarchies in which simple systems are synthesized into more complex systems. Consequently, each system has a set of boundaries that indicates some degree of differentiation between what is included and what is excluded from the system.

Von Bertalanffy (1955) and Litterer (1969) established several properties of systems. Properties of systems give us insight in how they work and function. The following list shows some of the most generally accepted system properties:

1. Holism. The system as a whole determines how the parts behave within the system rather than the sum of the component parts alone.
2. Goal seeking. System interaction must result in some goal or final state.
3. Regulation. Objects within a system must be regulated in some fashion so that its goals can be realized and corrected.
4. Differentiation. In complex systems, specialized units perform specialized functions.
5. Hierarchy. Systems are generally complex wholes made up of smaller subsystems. This nesting of systems within other systems is what is implied by hierarchy.
6. Transformation. All systems, if they are to attain their goal, must transform inputs into outputs.
7. Entropy. This is the amount of disorder or randomness present in any system.
8. Negative entropy. This is amount of new energy that is generated from inside or outside the system that brings orderliness to the system
9. Requisite variety. To be effective and cope with environmental demands, a system must be at least as complex as its environment
10. Interrelationship and interdependence. Related and dependent elements constitute a system.
11. Equifinality and multifinality. Open systems have equally valid alternative ways of attaining the same objectives.

1.4 A REPRESENTATION OF TOYOTA'S TPS SYSTEM PROPERTIES

The properties of Toyota's management system can be developed using systems theory. Figure 1.8 shows how each of Toyota's TPS properties correlates to a property of systems theory, but it is not exhaustive, and these types of system properties cannot be exclusive to those listed. Instead, it is a representation of how TPS can be illustrated as a system. Importantly, TPS cannot survive if these properties do not exist or function properly within the organization.

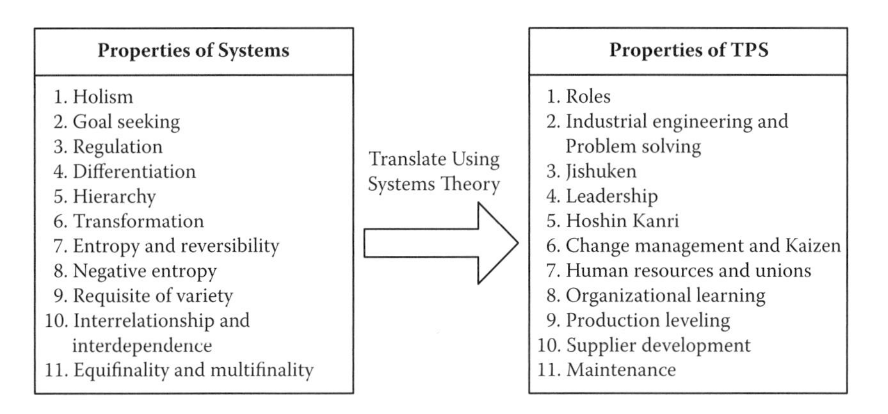

FIGURE 1.8
System properties theory and the translation to TPS.

1.4.1 Holism and Roles

Roles are defined as the rights and obligations (i.e., the relationship) each employee has to one another. Without roles, teamwork cannot be applied, which means TPS as a whole cannot exist. The holism property of TPS means that no single employee, work group, or department can achieve TPS without the entire organization working together. Understanding and applying teamwork is fundamental to TPS. Without teamwork, TPS cannot exist. Everyone has to be willing to use *one system* and work toward *one system* in achieving TPS.

1.4.2 Goal Seeking, Industrial Engineering, and Problem Solving

Goal seeking is described as a system interaction to achieve a desired function or objective. The identity of TPS is industrial engineering and systematic problem solving. Both of these concepts relate to how new ideas are tested when investigating workplace problems. While most outsiders would view that a methodology is not an end but a means to achieve a goal, TPS views that the means is more important than the end. In the context of a management system, there are many ways to apply continuous improvement. Toyota is successful because it shares a common way of thinking when applying continuous improvement.

1.4.3 Regulation and Jishuken

Regulation defines how a system self-regulates or controls its functions, characteristics, and objectives. At Toyota, the strength of TPS is defined by the number of people willing to use one system (i.e., holism) to improve the workplace. If employees share different opinions, ideas, or interpretations of TPS, fewer employees can work in alignment toward that system. A way to regulate TPS thinking is through *jishukens*. Jishukens are management-directed kaizen activities intended to improve the application of TPS thinking voluntarily. Jishukens regulate TPS thinking because they provide a way for managers to strengthen their understanding of TPS while working together with other managers. Toyota's belief is that if management shares the same type of understanding about TPS it can be applied consistently throughout the organization.

1.4.4 Differentiation and Leadership

The leadership property of TPS is the only system property that can actually start TPS. While many organizations are attempting to start Lean without leadership's involvement, TPS requires leadership participation and involvement. The unique and differentiated role of leadership is to teach and coach TPS. It is only property of TPS that cannot be switched, modified, or substituted with other organizational functions. This property implies that TPS is not activity that starts on the shop floor or is an activity that only involves a select group of employees. TPS is a management system that promotes companywide participation through coordination by leadership.

1.4.5 Hierarchy and Hoshin Kanri

Hoshin kanri or strategic planning is a system property that relates to how TPS is organized at various levels of hierarchies throughout the organization. The hierarchy system property is important for TPS implementation because it provides the mechanism for employee involvement at each level of the organization. Without a hierarchy, an organism cannot prioritize and coordinate its functions while adapting to its environment. Hoshin kanri provides the hierarchal structure of TPS by aligning the smaller components of the organization to the larger ones in meeting companywide targets and goals. Simply, hoshin kanri (strategic planning) provides a systematic way to coordinate organizational support.

1.4.6 Transformation, Change Management, and Kaizen

The transformational system property states that all systems should have a way to transform inputs into outputs. For TPS, that property can best be explained by how Toyota creates an open environment in which employees feel comfortable and confident to improve their work. Without an open climate, TPS cannot evolve, change, or develop. Toyota's change management practices as they relate to kaizen mean that employees can take a small risk without fear of losing their job. This property is extremely significant in TPS because no inputs or outputs can go through a transformation unless the supporting environment allows it. TPS requires an open climate in which employees feel confident to expose problems, make improvements in their work, and test ideas.

1.4.7 Entropy and Human Resources

Entropy and negative entropy describe the amount of disorderliness and orderliness in a system over time, respectively. One of the largest sources of entropy in TPS is the decaying nature of employee motivation. Initially, employees are motivated to contribute their ideas, but over time employee participation slows. The reason is that it is human nature to expend less effort to meet desired outcomes rather than to increase effort over time. Toyota manages entropy (i.e., the loss of employee engagement) through the human resource function.

Another source of TPS entropy is the loss of management effectiveness when the relationship between management and employees is unbalanced. Most Lean practitioners are not aware that many of Toyota's human resources practices and functions grew out of organized labor. Entropy can occur between management and employees when there is a loss of trust and structure in the relationship. While some argue that TPS is a autonomous environment between workers and management, this system property would suggest that TPS is more targeted and structured.

1.4.8 Negative Entropy and Organizational Learning

Negative entropy is when energy is added to increase the amount of order in the system. An illustration of negative entropy in TPS is organizational learning. Organizational learning helps TPS by encouraging employees to fulfill their full potential. Organizational learning is the main mechanism for improving and developing TPS. Organizational learning improves how TPS knowledge is created, transferred, and most importantly how it is shared. If this system property does not exist, TPS will not be able to improve, adapt, or prepare future generations.

Managing entropy is essential for sustaining TPS. Human resources, labor management relations, and organization learning show that the human side is essential for maintaining TPS. This is possibly why so many organizations struggle to sustain their Lean implementation efforts. Toyota utilizes the human resource function to minimize entropy, which is the loss of employee engagement. TPS is weakened by fewer employees willing to achievement it. The employee management relationship is another property of TPS that aims to minimize entropy due to losses associated with poor labor management relations. Both the company and the employees must develop a clear understanding of each other's perspectives to achieve mutual goals. If not, the organization will be split and unable to cultivate the necessary energy to prosper. Organizational learning is used as the primary driver to increase negative entropy. Employees are the greatest asset of the company because none of the systems and operations in the company can work without people.

1.4.9 Requisite of Variety and Production Leveling

A requisite of variety means for a system to cope with its environment it has to be at least as complicated as its environment. The TPS system property that best describes this property is production leveling. Production leveling is the ability of the organization to produce to a changing demand. If the organization cannot meet the various fluctuations of the market and is unable to adapt to its complexity, the organization will not survive. Production leveling is important to TPS because it drives the complexity needed throughout the organization to adapt to its environment.

1.4.10 Interrelationships and Supplier Development

Interrelationships and interdependence can best be described as the relationships between various systems. In the construction of automobiles, 70% of the cost is associated with suppliers. This means that TPS has to have a strong connection to the supply base to leverage supplier capability. The best way to leverage supplier capability is to replicate the organization's competitive strengths and capabilities. Importantly, there are some characteristics of TPS that are transferable and others that Toyota chooses not to replicate at suppliers. Understanding those aspects of TPS that are interdependent is significant for expansion.

1.4.11 Equifinality, Multifinality, and Maintenance

Equifinality and multifinality mean that the system has both alternate and equally effective ways of meeting its intended functions, respectively. The TPS property for equifinality and multifinality relates to how TPS is applied to both value-added and non-value-added work environments. In both of these environments, TPS should be equally effective. In value-added work environments, it is clear how industrial engineering can be used to find and eliminate waste. However, when the work environment does not contain value-added processes, only waste, is TPS applicable? An example that may challenge traditional TPS thinking is facilities maintenance. Facilities maintenance does not create or add value seen by the consumer, yet without maintenance equipment is likely to break down and fail. This property states that work areas like maintenance should have equal and alternate ways to satisfy TPS. While TPS may look different in these diverse environments, the management system should remain the same.

1.5 SUMMARY

Viewing TPS as a system is a helpful way to understand the big picture of Toyota's management system. Systems thinking explains how the pieces of a larger whole work and function. The system properties of TPS show how each piece of the system has a unique characteristic, function, and objective. Establishing a system without satisfying these properties poses trouble for effectiveness of the system and its ability to achieve its desired outcomes.

The best way to understand how Toyota achieves TPS is through a systematic study and evaluation of the eleven system properties. System properties can be described and evaluated through the use of models and frameworks. Organizational theory is helpful in understanding the system properties of TPS because it can explain how things work based on some general ideas or principles. While some practitioners may find it difficult to drop interesting details to explore theoretical concepts, abstraction gives managers an increased ability to process more information or to process information more quickly. Cognitive psychologists tell us that humans have the capacity to think about, roughly, seven pieces of information (plus or minus two) at one time (Miller, 1956). Concepts make it possible to communicate knowledge to others and allow us to relate large bodies of knowledge to each other.

Systems theory is better suited to raise important questions about how TPS can function in a unique organization rather than to provide ready-made answers to how Toyota solves problems in its unique business condition and climate. The premise of this book is to use systems theory to describe the properties of Toyota's management system so that Lean can be homegrown in any company. To accomplish this, one theory cannot be used, but several from varying perspectives are needed to help understand organizational patterns. It is hoped that the following chapters can provide some practical guidance while examining the unpredictable nature of human systems in organizations.

REFERENCES

Ackoff, R. (1981) *Creating the Corporate Future*, Wiley, New York.

Boulding, K. (1964) General systems as a point of view, in Mesarovic, J., *Views on General Systems Theory*, Wiley, New York.

Bowler, D. (1981) *General Systems Thinking*, North-Holland, New York.

Hegel, G. (1929) *The Science of Logic*, trans. Johnston, W., and Struthers, L., Allen and Unwin, London.

Litterer, J. (1969) *Organizations*, Wiley, New York.

Miller, G. (1956) The magical number seven, plus or minus two: Some limits on our capacity for processing information. *Psychological Review*, Vol. 63, 81–97.

von Bertalanffy, L. (1955) General systems theory. *Main Currents in Modern Thought*, 11, 71–75.

2

The Property of Holism in TPS

2.1 SYSTEM PROPERTY: HOLISM

One of the first tenets and fundamental properties of systems theory is the principle of holism. The principle of holism implies that the state and properties of a system cannot be determined by studying its parts alone. Any investigation of a complex system must involve a complete understanding of the pieces and how they work together. Holism means that studying or describing a system rarely takes place in isolation (Duhem, 1906). When pieces assemble to form a system, emergent properties develop to create something new that does not exist at the component level. Emergent properties of systems depend on the number of component interactions and how they are organized (Figure 2.1).

2.2 HOLISM AND THE TOYOTA PRODUCTION SYSTEM

The most obvious example of holism in the Toyota Production System (TPS) is the concept of teamwork. Teamwork occurs when two or more individuals come together to create a result that cannot be accomplished independently. Without teamwork, none of the organization's systems can function as one system or move in one direction. Teamwork allows the emergent properties of trust, cooperation, and consistency to occur in TPS. Trust allows each individual to work in a free state when exchanging ideas and effort with the company without fear of losing one's job. Cooperation is necessary in teams because without it, individuals can serve their own needs without meeting the needs of the whole. When there is trust and cooperation, teams can function more consistently because

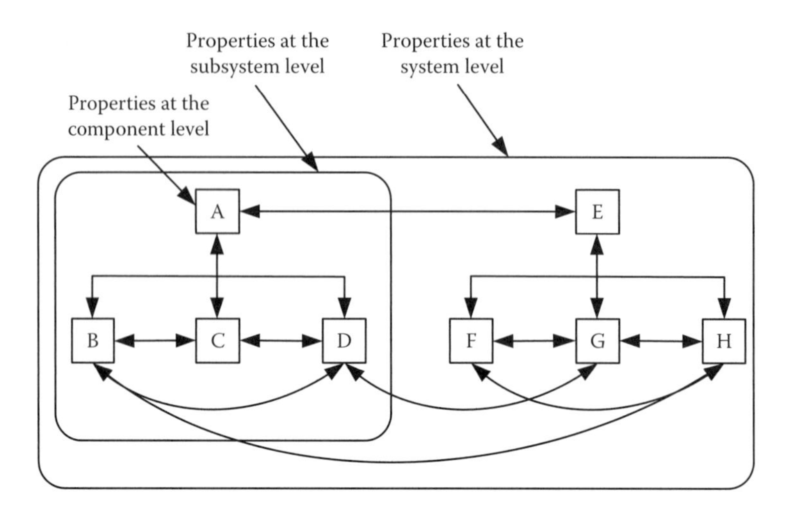

FIGURE 2.1
Illustration of a system: each level of a system has different and emerging properties.

political forces are less of a dominant factor in swaying a system's ability to meet its primary objective.

2.3 THE ROLE CONCEPT

A central theme in teamwork is the idea of roles. In any social system, individuals have obligations toward a system in their environment. When a person joins an organization, the person becomes a member, which entitles the person to all the rights and privileges of that membership. Consequently, being a member also means that the position must abide by certain rules. Everyone in the organization shares these expectations and beliefs, which make up a system of roles as shown in Figure 2.2.

All members are interconnected, and each member influences the other. Each role has a set of significant relationships that Katz and Kahn (1978) describes as the role sender and the role receiver. The role sender is the position or group that communicates the expectations of the role to the occupant (i.e., role receiver). The role receiver also has a set of obligations that makes the occupant a role sender.

In organizations, work roles are defined as the integration between the individual and the company. The interactions between the individual and the organization are a central concept in role theory that relates to worker

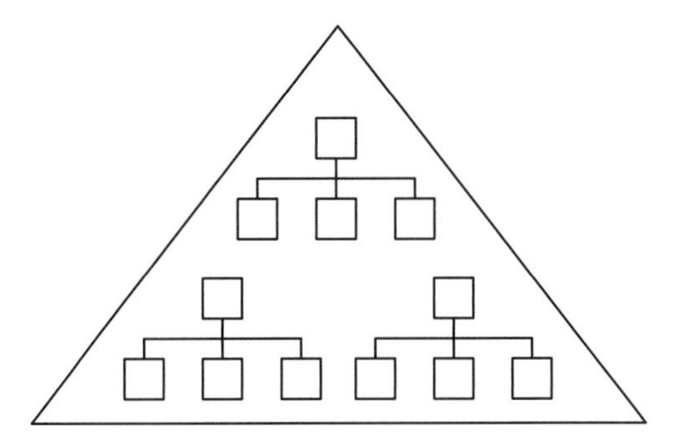

FIGURE 2.2
Organization as a system of roles.

satisfaction (Figure 2.3). If the organization does not communicate the role clearly or if the needs of the individual are not met while fulfilling the role, worker satisfaction and goal achievement of the organization suffer (Schein, 1978).

Biddle (1979) defined work roles as characteristic of one or more persons in a context. This definition is dependent on four terms: the behavior, the person, the context, and the characteristics (Figure 2.4). Roles have been described as behaviors and actions that people do and should total the behaviors of all human beings who perform that role in that context. Roles can be applied to many persons but are not limited or thought of being applied to a single group or individual. Roles are limited to the context for which they are applied and do not represent the total set of all behaviors of

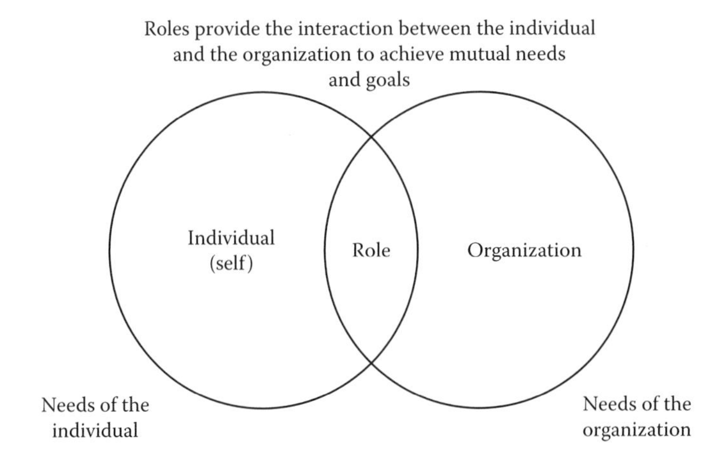

FIGURE 2.3
Interaction of the individual and the organization through roles.

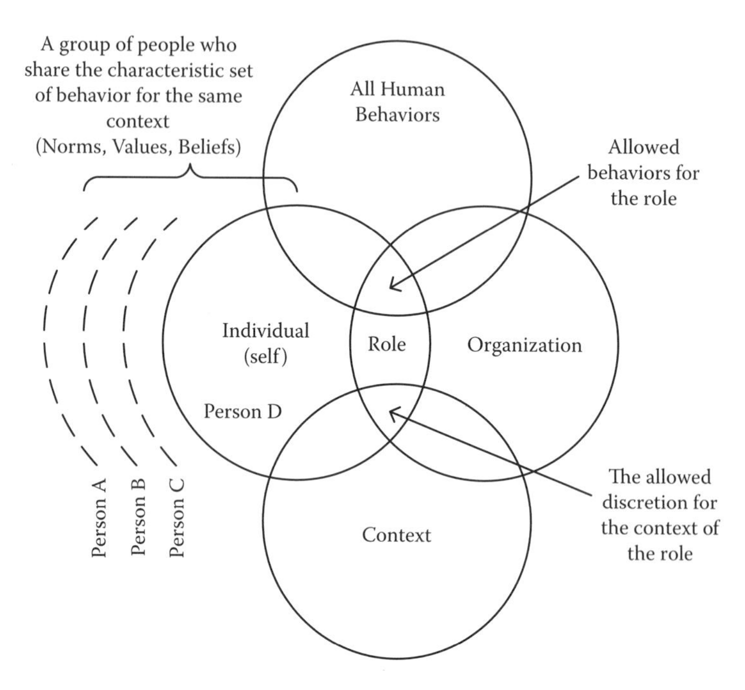

FIGURE 2.4
Work roles in organizations: Biddle's (1979) four properties.

human beings. Thus, roles contain contextual boundaries and are largely dependent on the situation and the encounter. Roles will change radically from context to context; however, role context refers to a class of places or occasions. Last, roles consist of behaviors that are characteristic of a set of persons and context (Biddle, 1979).

2.4 THE DIFFERENCE BETWEEN ROLES AND JOB DESCRIPTIONS

Roles and job descriptions are often used to describe work in organizations. Job descriptions are used by organizations to list the required tasks, functions, or responsibilities of a position. Job descriptions describe the static aspects of the job and assume that tasks are independent. The quality of work and performance of the position do not involve a large number of complex factors or any discretionary aspects to complete the job. In this manner, job descriptions state what should be completed and how it should be completed assuming there are no problems. No task can be completed without an employee exercising discretion. If there are no limits to the discretion, users will not be concerned how they carry out the task or if they infringe on the territory of another job (Brown and Jacques, 1965). This is why job descriptions are not used by management: They seldom represent an actual guide for users to make decisions (Dayal, 1969).

Roles, on the other hand, are dynamic. They define an allowed set of behaviors for a specified context for the job and the discretion for carrying out the job. Generally, roles are more detailed because they have to describe a range of conditions (i.e., contextual situations) and how to respond to them. Roles also describe the types of behavior that is acceptable for the position in a set of circumstances. Not all behaviors are acceptable, even if the context changes for the role.

2.5 ROLE PRESCRIPTIONS

The theoretical starting point of role development begins with a set of role expectations held by role senders about a focal person or role receiver (Katz and Kahn, 1978). These role expectations are often referred to as the role prescription and provide guides and standards on how the role should exist within the organization. Role prescriptions are important because they provide a means to manage the expectations of the company and its employees. Role prescriptions also describe obligations (the behaviors expected of the person in the role) and rights (those behaviors that others are expected to direct toward him or her). Thus, role prescriptions provide the norms, rules, and contextual cues of how to act within the organization. Last, role prescriptions are the most potent factor in the control of human behavior. In fact, much of social behavior that occurs is affected by role prescriptions (Katz and Khan, 1978; Biddle, 1979).

2.6 ROLE THEORY AND ANALYSIS

Role theory and analysis involve an important area of study in understanding the nature of human behavior in organizations. For over 30 years, role theory has been used as a technique to develop and manage the interdependencies of work relationships that are required by employees to perform required tasks effectively and efficiently (Katz and Kahn, 1978; Levine et al., 1983). The assignment of work roles is therefore a vital function in the achievement of organizational goals and the understanding of an organization's social network of planned, expected, and required task activities associated with the division of labor (Wickham and Parker, 2007).

2.6.1 Role Components and Features

Roles can be further understood by analyzing and dissecting its various components and interdependencies. For example, *role breadth* is the range of characteristics that appear within the role. All roles will contain a range of context and behaviors. *Role difficulty* is the degree of skill and energy required to perform the role. *Role coherence* is the degree to which the components of a role fit together. *Role conflict* is when expectations are contradictory, which often leads to job dissatisfaction. Role conflict can occur when a person's values do not match the organization's or the organization is asking those in the role to perform two different tasks that are in opposition. *Role ambiguity* occurs when the role has been defined unclearly, and there are no clues for a person to understand how to function within the role. *Role differentiation* (i.e., role proximity) is when two or more roles are said to be differentiated if they have few behavior elements in common (Gordon, 1972). *Role integration* refers to how well roles fit together in a system. *Role interdependence* refers to the degree to which roles are mutually supportive or hindering. Figure 2.5 illustrates these various role features and components (Katz and Kahn, 1978; Kahn et al., 1964; Biddle, 1979; Pareek, 1976).

2.7 THE EMERGENT PROPERTIES OF TPS: TEAMWORK

The heart of TPS is teamwork. To ensure the success of the company, each team member has the responsibility to work together, communicate honestly, and share ideas with the team (*Team Member Handbook*, 2006). Toyota believes that through coordination and collaboration the contribution of the team is greater than the sum of its members (*Toyota Way*, 2001). One of the successful characteristics of teams at Toyota is how they are organized and structured. Toyota utilizes a dense management structure

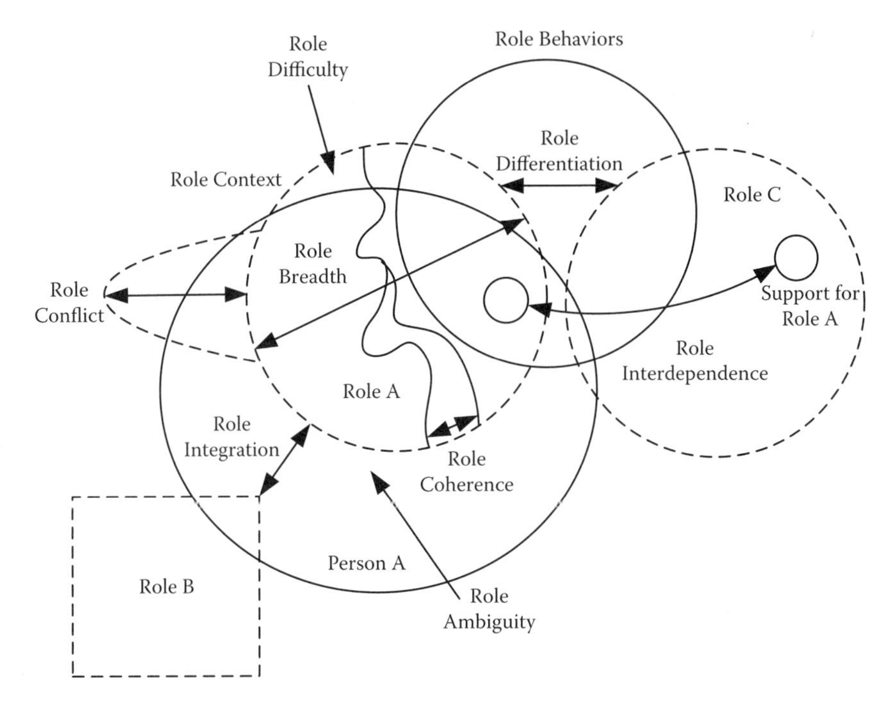

FIGURE 2.5
Role features and components.

throughout the hierarchy with a focus on small work groups. Figure 2.6 illustrates a typical organizational structure at Toyota.

Interestingly, many researchers and authors incorrectly describe Toyota's organizational structure as flat. In Figure 2.6, Toyota has seven levels of management and only two levels of direct labor. Toyota's teams are often described in articles and books as autonomous, self-managing, or self-directing groups that have flexibility to make decisions for their work. Observations like these give the impression that management does not manage, and teams can establish their own targets and goals. In TPS, management manages. Management sets targets and goals, while the means to achieve them flow upward from team members. Management approves trials to allow testing of ideas (which are encouraged), but approval for change lies with management.

A better representation to describe Toyota's team concept is how its teams are organized. The basic composition of a team is illustrated next. Each team is led by a leader a level higher than the members, and the team consists of working members. At Toyota, the leader of the team has the role to

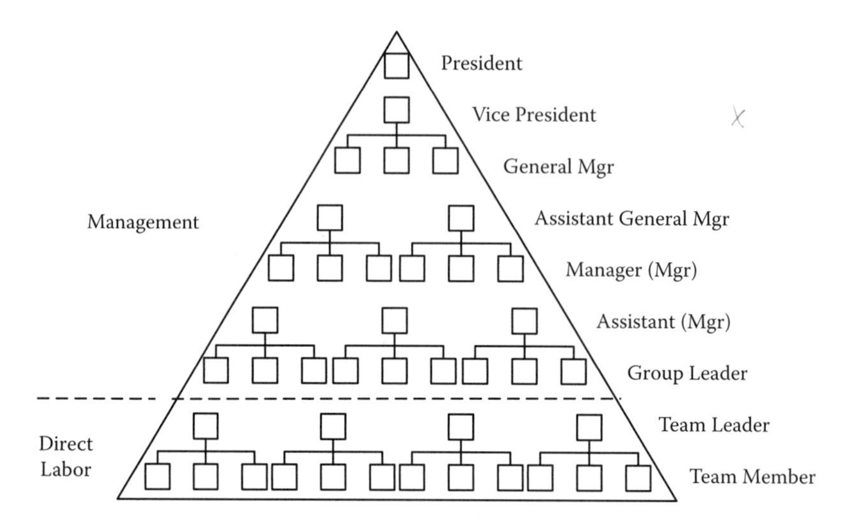

FIGURE 2.6
Toyota's organizational structure: typical representation for vehicle operations.

- Allocate team resources
- Set targets and goals
- Identify member contributions
- Elicit ideas and opinion from the team
- Provide approval

The working member of the team has the role to

- Ensure everyone is a working member (including the leader)
- Make contributions through participation
- Support the team's decisions as if their own
- Rotate through support roles vital to the performance of the team, such as planning, tracking, distributing, documenting, and the like

Role clarity improves team performance and task ambiguity. Without clarity, members are more likely to waste time and energy negotiating roles or protecting turf. When roles are clarified, members are more likely to collaborate to achieve team goals. When roles are not defined clearly, even the simplest organizational functions can be a challenge.

Consider the case of implementing standardized work. Standardized work is a tool that is used to perform work analysis. The output of the tool provides work allocation possibilities, layout information, and a record of the job. Many organizations begin with standardized work as a first step in implementing Lean. The belief is that documenting a job should occur before improving a job. At Toyota, it is clear who writes and approves standardized work. Organizations that do not have standardized work often select an engineer or Lean coordinator to write standardized work.

This approach to Lean is common; however, it is what the organization does informally that causes big problems for managing something as simple as standardized work. After a while, the organization tries to get team members to write standardized work instead of the engineers or coordinators. To boost morale and to raise excitement among team members, leadership gives various speeches. Phrases and words like *subject matter expert* and *bottom-up management* are used to describe how team members can be involved in standardized work. While it is not clear to what extent workers are to be involved in standardized work, employees get the expectation that they own the job rather than management.

In the context of roles, this example poses many problems. Supervisors experience role conflict because they are formally in control but informally not. Because leadership has articulated a definition of Lean that can be interpreted in many ways, undesirable expectations develop on the shop floor. The Lean coordinator or engineer has role coherence issues because it is not the typical function of administrative personnel to write and approve standardized work. In this scenario, the engineers and coordinators do not have the linkage to operations that make them knowledgeable and efficient in work analysis. Team members have role integration issues because they do not really know how to interact with management given these new and ambiguous responsibilities. Simply, it is not clear to what extent team members should follow, change, or approve standardized work.

Unfortunately, when roles are not clear, teamwork is difficult to achieve. When teamwork does not exist, each individual can act alone with no real obligations to another member. The end results are many individuals uninterested in helping others unless it benefits them. Figure 2.7 illustrates the example.

At Toyota, roles are clearly defined with the concept of teamwork in mind (Figure 2.8). For example, team members follow standardized work, the team leader writes standardized work, and the group leader approves

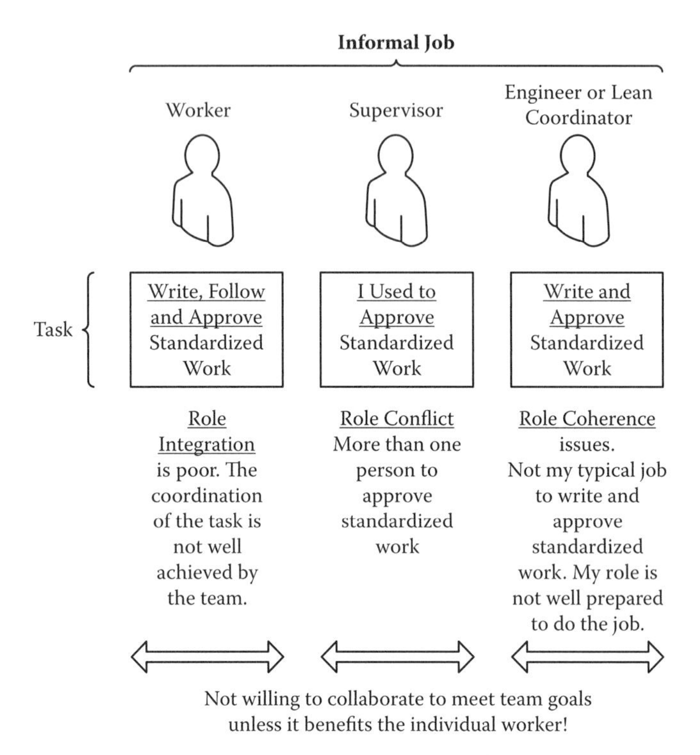

FIGURE 2.7
Typical approach for implementing and developing standardized work.

standardized work. Because the development of standardized work has been integrated into the roles of the organization, teamwork can be applied in a natural setting. Teamwork is reinforced in the organization because each member understands his or her part in how the organization functions. TPS uses roles to create the emergent property of teamwork so that members can work in a system with mutual goals.

Toyota also utilizes roles to define the interaction of its members when they are responsible for various organizational functions like safety, quality, productivity, cost, maintenance, and human resource development. In this manner, roles are defined vertically along the hierarchy and horizontally across the organization (i.e., matrix organization). Toyota views that roles should be blind to departmental and hierarchical boundaries so that the most direct and efficient approach can be applied using teamwork. Figure 2.9 conceptualizes how Toyota would develop roles and system interdependencies.

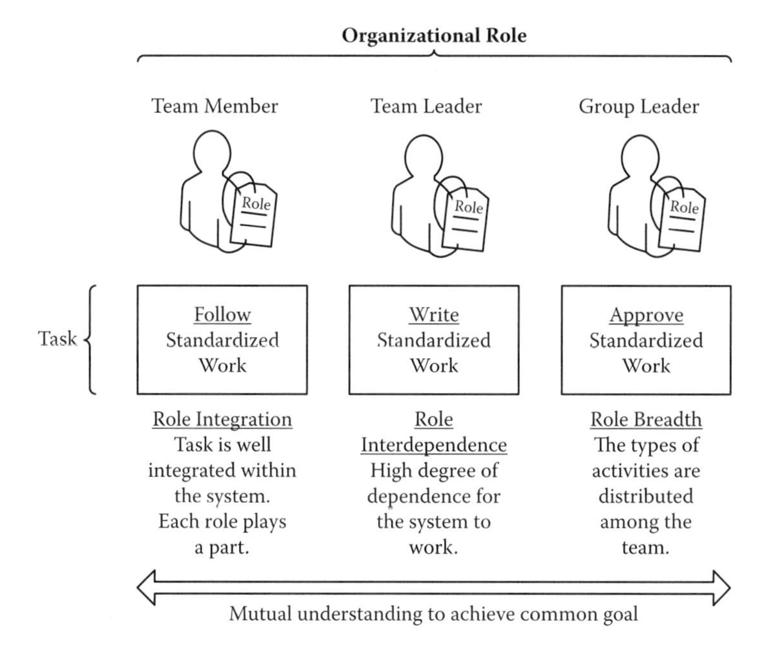

FIGURE 2.8
Example of teamwork when developing standardized work at Toyota.

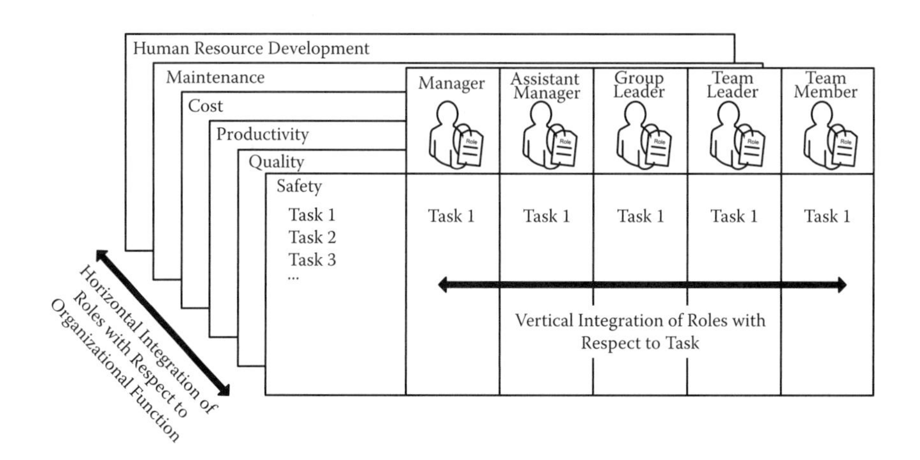

FIGURE 2.9
Role development approach for a system of roles: Toyota representation.

2.8 THE EMERGENT PROPERTIES OF TPS: COLLABORATION AND TRUST AMONG MEMBERS

Another emergent property associated with teamwork is the ability to build collaboration and trust among members. Research shows that the most significant component of a successful team is trust (Thacker and Yost, 2002; Erdem and Ozen, 2003). If trust is not present in a team, no other components can keep a team together. The two most common types of trust are cognitive and affective. Cognitive trust is concerned with the performance characteristics between its members, capability, and competence. Affective trust is psychological. Affective trust relates to emotional bonds and the sharing of ideas and feelings (Erdem and Ozen, 2003).

One of the most effective ways to build cognitive trust in organizations is the use of roles. Roles create a shared responsibility among members and themselves. Roles create an environment in which there is an expectation that a person will perform a task correctly and on time. If teams are allowed to stay together (which occurs in Toyota's organizational structure), affective trust can be built over time. The trust property in TPS means that teams can focus on achieving the goals of the organization without having to exert effort on group dysfunction (Figure 2.10).

A technique used to enhance affective and cognitive trust in organizations is the use of small teams. Small teams create a mechanism within the organization that makes it easier to nurture and build trust among members. When teams increase in size, information management and relationship building becomes more difficult to manage. Traditionally, large teams are less likely to share knowledge, learn from one another, help one another, or share resources.

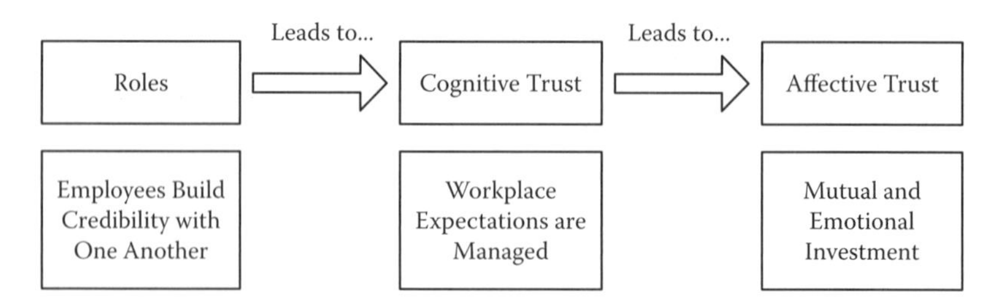

FIGURE 2.10
Toyota's trust-building approach using roles.

Other advantages of small teams include that they

- Allow more opportunities for social learning (Bandura, 1986) and hands-on teaching
- Improve communication, information exchange, and coordination among members (Martin and Bal, 2006; Dahlin et al., 2005; Keller, 2001)
- Can improve cohesion among team members (Barrick et al., 1998)
- Make it easier to apply behavior norms and settle within-team conflict (Pelled et al., 1999)
- Are more adaptable to changing environments (DeRue et al., 2008)

Other research showed that effective teams have a lot to do with how they view themselves as being alike. Researchers argue that effective teams should use people from different disciplines to build on each other's strengths (Gitlin et al., 1994). Other research shows it is hard for teams to get anything done when members become too specialized and diverse (Gratton and Erickson, 2007). Overall, a balance is needed to encourage diversity in thinking but not to the point that communications are hampered by opposing views.

2.9 THE EMERGENT PROPERTIES OF TPS: WORKPLACE CONSISTENCY

One of the most beneficial outcomes of roles is the tendency to make work more consistent. When roles are established, each employee has a defined way to achieve organizational outcomes. Ideally, when a new employee is hired, the employee is trained by the organization to fulfill a role. Roles are important to organizations because they provide new employees a record of success for doing their job. When new employees are not trained to the organization's role (or existing employees not required to fulfill their organizational role), employees often resort to their old ways of doing things. Everyone has a natural tendency to revert to experiences that have proven to be successful in previous jobs. This is problematic when everyone in the company is working to another system rather than that of the organization. Unfortunately, American business is good at rewarding individuals who can achieve results without working to the company's system. The

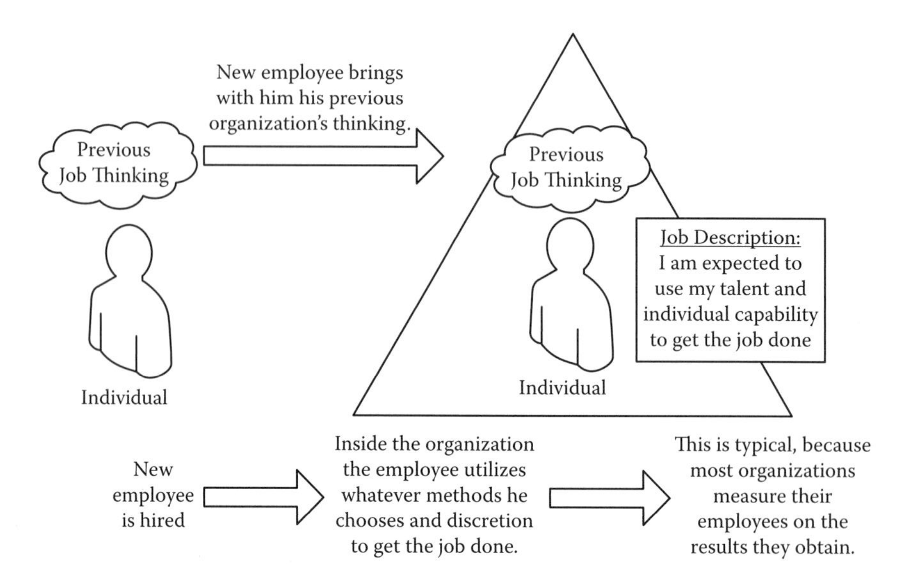

FIGURE 2.11
Results-focused organizations that use job descriptions.

thinking is that as long as an employee can get a good result, it is acceptable to violate the company's procedures or practices (Figure 2.11). In this context, employees are acting as individuals, not conforming to a role, and will use whatever discretion necessary to get results.

While this individualistic approach may seem to yield short-term results, the organization will suffer. Each individual will struggle to emerge as the new dominate and accept behavior for doing business with little concern for the company or the team. Each person is rewarded individually, but not for how the role is fulfilled. In the long term, knowledge about the role is never shared because the reward system did not support it. In this case, the job can never be improved formally because job knowledge means job security. Organizations fall into this trap by connecting results to individuals. In many cases, even the most successful employees do not fully understand how they achieved the results or can repeat the result if asked. Organizations connect results to individuals because they believe that people are individually talented and strong (Figure 2.12).

At Toyota, roles are connected to work methods, not individuals (Figure 2.13). As employees enter Toyota, they are taught to fulfill roles that contain standardized work methods that have proven to be successful. A successful work method is one that is repeatable and predictable and achieves the desired result. As work methods improve, the role is updated.

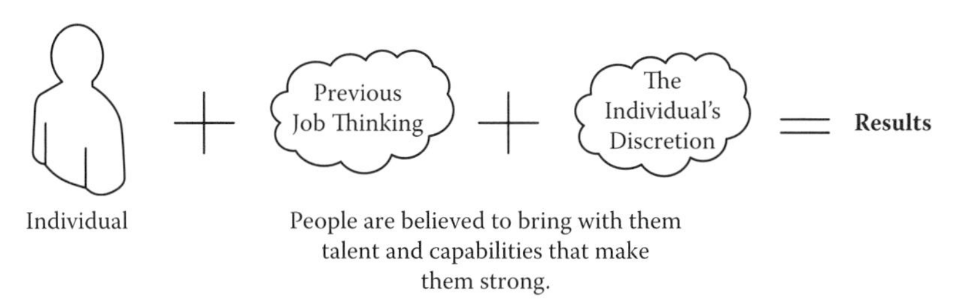

FIGURE 2.12
How most organizations view individuals and results.

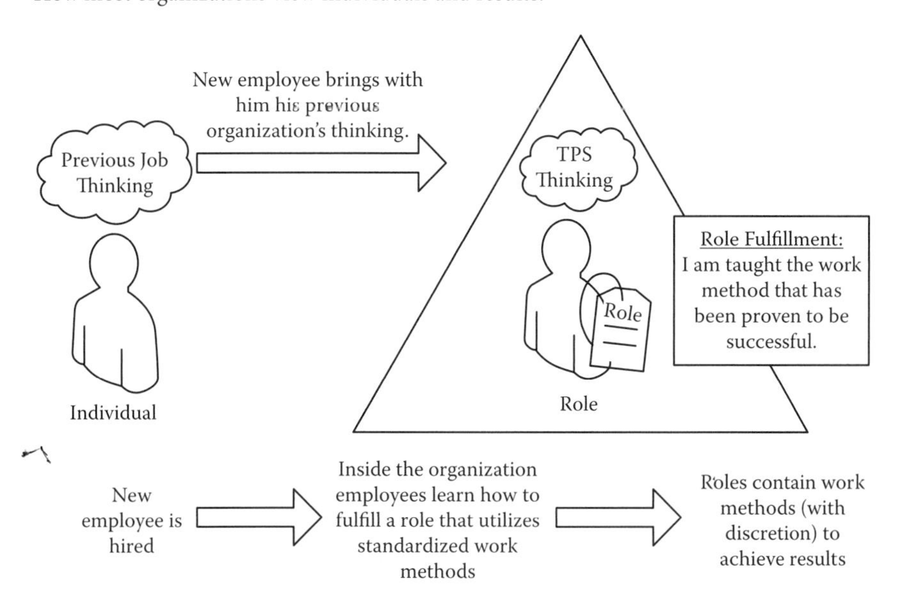

FIGURE 2.13
Process-focused organizations that utilize roles.

The discretion within the role creates an environment that is more consistent and predictable. Instead of employees referencing the way they performed the job at their last company, everyone in the organization references TPS when making decisions.

When roles are connected to work methods, the chance of failing is less when new employees use their effort and energy to meet organizational outcomes. Roles are considered to be developed when work methods are predictable and can be repeated by anyone. Without roles, work methods are likely to be developed without a formal coding, what Polanyi (1967) called tacit knowledge. Tacit knowledge is knowing what decision to make

Results are connected to roles and the work methods they perform

A standardized method is:

1. A tested idea that has shown to lead to success
2. The method offers predictability
3. The method can be repeated by anyone
4. Low risk for failure

FIGURE 2.14
Toyota's role concept: methods and results.

or how to do something without it being clearly coded or articulated (Polanyi, 1967). Toyota utilizes tools such as standardized work to decode knowledge into usable form that can be more easily shared, what Nonaka and Takeuchi (1995) referred to as externalization or explicit knowledge. Roles encourage the externalization of knowledge, which makes the workplace behavior more consistent (Figure 2.14).

2.10 SUMMARY

One of the most dominant and emerging properties of TPS is teamwork. Organizations are composed of complex social systems that can be represented as a system of roles. Role theory suggests that the needs of the organization and the individual must be evaluated to reconcile mutual goal achievement. Toyota utilizes roles to achieve various organizational outcomes in the support and development of TPS. Roles are used in TPS to establish a natural setting for teamwork in which employees have a mutual understanding of each other's rights and obligations in reaching common goals. Roles are also used to create a work environment in which employees can work with trust and collaboration among members. Roles help make the workplace more consistent by using work methods and procedures that have proven to be successful. When work methods are tied to roles, work methods can be improved and more easily shared. Every Toyota employee is expected to use standardized work methods in the employee's role to achieve results.

REFERENCES

Bandura, A. (1986) *Social Foundations of Thought and Action: A Social Cognitive Theory*, Prentice Hall, Englewood Cliffs, NJ.

Barrick, M. R., Stewart, G. L., Neubert, M. J., and Mount, M. K. (1998) Relating member ability and personality to work-team processes and team effectiveness. *Journal of Applied Psychology*, Vol. 83, 377–391.

Biddle, B. (1979) *Role Theory: Expectations, Identities and Behaviors*, Academic Press, New York.

Brown, W. and Jacques, E. (1965) Glacier Project Papers, Heinemann, London.

Dahlin, K., Weingart, L., and Hinds, P. (2005) Team diversity and information use. *Academy of Management Journal*, Vol. 48, 1107–1123.

Dayal, I. (1969) Role analysis technique in job descriptions. *California Management Review*, Vol. 11, No. 4, 47–50.

DeRue, D. S., Hollenbeck, J. R., Johnson, M. D., Ilgen, D. R., and Jundt, D. K. (2008) How different team downsizing approaches influence team-level adaptation and performance. *Academy of Management Journal*, Vol. 51, 182–196.

Duhem, P. (1906) *La Theoric Physique, Son Object et Sa Structure*, Paris.

Erdem, F. and Ozen, J. (2003) Cognitive and affective dimensions of trust in developing team performance. *Team Performance Management: An International Journal*, Vol. 9, No. 5/6, 131–135.

Gitlin, L., Lyons, K., and Kolodner, E. (1994) A model to build collaborate research or educational teams of health professions in gerontology. *Education Gerontology*, Vol. 20, 15–34.

Gordon, L. (1972) Role differentiation. *American Sociological Review*, Vol. 37 (August), 424–434.

Gratton, L., and Erickson, T. (2007) Eight ways to build collaborative teams. *Harvard Business Review*, Vol. 102 (November), 100–109.

Kahn, R., Wolfe, D., Quinn, R., Snoek, J., and Rosenthal, R. (1964) *Organizational Stress: Studies in Role Conflict and Ambiguity*, Wiley, New York.

Katz, D. and Kahn, R. (1978) *The Social Psychology of Organizations*, 2nd ed., Wiley, New York.

Keller, R. (2001) Cross-functional project groups in research and new product development: Diversity, communications, job stress, and outcomes. *Academy of Management Journal*, Vol. 44, 547–555.

Levine, E., Ash, R., Hall, H., and Sistrunk, F. (1983) Evaluation of job analysis methods by experienced job analysts. *The Academy of Management Journal*, Vol. 26, No. 2, 339–348.

Martin, A. and Bal, V. 2006. *The State of Teams: CCL Research Report*, Center for Creative Leadership, Greensboro, NC.

Nonaka, I. and Takeuchi, H. (1995) *The Knowledge-Creating Company*, Oxford Press, New York.

Pareek, U. (1976) *Interrole Exploration, Annual Handbook for Group Facilitation*, University Associates, La Jolla, CA.

Pelled, L. H., Eisenhardt, K. M., and Xin, K. R.(1999) Exploring the black box: An analysis of work group diversity, conflict, and performance. *Administrative Science Quarterly*, Vol. 44, 1–28.

Polanyi, M. (1967) *The Tacit Dimension*, Anchor Books, New York.

Schein, E. (1978) *Career Dynamics: Matching Individual and Organizational Needs*, Addison-Wesley, Reading, MA.

Team Member Handbook (2006) Toyota Motor Manufacturing Kentucky, Inc.

Thacker, R. and Yost, C. (2002) Training students to become effective workplace team leaders. *Team Performance Management: An International Journal*, Vol. 8, No. 3/4, 89–94.

Toyota Way (2001) Toyota Motor Corporation.

Wickham, M. and Parker, M. (2007) Reconceptualising organizational role theory for contemporary organizational contexts. *Journal of Managerial Psychology, Bradford*, Vol. 22, No. 5, 440.

3

The Property of Goal Seeking in TPS, Part 1

3.1 SYSTEM PROPERTY: GOAL SEEKING

Systems are designed to serve goals. The goal makes up the identity of the system, and it is used to judged its performance (Putt, 1978). Systems engage in purposeful behavior toward the attainment of their goals in reaching a final state. Goals influence how a system operates, which gives rise to the efficiency of the system itself (Gigch, 1978). A central interest in efficiency is how the system accomplishes its goals in different situations. Thus, a system is defined by the way it behaves in reaching its goals (Ackoff, 1971).

A characteristic of goal-seeking systems is the ability to accomplish the same thing in different ways and in different conditions (Ackoff, 1971). Systems can exhibit alternate choices in reaching their goals (Gigch, 1978) and adapt to their environment when conditions change. A key concept in goal-seeking systems is that they have a feedback loop to receive and register signals that indicate the behavior is working toward its goals.

A multigoal-seeking system is one that is goal seeking in two or more final different states. A prerequisite in multigoal-seeking systems is that they must have memory to store and retrieve information. Memory helps multigoal-seeking systems to increase their efficiency when producing outcomes related to various goals (Ackoff, 1971). A final property of multigoal-seeking systems is that they must have a way to prioritize their goals in a hierarchy.

3.2 GOAL-SEEKING PROPERTIES OF THE TOYOTA PRODUCTION SYSTEM

The goal-seeking properties of the Toyota Production System (TPS) relate to efficiency. TPS is a management system whose goal is to improve workplace productivity. TPS is a strategy applied to organizations to improve how they work, function, and meet intended goals. TPS is not a goal in itself but a way to improve the systems of the organization to meet its desired outcomes. The technique that Toyota uses to improve organizational effectiveness is industrial engineering (IE; Figure 3.1). IE is the identity of TPS; however, IE has evolved over the years. Interestingly, many organizations are defining their version of TPS by the type and style of IE that is applied today. Not all forms of IE are considered equal, and there is not enough agreement or consistency in the profession that allows it to be applied the same way every time.

Another goal-seeking property of TPS is the manner by which it accomplishes the goal of efficiency. The sequences of behavior that drive TPS to meet its conforming goals of efficiency are problem solving. Toyota uses problem solving as a mechanism to improve and correct organizational systems. In this way, TPS acts as a feedback loop to improve all of the company's systems. Much like the style of IE that shapes the identity of TPS, so does the type of problem solving that shapes the feedback loop for improving organizational systems (Figure 3.2). Not all problem-solving methodologies are the same, and they all do not act as effective feedback systems for getting the organization back on track.

Due to the depth of these two topics, a chapter has been devoted to each. This chapter is dedicated to understanding the style of IE that Toyota utilizes to identify TPS. The next chapter reviews various types of human problem-solving methodologies and the specific form that Toyota has created to appeal to certain users.

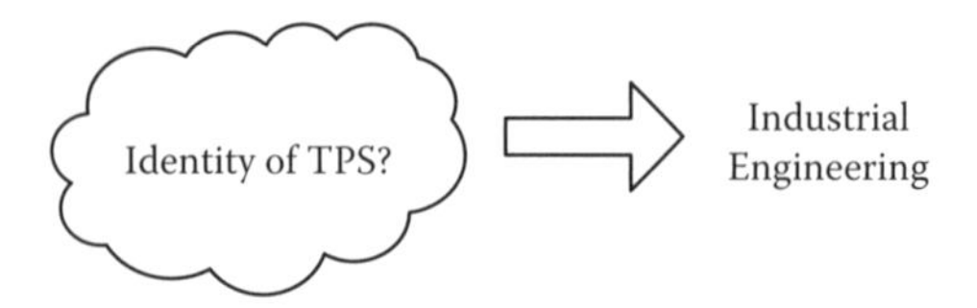

FIGURE 3.1
Goal-seeking property of TPS: industrial engineering identity.

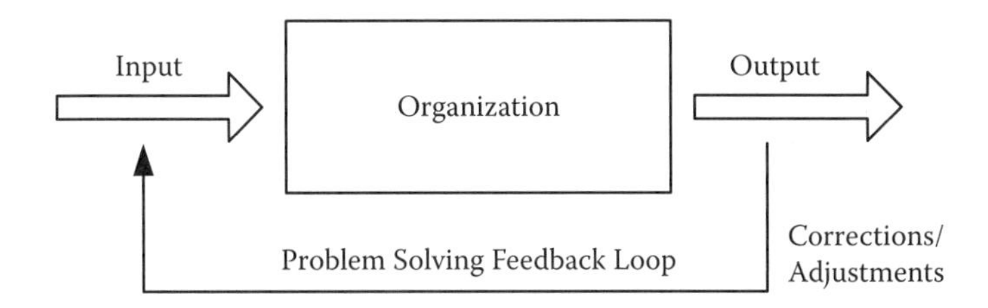

FIGURE 3.2
Goal-seeking property of TPS: the problem-solving feedback loop.

3.3 BACKGROUND: TPS AND THE INDUSTRIAL ENGINEERING CONNECTION

"Catch up with America in three years. Otherwise, the automobile industry of Japan will not survive," Kiichiro Toyoda, then president of Toyota Motor Company, said on August 15, 1945. "To accomplish this mission, we had to know America and learn American ways," Taiichi Ohno relates in *Toyota Production System* (1988a).

Ohno believed that the quickest way to catch up with America was to import American production management techniques and business management practices. Toyota studied IE, which by Ohno's accounts best describes TPS, a companywide system tied directly to management to lower cost and raise productivity systematically (Ohno, 1988a).

One of Ohno's close friends and consultant at the time was Sheigo Shingo. Shingo also viewed TPS as a form of IE to address plant improvement. He believed that management should possess a set of fundamentals closely related to IE as a way to spread and teach the TPS.

In Shigeo Shingo's book, *A Study of the Toyota Production System*, he pledged his life's work to the study of Taylor's principles of scientific management. Shingo believed that TPS is a system made up of principles that can be applied through practical implementation. If management cannot understand how to attack the rationalization of the current system through scientific study, then it cannot be expected to improve or change (Shingo, 2005).

3.4 THE ROLE OF INDUSTRIAL ENGINEERING

An industrial engineer is one who is concerned with the design, installation, and improvement of systems. Industrial engineers utilize specialized knowledge and skills in mathematical, physical, and social sciences, combined with the principles and methods of engineering analysis (Salvendy, 2001). Over the years, IE has drawn on mechanical engineering, economics, labor psychology, philosophy, and accountancy in an effort to bring together people, machines, materials, and information (Saunders, 1982). If industrial engineers had to focus on one aspect of their field, it would be productivity or productivity improvement. Simply, industrial engineers focus on the total elimination of waste by increasing efficiency through cost reduction (Going, 1911).

Industrial engineering covers not only the technical aspects of systems but also systems relating to management (Figure 3.3). Anderson proposed that IE is one of the primary drivers for linking the needs of the employers to the needs of the employees (Anderson, 1928). Employers want industrial peace, reduction of cost, higher efficiency, and improvement in quality. Employees want steady work, higher wages, better personal relations with their supervisor, and good working conditions. By utilizing IE techniques, management can develop, evaluate, and improve the wants of both groups (Anderson, 1928).

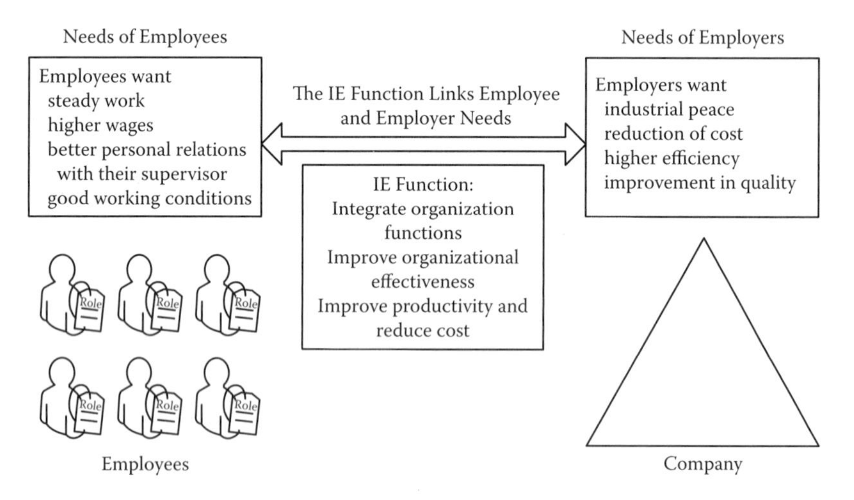

FIGURE 3.3
Representation of the industrial engineering function.

3.5 INDUSTRIAL ENGINEERING THEORY

The theory of IE states that everything should be treated as a process: studied, disassembled, changed, and evaluated to improve its performance. The laws of IE state that a system cannot be improved unless four prior laws are satisfied (Figure 3.4). The first law of IE states that results should only be improved when processes are improved. When a system is improved but it cannot be determined what changed in the system, the outcome is by chance. A good process is when what makes results repeatable, consistent, and predictable is known. Industrial engineers want to link results with the process to achieve consistency before output.

The second law states that before a process can be improved the process must first be stable. Most organizations want to improve a process before stabilizing the process. The belief is that the improved process is the quickest way to the new result. Unfortunately, jumping from one point of instability to another is unreliable. Industrial engineers want to move from a known state to another known state. If the current state is not stable,

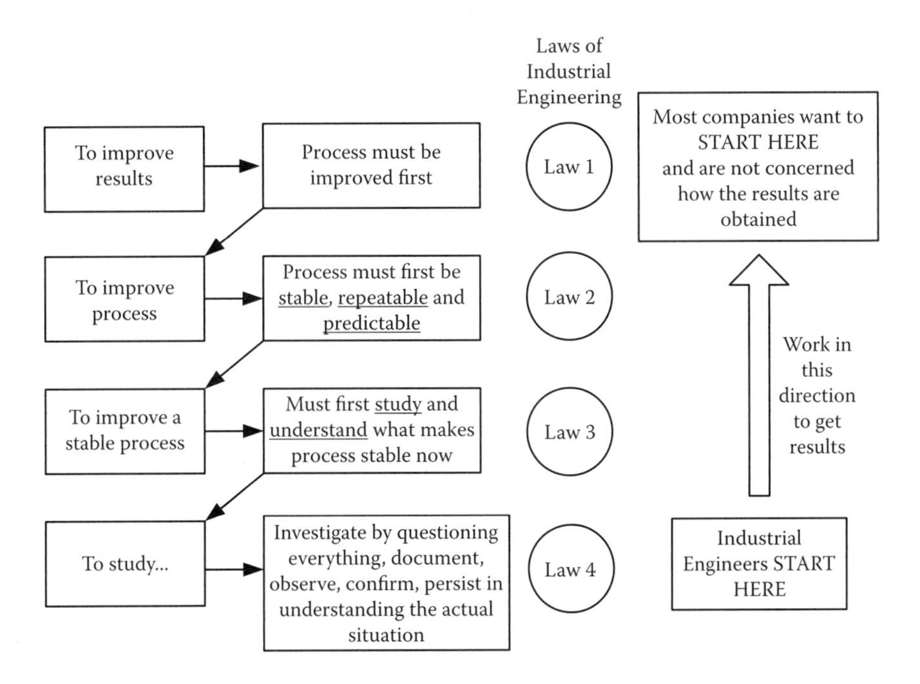

FIGURE 3.4
The four laws of industrial engineering.

repeatable, and predictable, it makes little difference switching to another state that is not reliable, repeatable, or predictable.

The third IE law states that to improve a stable process, it must be known what makes the process currently stable. To ensure a stable system, process variables have to be identified and controlled. Otherwise, the system is floating, and its ability to maintain stability is by chance. Industrial engineers want reliable systems and want to know the factors in the process that make the system reliable.

The fourth law relates to the study habits of the IE profession. To study a process, everything must be questioned about the process. The process must be investigated thoroughly by documenting, observing, and confirming. The industrial engineer must be patient and persistent in understanding the actual situation. By going slow, step by step, the insights of a process can be understood.

3.6 LITERATURE REVIEW OF INDUSTRIAL ENGINEERING METHODS

3.6.1 The Scientific Management Era

One of the earliest contributions to the field of IE and to the industrial efficiency movement in the early 1900s was by Frederick Taylor with his invention of scientific management (Taylor, 1911). Scientific management is the saving of energy, materials, and time or in other words, the elimination of waste through studying, recording, and analyzing work. The Gilbreths (Frank and Lillian) were also advocates of scientific management and were concerned with how to raise productivity properly without degradation of an employee's health and well-being (Gilbreth, 1973). Scientific management was never a manner of how much a person can do under a short burst of speed but instead at a safe and comfortable working speed that can be done day after day (Mogensen, 1935b). Unfortunately, charlatans attempting to break into the field of time and motion study portrayed scientific management poorly. Consultants who had neither proper training nor interest except for a quick financial benefit did not follow Taylor's true teachings of scientific management. Scientific management was intended to secure the maximum prosperity of the employer coupled with the maximum prosperity for the employee. The long-term prosperity of the employer cannot exist unless it is accompanied by the prosperity of the employee (Gilbreth, 1973). Antiscientific management extremists do not realize that Taylor encouraged shop floor employees to be involved in decision making. In 1935, Mogensen suggested that industrial engineers (then referred to as efficiency experts) should solicit suggestions and ideas from those working directly with the operation. Mogensen (1935a) went on to say the following: "While it is assumed that workmen and the foreman are incapable of such suggestions, some of the most valuable ideas have come from this source."

3.6.2 Skill Sets of Industrial Engineers in the Era of Scientific Management

A list of the efficiency tools and concepts used by IE during the era of scientific management is shown in this section. Industrial engineers were mostly concerned with defining processes, identifying problems, and establishing standard operations (Harrington, 1911). Some of the most practical techniques they employed were direct observation and work sampling (Staley and Delloff, 1963). Combined with a questioning attitude, industrial engineers could obtain facts to make productivity improvements quickly and easily. The industrial engineer was also proficient with the use of charting. By breaking down processes into smaller units, the industrial engineer could analyze work flow by examining process steps visually. Last, the industrial engineer was concerned with running trials to test new productivity ideas. By testing factors one at a time and by sequentially changing those parameters based on previous trials, the industrial engineer could speed up decision making while focusing on improvement. Various skill sets of an industrial engineer in the era of scientific management are discussed next.

3.6.2.1 Skill Set: Scientific Method

The scientific method is the basis of scientific management. It is a way of outlining a series of small steps used for inquiry and learning based on the unbiased evaluation of data (Morse, 1956). When steps are arranged in a sequence, decision-making processes can be evaluated for effectiveness. The scientific method is more commonly known as a form of problem solving. An example of a scientific method is as follows:

1. Observe a problem or phenomenon
2. Review existing research and theory (when relevant)
3. Make hypotheses or research questions
4. Determine an appropriate methodology/research design
5. Collect relevant data
6. Analyze and interpret the results
7. Replicate the study (if necessary)

3.6.2.2 Skill Set: Importance of Direct Observation

Direct observation is carried out to ensure that all pertinent facts are collected and that each fact is checked for accuracy. Logic or deduction cannot enter the analysis until observed facts are obtained (Mogensen, 1935b).

3.6.2.3 Skill Set: Work Sampling

Work sampling is also known as the ratio delay technique. A small number of chance occurrences tend to follow the same distribution pattern as that in the entire population of occurrences. Work sampling is an excellent and inexpensive tool for making accurate studies and predictions about the effectiveness of a process (Staley and Delloff, 1963).

3.6.2.4 Skill Set: Questioning Attitude

A questioning attitude is essential for developing critical thinking skills. A questioning attitude is the first step to continuous improvement and waste elimination (Tippett, 1953; Ireson and Grant, 1971).

Tippett's (1953) list of common productivity areas has many similarities to Ohno's (1988a) seven wastes.

1. Delays in routing work to and from the operators
2. Excessive personal time (variation)
3. Lost time in setting up or in other work preparation
4. Insufficient work to do (waiting)
5. Bottlenecks
6. Obsolete methods (using the wrong work method)
7. Unbalanced work loading
8. Defects in process

3.6.2.5 Skill Set: Standardization

Standardization is one the basic tools for eliminating waste. A standard is simply a carefully thought out method of performing a function. The idea of perfection is not involved in standardization. The standard method is the best method that can be devised at the time the standard is drawn. Improvement in standards are wanted and adopted whenever and wherever they are found. Safeguards protect standards from change or changing just for the sake of changing. Standardization in this way is a constant invitation to experimentation and improvement (Cooke, 1910; Gilbreth, 1973).

3.6.2.6 Skill Set: Standards Engineer

The standards engineer is the operational starting point for standardization. It is a position that was established in the early 1900s to advance standardization throughout a company. A good standards engineer is not necessarily a good engineer but one that is strong in the area of human relations. The standards engineer must have the ability to understand people as individuals and in groups and have the ability to cope and be patient with the common human characteristics of resistance to change and resentment of criticism. The standards engineer is encouraged to cut across organizational lines to make standardization companywide and not departmentalized (Rayfield, 1964).

3.6.2.7 Skill Set: Systems Thinking

One of the most well-known techniques of industrial engineers in the scientific era is the idea of systems and systems engineering. Systems thinking helps engineers focus on the entire system rather than its parts. An effective industrial engineer cannot improve a system by analyzing the parts of the system in isolation from the whole. The whole determines the nature of the parts (Kadota, 1982; Gottlieb, 1971).

3.6.2.8 Skill Set: The Process Flowchart

Charts are graphical representations of work that has been broken down into basic components or units. The process flowchart was credited by Frank Gilbreth to record in detail operations that cannot be understood through direct human observation (Staley and Delloff, 1963; Mogensen, 1935a).

After charting the process, ask the following:

1. Can the operation be eliminated?
2. Can it be combined?
3. Can we combine the sequence of operations?
4. Can it be simplified?

3.6.2.9 Skill Set: Work Distribution Charts

More formally known as line balance charts, work distribution charts are used in the factory and the office (Staley and Delloff, 1963). Evaluate areas that have

1. Tasks that consume the most time
2. People working on jobs below or above their skill
3. People who are doing too many different things
4. Tasks in which everyone has some part
5. Employees who require overtime to complete their duties

3.6.2.10 Skill Set: Time Study

A time study is one of the most common methods for setting standards. Time study is the division of operations into motion components or elements. Industrial engineers use time and motion study to determine a representative time for each element in a job (Staley and Delloff, 1963).

3.6.2.11 Skill Set: Testing, Adaptive One Factor at a Time

Testing one variable at a time is the quickest way to gain information about a system when making improvement (AOFAT, adaptive one factor at a time). By adjusting the experiments along the way, investigators can find out more rapidly whether a factor has any effect. When performance improvement is the primary purpose of the experimental effort, testing one factor at a time is the best choice (Friedman and Savage, 1947; Daniel, 1973).

3.6.3 Contemporary Industrial Engineering

Today, IE has become a more integral part of the organization. With the invention of the high-speed computer in the 1960s, IE has evolved into a hard discipline: Data can be recalled at any time, and decision making can be improved through the use of models and simulations (Saunders, 1982). Computers have given industrial engineers the ability to analyze and optimize complex systems throughout the organization (Katzell et al., 1977). The field has also become more specialized over the years, much like mechanical engineering in the earlier twentieth century. IE offers several subspecialties, such as human factors, job design, labor psychology, and systems engineering. Today, it is not uncommon for industrial engineers to work on planning systems, supply chains, accounting systems, and organizational polices.

One of the most significant changes in the IE profession has been its role in change management (Zandin, 2001). The industrial engineer is a strategist, senior champion, sponsor, educational activist (i.e., trainer), and a change master. One of the main reasons why the IE function has become more of a driver for change is the growth of service functions within modern industry. Because industrial engineers have skills to analyze social-technical systems, they can help improve the fit between technology and the worker (Salvendy, 2001).

3.6.4 Skill Sets of Contemporary Industrial Engineering

The skill sets of the modern industrial engineer are much different from the days of scientific management. While the objective remains the same—to improve efficiency through cost reduction—the approach has changed dramatically. Most modern IE skill sets emphasize rapid organizational change instead of spending time stabilizing and documenting current operations. Techniques such as process design and reengineering (PDR) result in radical change by focusing on end-to-end processes. PDR does not study existing systems mainly because it is assumed that nothing can be gained from analyzing a broken system. PDR advocates radical innovation, which requires a clean slate change. Documenting existing processes often limits the vision of the design team when trying to apply PDR (Taylor et al., 2001).

Other tools used by contemporary industrial engineers include statistical and numerical methods such as experimental designs (EDs) and design of experiments (DOE). These techniques allow industrial engineers to understand the complexities of the business and the interacting factors acting on and within the organization before leaping toward a new state

(Czitrom, 1999). Today, DOEs are packaged with structured initiatives for business improvement known as Six Sigma and Lean Sigma (Pyzdek and Keller, 2009; Wedgwood, 2007). Six Sigma is a systematic method for improvement that relies on statistics to make dramatic reduction in defect rates (Tanco et al., 2009). Initially established by Motorola in 1987, Six Sigma has been extremely popularized as a new form of business management strategy (Jugulum and Samuel, 2008). Six sigma often involves large masses of data and concerns itself with percentages and averages or the presentation of data in tables and charts (Bowker and Lieberman, 1971).

Lean Sigma is another improvement methodology that is being employed by industrial engineers. Proponents of Lean Sigma suggest that by integrating statistical methods with the ideas of work simplification, a common language can be developed to help organizations be responsive to changing markets while eliminating defects (Wedgwood, 2007). Jugulum and Samuel suggested that the key to Lean Sigma is through integration. Six Sigma provides the detailed statistical study to optimize projects, while Lean is usually implemented through a series of short, focused kaizen blitzes. Lean Sigma allows organizations a comprehensive framework that enables them to have the benefits of both worlds, a mile wide and a mile deep (Jugulum and Samuel, 2008). Other contemporary IE techniques are discussed next.

3.6.4.1 Skill Set: Systems Engineering and Optimization

Systems engineering is derivative of the systems approach that specializes in optimizing mathematical models using numerical methods. Systems engineering divides the total system into smaller subsystems, specifies the input-output requirements of each subsystem and each smaller component, and determines the method of interconnection to accomplish the overall objective of the system. Systems optimization utilizes a variety of different mathematical models, including the study of systems that are continuous, discrete, lumped or distributed, linear or nonlinear, constant or time varying, deterministic or structured, and behaviors. (Mar, 1994; von Bertalanffy, 1968; Martens and Allen, 1969; Castro et al., 2010)

3.6.4.2 Skill Set: Process Design and Reengineering

PDR is a systematic discipline for achieving dramatic performance improvements by fundamentally reexamining, rethinking, and redesigning the processes that an organization uses to carry out its mission. Reengineering processes are usually described in terms of the beginning and end states and activities in between (Taylor et al., 2001; Lee and Dale, 1998; General Accounting Office, 1995; Manganelli and Klein, 1994; Lee, 1996).

3.6.4.3 Skill Set: Experimental Design, Design of Experiments, Taguchi Methods

Experimental design enables industrial engineers to study the effects of several variables affecting the response or output of a process using statistics. Taguchi's approach to DOE was based on orthogonal designs to simplify and accelerate testing. Replications and randomization are required for an estimate of error to determine the basis for decision making on the importance of factors contributing to the response variables (Anderson and McLean, 1974; Tanco et al., 2009; Montgomery, 2005).

3.6.4.4 Skill Set: Six Sigma

Industrial engineers specializing in statistical improvement utilize a variety of statistical and quality tools such as the following (Tanco et al., 2009; Pyzdek and Keller, 2009):

1. Affinity diagrams
2. Multivariate charts
3. FMEA (failure mode and effects analysis)
4. DMAIC (define, measure, analyze, improve, control)
5. Analysis of variance (ANOVA)
6. Regression analysis
7. TRIZ (theory of inventive problem solving)*

3.6.4.5 Skill Set: Lean Sigma

Lean Sigma is a business improvement strategy based on combining the statistical tools of Six Sigma and the waste reduction methodologies of Lean (Wedgwood, 2007; Jugulum and Samuel, 2008).

Integrating the tools of six sigma and Lean includes

---

* TRIZ was developed by Soviet scientist and writer Genrich Saulovich Altshuller and colleagues beginning in 1946.

1. Chi-square analysis and 5S
2. DOE and kanban
3. FMEA and value stream mapping

3.6.5 Literature Review Summary

Over the years, the types of skills used by industrial engineers have changed significantly. Industrial engineers in the early nineteenth century applied skills that were less specialized than industrial engineers in the twenty-first century. Contemporary industrial engineers emphasize system optimization, advanced computational mathematics, and rapid overhaul within the organization. Industrial engineers in the era of scientific management were more concerned about applying practical skills, things that could be accomplished without a computer. While most views of scientific management tend to be described as extreme Taylorism, earlier discussions in this book pointed out that scientific management was not completely top-down directed.

Improvement initiatives such as Lean manufacturing, Six Sigma, and even Lean Sigma, have followed contemporary IE. The literature shows that kaizen specialists or master black belts should be capable of performing value engineering in product design (Bicheno, 2000), environmental scanning (Jackson, 2006), cellular manufacturing, production flow analysis, and supply chain infrastructure analysis (Askin and Goldberg, 2002; Srinivasan, 2004). In short, contemporary IE is concerned with optimizing complex systems and structures that require specialized skills (Figure 3.5).

3.7 WHAT TYPE OF INDUSTRIAL ENGINEERING IDENTITY IS TPS?

The identity of TPS cannot be defined without understanding what makes TPS strong. In Chapter 2, teamwork was shown as an emergent property of TPS. The work of Besser suggested that Toyota has three types of team classifications: the small work group team, the company team, and the corporate team (Besser, 1991). A conceptual illustration is shown in Figure 3.6. The strength of a team is defined by the number of members willing to contribute toward the goals of the team. The strength of TPS is also determined by the number of people (i.e., group, company, or

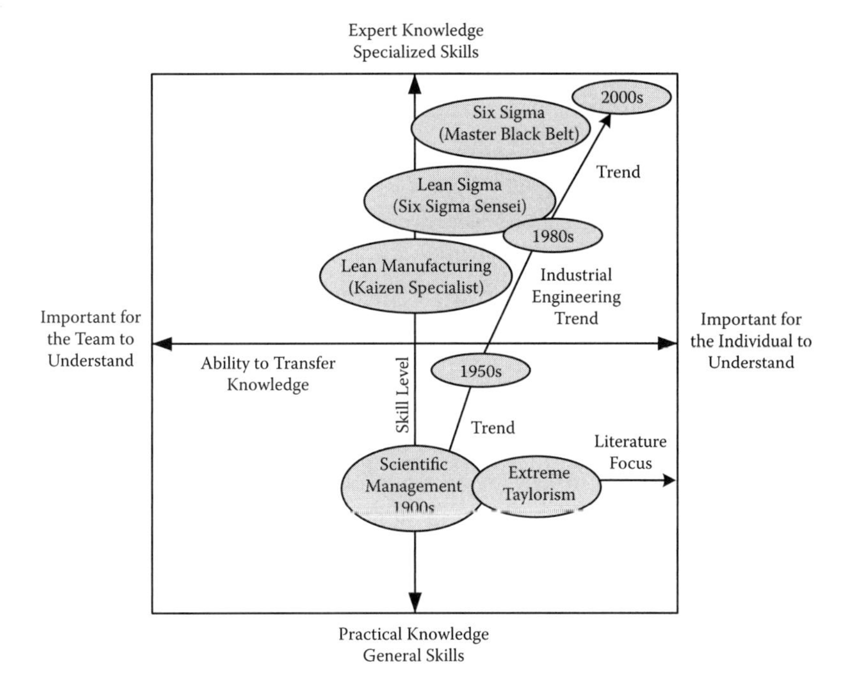

FIGURE 3.5
Comparison of skill trends for IE, Lean, Six Sigma and Lean sigma.

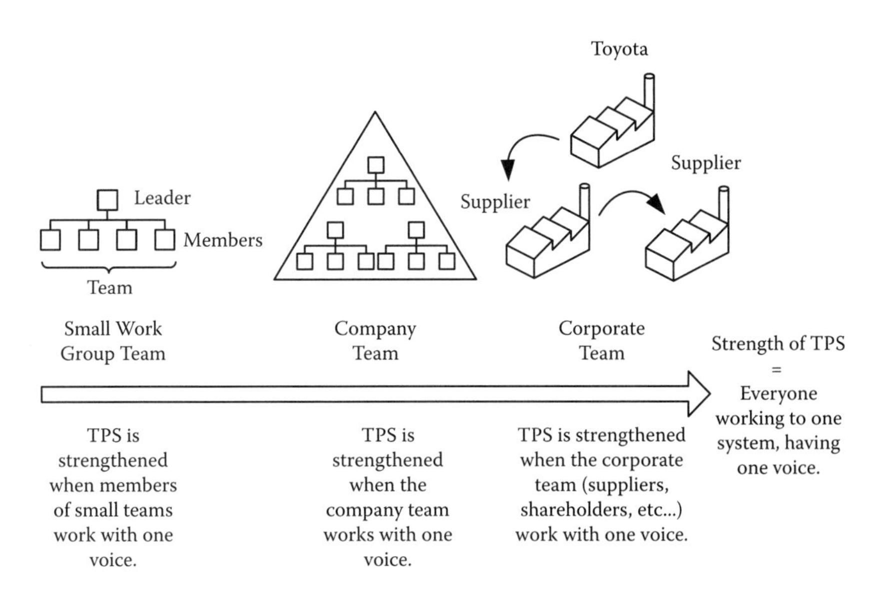

FIGURE 3.6
Representation of the types of teams at Toyota (extended from Besser's concept of teamwork at Toyota).

enterprise) open and willing to work toward its achievement. For TPS to be strong, it requires a large number of people to use its systems, processes, procedures, and business practices. If members are unwilling to use the system or work within its parameters, the system (i.e., the team) is weakened and less effective. Members need to agree that there is only one best way to accomplish team goals and are willing to use the system or way (i.e., the Toyota Way) to do it.

The type of IE that makes up the identity of TPS is largely dependent on the number of people who can use it. If the identity of TPS is contemporary IE, like most views on Lean manufacturing, only the elite can be expected to apply TPS. In this context, the strength of TPS is limited to a small group. While these groups can apply sophisticated techniques that are effective for improving the organization, other companies can also purchase or hire these types of people to perform IE. In this case, Lean does not provide a unique competitive advantage (Figure 3.7). The improvement effect that the organization sees or feels is the addition of the

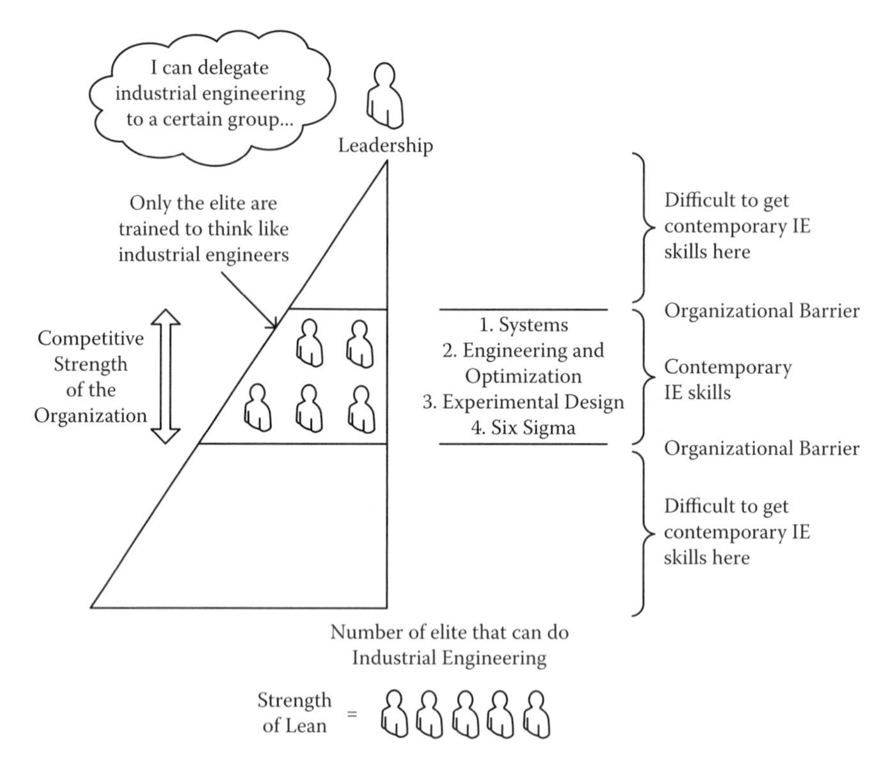

FIGURE 3.7
Strength of Lean manufacturing for most organizations.

IE function to the organization. It should be noted that Toyota already has a production engineering/IE department, but utilizes TPS to get the full power of the workforce involved.

The identity of TPS closely resembles the era of scientific management. In the 1950s, Taiichi Ohno initiated a new type of production system (i.e., TPS) with help from Shigeo Shingo to encourage employees to use a reliable way to prove their ideas. Ohno believed that if employees could test their ideas immediately, they would be more willing to improve their jobs (Ohno, 1988b). Early nineteenth-century IE practices armed Ohno with a structured process to teach employees work analysis. The idea was to adapt IE so that TPS could be applied voluntarily. The end result was a system that enabled everyone in the organization to think like an engineer (Figure 3.8).

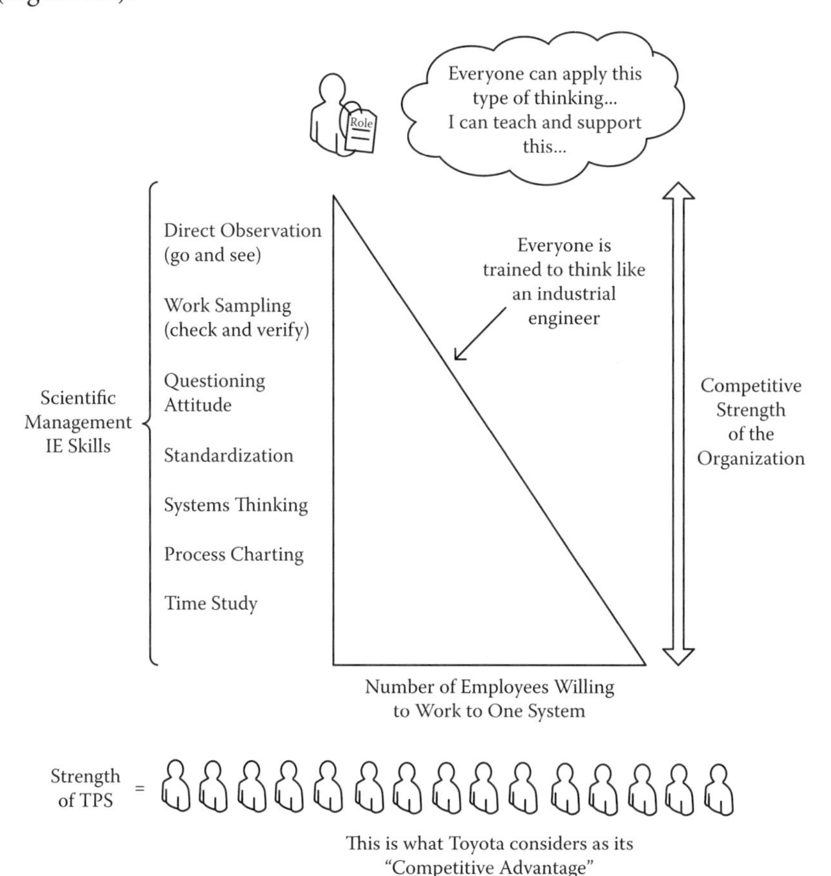

FIGURE 3.8
Strength of the Toyota Production System.

3.7.1 Differences between TPS, the Efficiency Expert, and Traditional Lean

Another way to describe the competitive advantage of TPS is to understand who is responsible for the result and the status of the result when IE is applied. In the 1920s, efficiency experts were assigned by management to reduce cost. They were responsible for obtaining the results using whatever methods possible. Efficiency experts performed IE by telling people what to do and how to do it. In this context, extreme Taylorism made the IE function extremely confrontational to employees in the workplace.

In the 1990s, Lean manufacturing and Six Sigma were recognized as successful alternatives for increasing a firm's competitive advantage. In these initiatives, the title of those in the IE profession was renamed kaizen specialists and master black belts. Surprisingly, most companies like to rename their IE programs using Japanese words but still perform IE. In these programs, the master black belt and the kaizen specialist are still responsible for the results. While they look and feel more cooperative than their earlier efficiency expert counterparts, they are still largely responsible for the outcome.

Modern Six Sigma and Lean programs are less confrontational than classical industrial engineering programs mainly because they try to perform IE with employees. Typically, kaizen specialists are assigned to work with a group of employees to improve their work areas. These activities are often known as kaizen events and typically last for a short time. Most companies apply these kaizen events as a way to implement or jump-start Lean. Interestingly, Toyota does not perform kaizen events. Toyota does have problem-solving and improvement activities, but they last until the job is completed, which can take anywhere from three days to three months.

Kaizen events have become extremely popular in the United States mainly because they use IE to make a drastic improvement in the workplace. Kaizen events have been known to increase employee participation by involving employees in the workplace. Management's presence in kaizen events intensifies kazien and draws attention to eliciting employee ideas. Artificially, employee engagement and involvement work in kaizen events. As long as the kaizen specialist stays with the group and performs IE to justify employee ideas, Lean works. The problem is when the kaizen specialist moves on to the next work area. When conditions at the workplace deteriorate, the group is not trained in how to hold standardization or know-how to return to the improved state without help. In this situation, the group has to wait until the kaizen specialist returns or until the next event. The issue is that IE skills were never adapted to the level of the group and transferred to the group. The problem is amplified when management does not take a leading role in teaching or coaching these skills, which means management cannot model the behavior that is expected from their employees. In this scenario, the strength of Lean is determined by the number of industrial engineers available to perform kaizen events. This is not a competitive advantage because any company can hire industrial engineers to perform IE with their employees.

On the other hand, TPS is an identity that everyone in the company can share (Figure 3.9). Toyota has adapted scientific management so that it can be applied and used by a larger group of employees in the company.

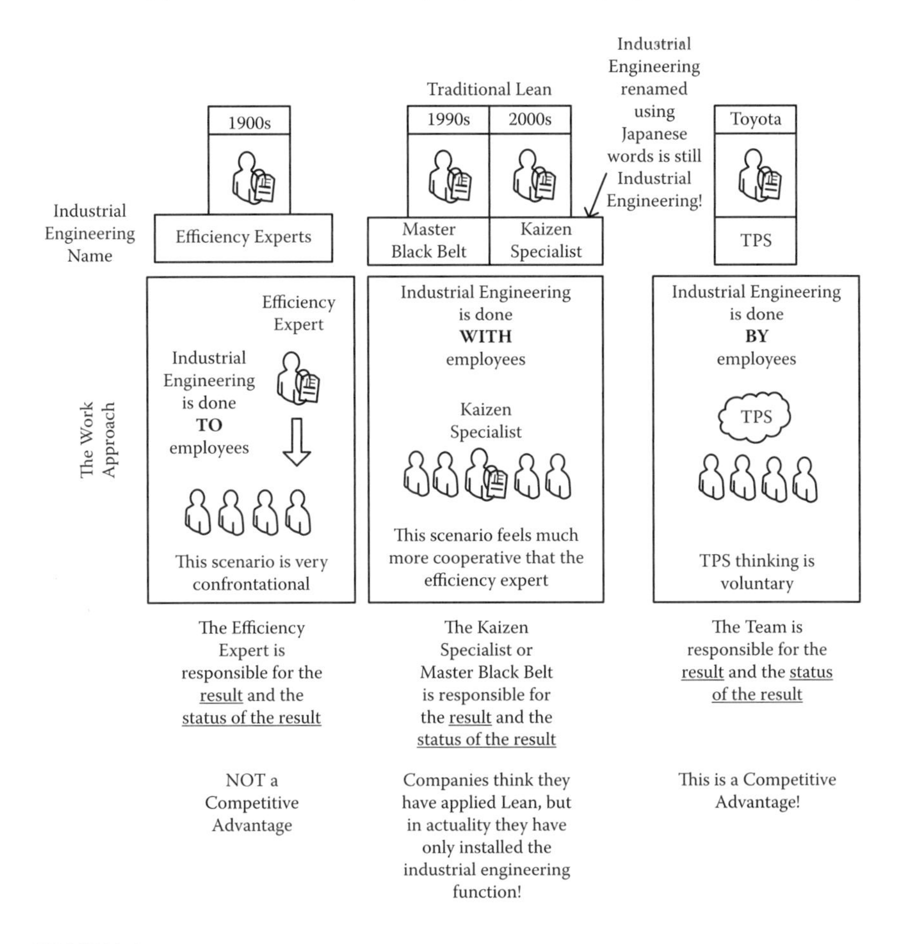

FIGURE 3.9
TPS compared to the efficiency expert and traditional Lean.

In this way, IE is done by the employees instead of done to the employees. Toyota's emphasis on developing and transferring skills throughout the organization is how TPS is able to recruit and attract so many members. In this way, results and means to achieve the results, belong to the group, not a kaizen specialist or master black belt. Because Toyota's leadership and management are also trained to apply these simplified techniques, IE can be applied and supported voluntarily by the employees themselves. Again, this is what Toyota considers as a plus factor or competitive advantage that cannot be easily copied or emulated by their competitors.

3.8 ADAPTING SCIENTIFIC MANAGEMENT TO TPS AND THE COMPARISON TO CONTEMPORARY IE

TPS is a form of IE that provides a reference point for decision making when employees voluntarily want to evaluate and improve their work. Following are examples that illustrate how TPS has adopted many of the practical skill sets of scientific management compared to contemporary IE. The easier it is for employees to understand TPS, the easier it becomes for employees to apply it on their own.

3.8.1 Standardization or Kaizen (Business Process Reengineering)?

Kaizen and IE tools such as business process reengineering (BPR) have become extremely popular in the Lean community. The idea of overhauling and improving existing systems in a short amount of time is appealing to executives. Most companies implementing Lean interpret kaizen as the first step in improving their systems. It is also a much easier sell to employees that they can throw out the things that do not like and change the things they want. Kaizen is sexy, appealing, and fashionable and brings a new opportunity for something better. The Lean community has capitalized on this idea of kaizen and has made it a priority in Lean implementation. In the Lean community, there is a rush to kaizen and short-term results.

According to the laws of IE, there cannot be kaizen (i.e., improvement) until processes are in control. Switching from an unknown state to another is still a move to the unknown. The Lean community has treated kaizen similar to BPR, meaning there is a belief that nothing can be gained by

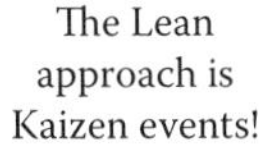

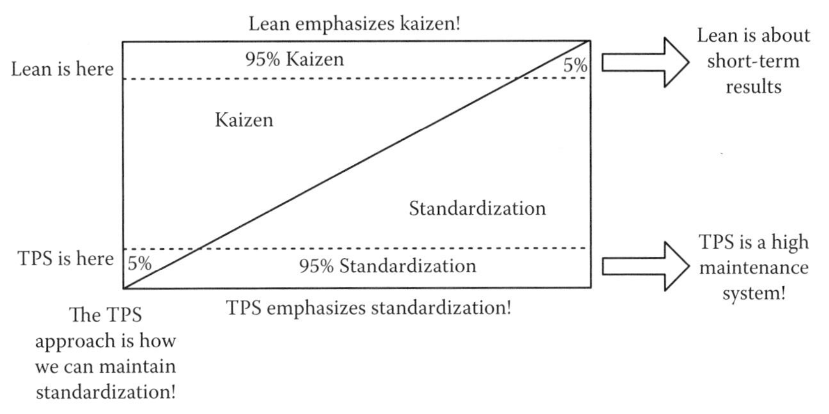

FIGURE 3.10
Comparing standardization and kaizen.

studying and stabilizing existing systems. Industrial engineers who use BPR (like kaizen) believe that the existing system is too broke to fix. Scientific management would disagree with this logic. The quickest and lowest risk to obtaining results is by stabilizing existing systems by standardizing practices that are already known to be successful. Standardization offers both short-term and long-term results. It is the reason why Toyota considers it one of its primary tools to eliminate waste in the workplace. Of TPS, 95% is about standardization, not kaizen (Figure 3.10). The reason why so many companies have difficulty implementing TPS-like systems is because there is a focus on kaizen and not standardization. Standardization is not sexy, fashionable, or appealing. Standardization requires effort, discipline, and patience. Toyota adopted scientific management because it explained how to study systems to find the factors that make processes stable. TPS is a high-maintenance system. The search for process variables never ends. TPS is about helping employees find process variables in their job so problems do not have to be repeated.

3.8.2 Knowledge by Observation (Genchi Genbutsu) or by Inference (Data)?

One of the most important aspects of scientific management is the concept that all decisions should be based on facts. The industrial engineer is trained to ensure that all pertinent facts are collected and checked for

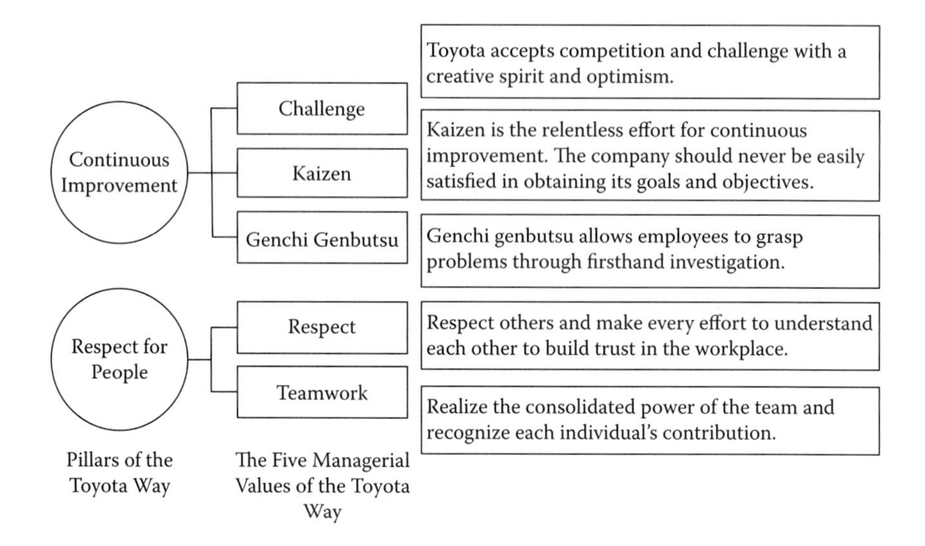

FIGURE 3.11
The Toyota Way.

accuracy. Engineers in the scientific era are trained to collect facts through direct observation.

Over the years, direct observation has been less emphasized in the IE profession with the invention of the high-speed computer. Contemporary IEs are less concerned about observing the workplace and more concerned about collecting large amounts of data. This disconnect from the shop floor greatly diminishes the ability of the industrial engineer to grasp what is really occurring.

A workplace value that Toyota adopted from scientific management that relates to workplace observation is genchi genbutsu. *Genchi genbutsu* means "go and see for yourself." Toyota would like for all employees to witness problems first hand to understand what is really occurring. Toyota believes that if employees can refrain from jumping to conclusions and not rely on other people's interpretation of the problem, decision making can be improved. While data are important, facts are more important. In 2001, Toyota reemphasized the importance of genchi genbutsu by including it as one of the five managerial values of the Toyota Way (Cho, 2001) (Figure 3.11).

3.8.3 Scientific Method or System Optimization?

One of the most popular tools in the implementation of Lean manufacturing is the use of value stream maps. A value stream map is a form of

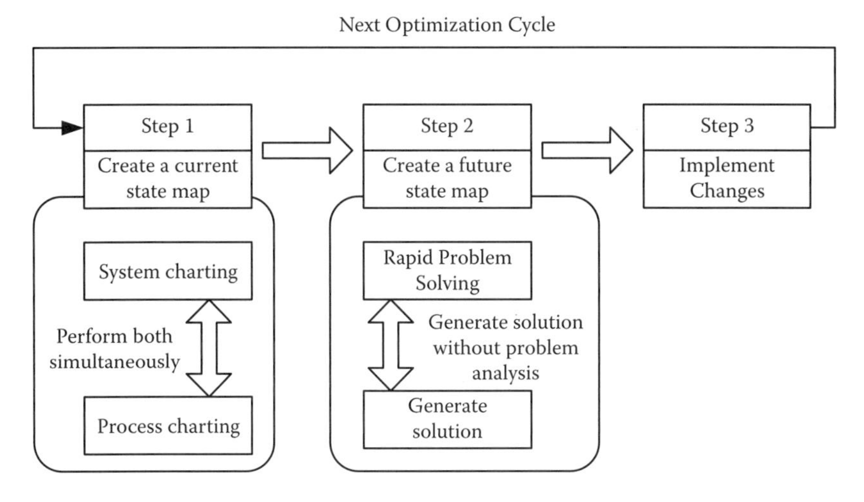

FIGURE 3.12
Optimization cycle of value stream mapping.

system optimization that encourages predictions about a future state to be made from analyzing the current state. Value stream mapping is iterative in nature, always searching and switching from an initial state to a better one. The technique combines process charting, system charting, and rapid problem solving. Figure 3.12 illustrates the optimization sequence.

The scientific method would propose that this form of IE has two critical flaws. The first problem relates to performing process analysis and system analysis simultaneously. Process study should always follow system study. If an industrial engineer works at the process level without understanding the problem on the system level, improvements could cause adverse affects at the system level. Also, attacking or solving problems at the process level do not necessarily improve system performance. In addition, information has a shelf life. The time spent in performing process analysis before system analysis is wasted because there is only one bottleneck at the system level. By the time the bottleneck is improved, all processes are likely to change, which makes all prior process analysis information obsolete. Process analysis is used to hone and refine issues at the micro level once the problem at the macro level has been identified.

The second problem relates to the idea of generating a solution (i.e., future state) without the scientific method. Decision analysis (the process of generating a solution or a hypothesis without problem analysis) can lead to biased decision making. Interestingly, Toyota does not create future state maps. If it did, it would violate its eight-step problem-solving methodology, meaning solutions would be developed without problem analysis. To develop a future state, Toyota would have to skip steps 1, 2, 3, and 4 in their 8-step problem-solving technique to reach step 5, which is develop a countermeasure. While this technique represents one of many in the field that has been adapted to help companies pursue Lean activities, Toyota follows the scientific method to encourage a thorough study of problems and their solutions. If employees do not have a structured process for developing and testing ideas, problems may return.

3.9 SUMMARY

This chapter proposed that the identity of TPS is a form of IE that can be applied voluntary by a wide range of employees within an organization. Toyota has emulated many of the practical techniques from scientific management in an effort to simplify workplace improvement. The strength of TPS is determined by the number of employees able to work toward its goals and objectives. Traditional Lean practices tend to take on a more contemporary version of IE, which makes its less applicable to team members, team leaders, senior management, and leadership. Organizations that do not have an IE department are able to replicate the IE function by establishing traditional Lean. In this context, kazien specialists and master black belts serve as industrial engineers who are largely responsible for the result and the outcome. The strength of these systems is dependent on the number of kaizen specialists and master black belts that exist inside the company. The competitive advantage of TPS is that all employees are trained to think like industrial engineers—specifically, techniques used in the scientific management era. Employees are skilled at applying IE techniques themselves independently of kaizen specialists or master black belts and are responsible for the results themselves.

REFERENCES

Ackoff, R., (1971) Toward a system of systems concepts, *Management Science*, Vol. 17, No. 11, 661–671.

Anderson, A. (1928) *Industrial Engineering and Factory Management*, Ronald Press, New York.

Anderson, V., and McLean, R. (1974) *Design of Experiments: A Realistic Approach*, Dekker, New York.

Askin, R., and Goldberg, J. (2002) *Design and Analysis of Lean Production Systems*, Wiley, New York.

Besser, T. (1991) Japanese management through the workers' perspective, dissertation, University of Kentucky, Lexington.

Bicheno, J. (2000) *The Lean Toolbox*, PICSIE Books, Buckingham, England.

Bowker, A., and Lieberman, G. (1971) Industrial statistics, in *Handbook of Industrial Engineering and Management*, eds. Ireson, W., and Grant, E. Englewood Cliffs, NJ: Prentice Hall.

Castro, M., Sebastian, R., and Quesada, J. (2010) A systems theory perspective of electronics in engineering education, IEEE EDUCON Education Engineering, the Future of Global Learning Engineering Education, April 14–16, Madrid, Spain.

Cho, F. (2001) *The Toyota Way*, Toyota Institute, Toyota Motor Corporation (unpublished).

Cooke, M. (1910) Academic and industrial efficiency. *Bulletin of the Carnegie Foundation for the Advancement of Teaching*, No. 5, 6.

Czitrom, V. (1999) One-factor-at-a-time versus designed experiments. *The American Statistician*, Vol. 53, No. 2, 126–131.

Daniel, C. (1973) One at a time plans. *Journal of the American Statistical Association*, Vol. 68, 353–360.

Friedman, M., and Savage, L. (1947) Planning experiments seeking maxima, in *Techniques of Statistical Analyses*, ed. Eisenhart, C., Hastay, M., and Wallis, W. New York: McGraw-Hill, pp. 365–372.

General Accounting Office (GAO) (1995) *Business Process Reengineering Assessment Guide*, GAO, Washington, DC.

Gigch, J. (1978) *Applied General Systems Theory*, Harper & Row, New York, NY.

Gilbreth, F. (1973) *Primer of Scientific Management*, Easton Hive, New York.

Going, C. (1911) *Principles of Industrial Engineering*, McGraw-Hill, London: Hill Publishing Co.

Gottlieb, B. (1971) The attitudes of organized labor toward industrial-engineering methods, in *Handbook of Industrial Engineering and Management*, ed. Ireson, W., and Grant, E. Englewood Cliffs, NJ: Prentice Hall.

Harrington, E., (1911) *The Twelve Principles of Efficiency*. The Engineering Magazine Company, New York.

Ireson, W., and Grant, E. (1971) *Handbook of Industrial Engineering and Management*, 2nd ed., Prentice Hall, Englewood Cliffs, NJ.

Jackson, T. (2006) *Hoshin Kanri for the Lean Enterprise*, Productivity Press, New York.

Jugulum, R., and Samuel, P. (2008) *Design for Lean Six Sigma: A Holistic Approach to Design and Innovation*, Wiley, Hoboken, NJ.

Kadota, T. (1982) Charting techniques, in *Handbook of Industrial Engineering*, ed. Salvendy, G. New York: Wiley.

Katzell, R., Bienstock, P., and Faerstein, P. (1977) *A Guide to Worker Productivity Experiments in the United States, 1971–1975*, New York University Press, New York.

Lee, C. (1996) Process reengineering at GTE: Milestones on journey not yet completed. *Strategy and Business*, No. 5, 58–67.

Lee, R., and Dale, B. (1998) Business process management: A review and evaluations. *Business Process Management Journal*, Vol. 4, No. 3, 214–225.

Manganelli, R., and Klein, M. (1994) *The Reengineering Handbook: A Step by Step Guide to Business Transformation*, Amacom, New York.

Mar, B. (1994) Systems engineering basics, *Journal of NCOSE*, Vol. 1, No. 1, 17–28.
Martens, H., and Allen, D. (1969) *Introduction to Systems Theory*, Merrill, Columbus, OH.
Mogensen, A. (1935a) From 59 steps to 22. *Factory Management and Maintenance*, Vol. 93, No. 3, 17–17.
Mogensen, A. (1935b) How to set up a program for motion economy. *Factory Management and Maintenance Plant Operation Library*, Vol. 93, No. 11.
Montgomery, D. (2005) *Design and Analysis of Experiments*, Wiley.
Morse, P. (1956) Statistic and operations research. *Operations Research*, February, p. 3.
Ohno, T. (1988a) *Toyota Production System: Beyond Large-Scale Production*, Productivity Press, New York.
Ohno, T. (1988b) *Workplace Management*, Productivity Press, New York.
Putt, A. (1978) *General systems theory applied to nursing*, Little Brown, Boston, MA.
Pyzdek, T., and Keller, P. (2009). *The Six Sigma Handbook*, 3rd ed., McGraw-Hill, New York.
Rayfield, J. (1964) Training a standards engineer. *Standards Engineering*, Vol. 14, No. 11, November, 1–8.
Salvendy, G. (2001) *Handbook of Industrial Engineering: Technology and Operations Management*. Wiley, New York.
Saunders, B. (1982) The industrial engineering profession, in *Handbook of Industrial Engineering*, ed. Salvendy, G. New York: Wiley.
Shingo, S. (2005) *A Study of the Toyota Production System: From an Industrial Engineering Viewpoint*, CRC Press, Taylor & Francis Group, New York.
Srinivasan, M. (2004) *Streamlined: 14 Principles for Building and Managing the Lean Supply Chain*, Thomson, Stamford, CT.
Staley, J., and Delloff, I. (1963) *Improving Individual Productivity*, American Management Association, New York.
Tanco, M., Ilzarbe, L., and Alvarez, M. (2009) Barriers faced by engineers when applying design of experiments. *TQM Journal*, Vol., 21, No. 6, 565–575.
Taylor, F. (1911) *Principles of Scientific Management*, Harper, New York.
Taylor, J., Dargan, T., and Wang, B. (2001) Process design and reengineering, in *Handbook of Industrial Engineering*, 3rd ed., ed. Salvendy, G. New York: Wiley.
Tippett, L. (1953) The ratio delay technique. *Time and Motion Study*, Vol. 2, No. 5, 10–19.
Wedgwood, I. (2007) *Lean Sigma: A Practitioners Guide*, Prentice Hall, Boston.
Zandin, K. (2001) *Maynard's Industrial Engineering Handbook*, 5th ed., McGraw-Hill, USA.

4

The Property of Goal Seeking in TPS, Part 2

4.1 PROBLEM SOLVING: THE IDENTITY OF THE TOYOTA PRODUCTION SYSTEM

Researchers, authors, and consultants are in a rush to understand the Toyota Production System (TPS); however, is studying Toyota as they are now really helpful? Much of what is known about TPS has been documented after Toyota came to America. Before Toyota had standardized work, kanban, and heijunka, what were they doing that led up to the system that is known today? Taiichi Ohno commented in his book that in 1937 he once heard that it took ten Japanese workers to do the work of one American worker (Ohno, 1988). He went on to say:

> Further more, the ratio was an average value. If we compared the automobile industry, one of America's most advance industries, the ratio would have been much more. But could an American really exert ten times more physical effort? Surely, Japanese people were wasting something. If we could eliminate waste, productivity should rise by a factor of ten. This idea marked the start of the present Toyota Production System.

Ohno knew that Toyota had to analyze its systems completely and thoroughly to improve efficiency. By improving efficiency, waste could be eliminated, and cost could be reduced. Ohno pursued efficiency improvements for each operator and for the plant as a whole. Ohno questioned everything that the company did to accomplish its goals and focused on the one place where a company makes and loses money: on the shop floor. Most important, Ohno's approach to raise efficiency was simple: distinguish

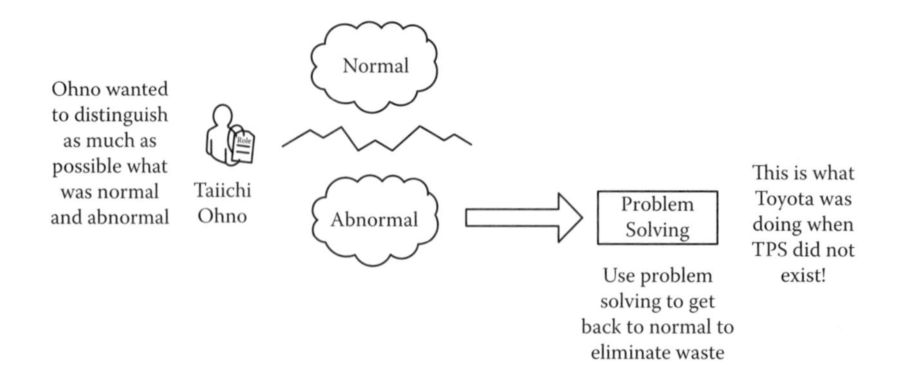

FIGURE 4.1
Ohno's path to TPS when TPS did not exist.

what is normal (what should be happening) and what is abnormal (what should not be happening) in a process. In most companies, there is no distinction between normal and abnormal. When errors are made in the workplace, it is not known until it is too late. Ohno believed that problems should be visible in the workplace so that problems can be identified when they occur. Ohno made every attempt to uncover the reason why problems happened and influenced many others like him to be diligent in their thinking when workplace conditions were abnormal. Ohno experimented, tested ideas, and analyzed operations from every angle. Ohno used the scientific method to improve the way people make decisions when correcting processes (Ohno, 1988) (Figure 4.1).

4.1.1 Problem Solving versus Kaizen (Improvement)

Many Lean practitioners misinterpret problem solving and kaizen (i.e., improvement). Problem solving is what Ohno wanted to do to bring processes from abnormal back to normal (Figure 4.2). Ohno wanted to use a scientific method (i.e., way of thinking) to help employees investigate and correct shop floor processes. Kaizen or continuous improvement occurs when a stable process is improved. In this context, kaizen means to raise the standard, whereas problem solving implies to achieve the standard (Figure 4.3). Ohno started TPS by establishing standards so that abnormal processes could be identified. Ohno used the scientific method to work toward solutions to correct abnormal processes.

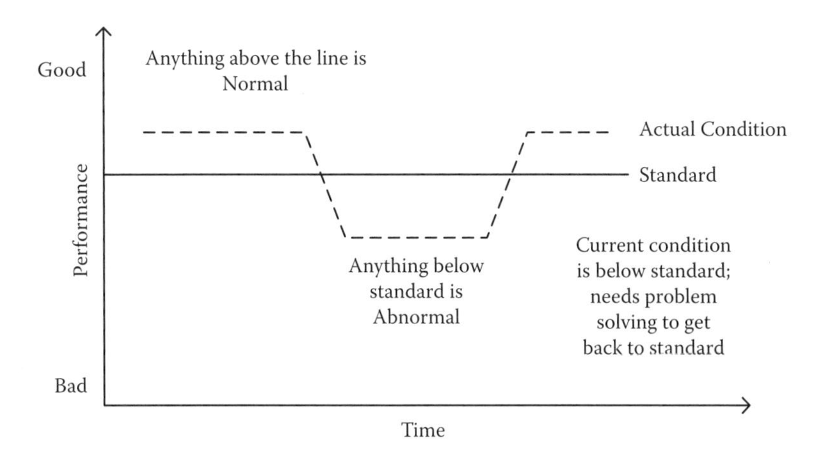

FIGURE 4.2
Example of problem solving: Actual condition is below standard.

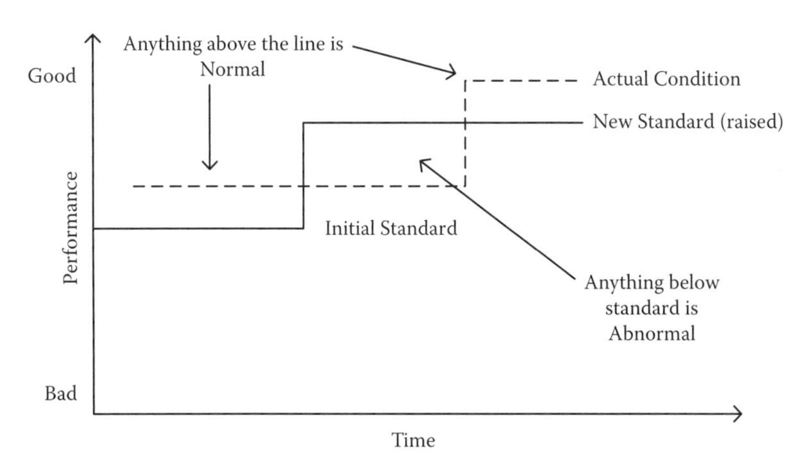

FIGURE 4.3
Example of kaizen or continuous improvement: Actual condition and standard are raised.

4.2 STRUCTURED PROBLEM SOLVING

Organizational excellence is dependent on being effective at solving problems. It has been shown by logic theorists and behavioral scientists that better problem solvers seem to have a sense of direction (Maier, 1930), are analytical in approach (Bloom and Broder, 1950), and are methodical (Fattu et al, 1954). A technique to improve problem solving has been the use of strategies, also known as problem-solving methodologies or processes (Bruner et al., 1956; Moore, 1995). Problem-solving methods (initially created and suggested by psychologists and management scientists) are an attempt to improve thinking, reasoning, and insight. One of the advantages of using structured methodologies is that they resemble heuristics, which are applicable to many different types of problems. While success is not guaranteed, such methods do increase the probability of solution generation (Thompson, 1991).

4.3 TOYOTA'S EIGHT-STEP PROBLEM-SOLVING METHODOLOGY

One of the most popular problem-solving methods in industry is Toyota's eight-step problem-solving process (Figure 4.4), sometimes referred to as A3 problem solving. A3 represents the size of a sheet of paper (11 × 17 inches), which before the widespread use of e-mail was the largest size of paper that could be faxed. The A3 process has been widely benchmarked, studied, and identified as a contributing factor in Toyota's success.

Outside observations of Toyota's process have generated much speculation and inquiry on how problem solving is applied and how it is embraced companywide. In 2004, the company renamed its eight-step process as the Toyota Business Practices (TBP). TBP represents a more generalized view of the eight-step process, applicable to all Toyota business units, including those outside the automotive industry, such as the housing division, electronic commerce, and marine and biotechnology groups.

Toyota's soft-sided philosophies have also shaped TBP and can be observed in their "Drive and Dedication" points, which are said to describe the company's heart and spirit that every team member should model when completing each of the eight steps. These motivational guidelines establish the accepted behaviors when carrying out eight-step problem solving. Research has shown that attitude and personal openness to problem solving are among the most powerful factors in problem-solving effectiveness (Heppner et al., 2004). The end result is highly structured behavior and a context that is shared for all organizational members when solving workplace problems (Figure 4.5).

Unfortunately, little is known about how Toyota's method is designed or what makes it unique compared to other classical problem-solving

Step 1 Clarify the problem	1. Clarify the "ultimate goal" of your responsibilities and work 2. Clarify the "ideal situation" of your work 3. Clarify the "current situation" of your work 4. Visualize the gap between current situation and the ideal situation
Step 2 Breakdown the problem	1. Break down the problem 2. Specify the point of cause by checking the process through Genchi Genbutsu
Step 3 Target setting	1. Make the commitment 2. Set measurable, concrete, and challenging targets
Step 4 Root cause analysis	1. Consider causes by imagining the actual situation where the problem occurs 2. Based on facts gathered through Genchi Genbutsu, keep asking, "Why?" 3. Specify the root cause
Step 5 Develop countermeasures	1. Develop as many potential countermeasures as possible 2. Narrow down the countermeasures to the most practical and effective 3. Build consensus with others 4. Create a clear and detailed action plan
Step 6 See countermeasures through	1. Quickly and as a team, implement countermeasures 2. Share progress by following the correct reporting, informing, and consulting 3. Never give up, and proceed to the next step quickly
Step 7 Monitor both results and processes	1. Evaluate the overall results and the processes used, then share the evaluation with involved members 2. Evaluate from the three key viewpoints: customers, stakeholders, and your own
Step 8 Standardize successful processes	1. Evaluate the overall results and the processes used, then share the evaluation with involved members 2. Evaluate from the three key viewpoints: customers, stakeholders, and your own

FIGURE 4.4
Toyota's eight-step problem-solving methodology.

methodologies. It is speculated that Toyota's problem-solving method may employ specific thinking styles that are only valid in solving well-structured problems or problems that exist in highly repetitive manufacturing environments. Some researchers propose that Deming's dynamic scientific inquiry model (Plan-Do-Check-Act) is what makes Toyota's method successful. Books written about Toyota's kaizen process suggest that Toyota promotes creative thinking in problem solving, which encourages its employees to practice divergent thinking. It could also be debated that the strength of their methodology is the strict logical

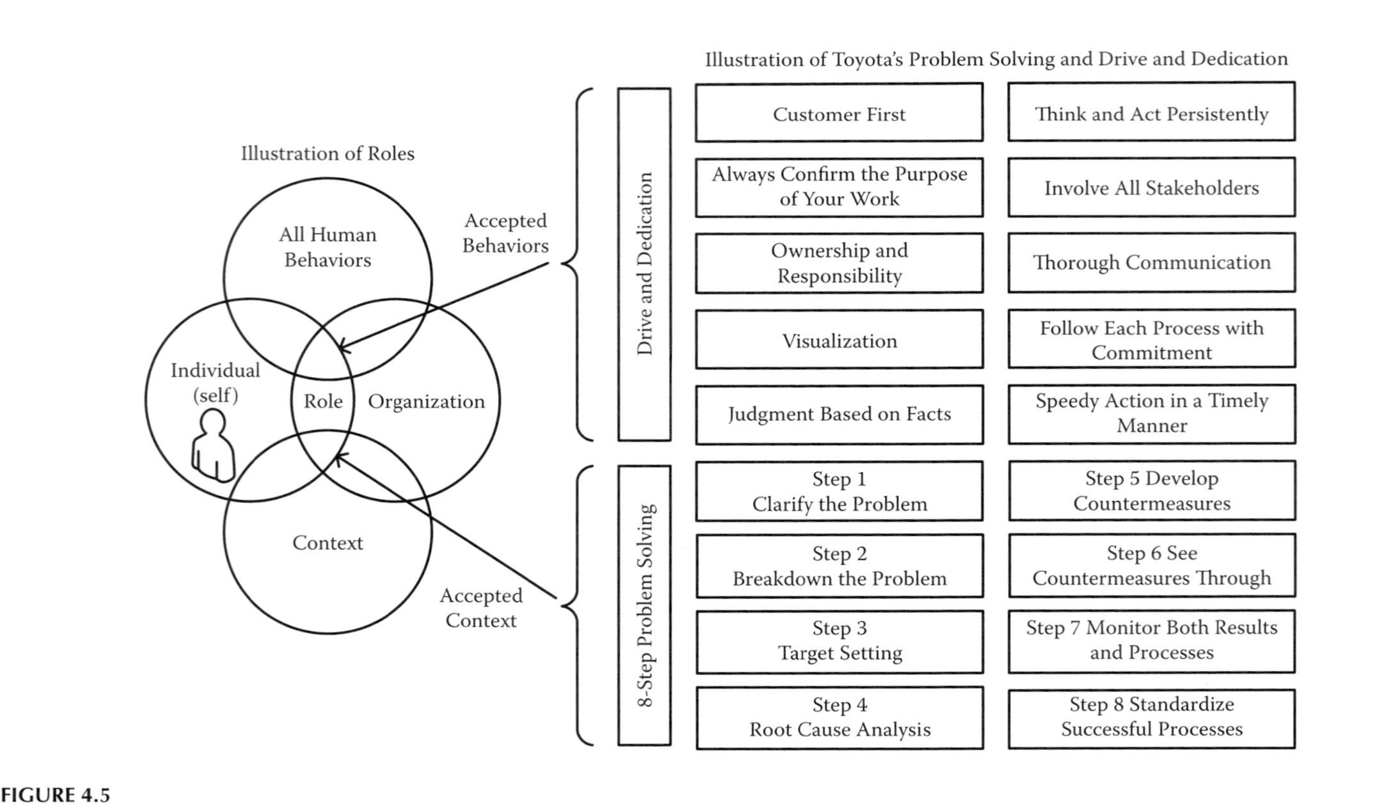

FIGURE 4.5
Accepted behaviors and context for Toyota's problem-solving process.

compliance of completing each step thoroughly without the distraction of bouncing around from idea to idea.

Most of the work in understanding problem solving has shown that no single method can fully describe the complexity of human behavior and thought. Therefore, various forms of problem-solving methods have been developed to serve the purpose of the organization. The wide varieties of methods suggest that different approaches could be used in different environments to serve different objectives. The next sections apply the theory of human problem solving to understand which characteristics of Toyota's method make it successful and efficient. This section attempts to unravel and break down Toyota's specific thinking styles that have been successful in guiding employees when solving problems.

4.4 HUMAN PROBLEM-SOLVING BACKGROUND

Human problem solving has been well studied and researched to understand the basic cognitive process of human thought (Robinson, 1987; Sternberg, 1997; Heppner et al., 2004). Problem solving has evolved to touch on many aspects, including scientific induction (Dewey, 1910); creative thinking (Wallas, 1926); invention (Rossman, 1931); general management (Kepner and Tregoe, 1965); productive thinking (Wertheimer, 1959); decision making (Simon, 1977); learning (Gagne and Brown, 1961); artificial intelligence (Schank, 1973); cybernetics (Wiener, 1948); and information-processing theory (Anderson et al., 1996). A common theme in problem-solving frameworks is that they all provide structured thinking for processing information that is essential in learning and the transfer of knowledge.

A significant factor in structured thinking is the amount of guidance a methodology should use to encourage productive thinking. Productive thinking is the process of learning and leading up to the solution by distinguishing what is relevant and what is not (Figure 4.6). Each problem has a unique space, known as the problem domain, in which the user works to reach the solution. The problem domain represents all the possible outcomes and information that is used to reach the solution. A person is said to have domain knowledge if the person has information that exists in the problem domain. Experts who have domain knowledge can eliminate or combine information that is needed to reach the solution.

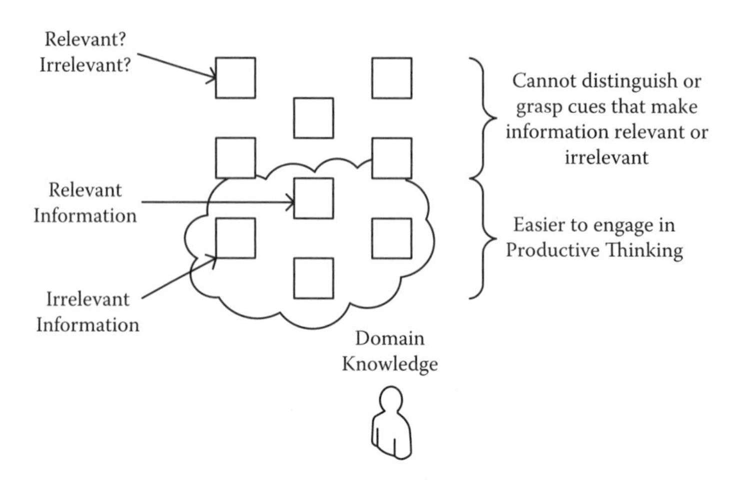

FIGURE 4.6
Representation of productive thinking.

Some problem-solving methodologies promote "discovery learning," which is a type of learning that appears to gain its effectiveness from the fact that it requires the individual to reinstate (and in a sense to practice) the concepts that will be used in solving new problems in the future (Katona, 1940). Discovery learning increases knowledge transfer because it allows the user to interact completely with the problem domain with little guidance. Discovery-based learning encourages users to learn from their mistakes while testing the relevancy of information as they work toward the solution. The main advantage of discovery learning is that when the solution is discovered by the user, the learning has a greater chance of being transferred to the user.

The drawback with discovery learning is that it is time consuming to test all possible scenarios in the problem space. Since none of the choices has been eliminated, the time to find the solution is largely dependent on the domain knowledge of the user. Discovery-based learning works best when it is used in team problem-solving environments (Figure 4.7). When the skill level of each member is the same, members can work cooperatively with each other to divide and conquer the problem domain. Organizations that promote diversity in thinking may not realize until it is too late that discovery-based learning is not intended when members have different levels of domain knowledge. In this case, expert members will quickly discount known information, which could create an undesirable participation dynamic within the team. Again, discovery-based learning should be used when there is a need to learn the entire problem domain or in teams that share similar domain knowledge.

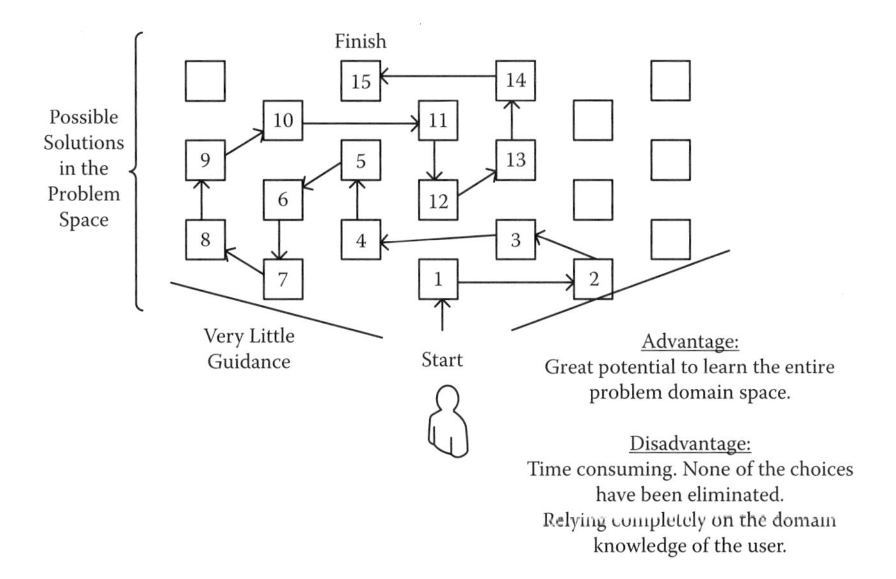

FIGURE 4.7
Representation of discovery-based learning.

Guided discovery learning (Figure 4.8) or inquiry-based learning is said to be more effective compared to discovery-based learning because it can lead to the problem solution by speeding up decision making by narrowing searches for solutions (Gagne and Brown, 1961). Thus, methodologies can discourage guessing and modify the learner's discovery so that the probability of errors is decreased, greatly improving the effectiveness of problem solving. In this way, knowledge transfer is still good since the discovery of the solution is accomplished by the user. The disadvantage of guided discovery learning is that the degree of narrowing is difficult to know for the user and type of problem. Expert problem solvers will generally prefer less narrowing and more flexibility, while novice problem solvers will need more help and guidance.

4.5 LITERATURE REVIEW OF PROBLEM-SOLVING METHODOLOGIES

Table 4.1 summarizes various problem-solving methodologies and frameworks used in structured thinking. Despite the wide variety of problem-solving frameworks, several major themes exist: scientific inquiry,

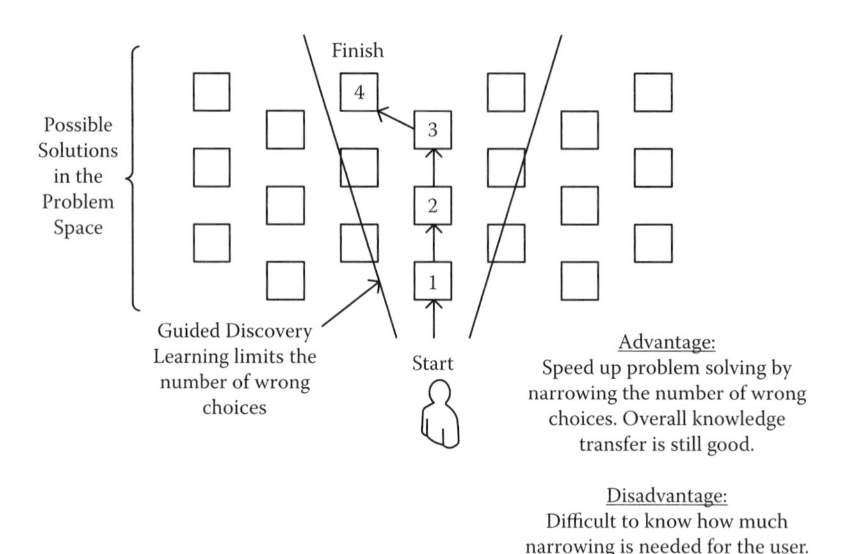

FIGURE 4.8
Representation of guided discovery learning.

creative thinking, heuristics and instructional theory. *Scientific inquiry,* as originally proposed by Dewey, has become extremely influential in structured thinking. Scientific inquiry emphasizes convergent thinking, which involves aiming for a single correct solution to a problem by generating and testing a hypothesis. Convergent thinking emphasizes speed, accuracy, and logic. Another unique thinking style is *creative thinking.* Creative thinking brings about divergent thinking, which involves the generation of multiple answers to a problem set. Researchers have connected divergent thinking as a type of flexible thinking that draws on ideas from across disciplines and fields of inquiry (Guilford,1967). *Heuristics* is concerned with the mechanics of the decision-making process and compares consequences of decisions. Various heuristics are used in generating solutions, such as induction, generalization, subgoal development, pattern recognition, reduction, specialization, working backward, diagrammatic reasoning, and extension (Simon, 1977; Polya, 1945). *Instructional theory* is the study of facilitation of human learning through instruction, often labeled as guided discovery (Gagne and Brown, 1961). It is a process not simply of applying learned rules but also of yielding new learning.

More general problem-solving frameworks are shown in Table 4.2. These frameworks vary in emphasis yet share many similarities with methodologies shown in Table 4.1. Table 4.3 summarizes various problem-solving

TABLE 4.1

Review of Human Problem-Solving Theoretical Frameworks

Scientific Inquiry			Creative Thinking			Heuristics		Two Phase	Instructional/ Educational
Dewey (1910)	**Shewhart (1939)**	**Deming (1986)**	**Wallas (1926)**	**Rossman (1931)**	**Osborn (1953)**	**Polya (1945)**	**Simon (1977)**	**Johnson (1962)**	**Gagne and Brown (1961)**
Suggestion	Specification	Plan	Preparation	Observation of the need	Orientation	Understand the problem	Identify all the possible alternatives	Preparation	Reception of the stimulus situation
Intellectualization	Production	Do	Incubation	Analysis of the need	Preparation	Devise a plan	Determine all the possible consequences of these alternatives	Solution	Concept formation or concept invention
Hypothesis	Inspection	Check	Intimation	A survey of all available information	Analysis	Carry out the plan	Evaluate all the possible consequences		Determining course of action
Reasoning		Act	Illumination or insight	A formulation of all objective solutions	Ideation	Look back (check results and problem)			Decision making
Verification			Verification	A critical analysis of the solutions	Incubation				Verification
				The invention (idea)	Illumination				
				Experimentation and final solution	Synthesis				
					Evaluation				

TABLE 4.2

Review of Human Problem-Solving Theoretical Frameworks

General Problem Solving								
Kepner and Tregoe (1965)	**Davis (1969)**	**Gregory (1967)**	**Hayes (1981)**	**Koberg and Bagnall (1981)**	**Lyles (1982)**	**Elias and David (1983)**	**Tyre (1989)**	**Hurson (2007)**
Situation appraisal	Problem statement	Deciding on objective	Find the problem	Accept the situation	Define the problem	Assess the situation	Problem definition	What's going on?
Problem analysis	Problem definition	Analyzing problems	Represent the problem	Analyze	Generate countermeasure	Identify the problem	Problem analysis	What's success?
Decision analysis	Search and formulate a hypotheses	Gathering data	Plan the solution	Define (the main issues and goals)	Select countermeasure	Define the goal	Generation and selection of solutions	What's the question?
Potential problem/ opportunity analysis	Verify the solution	Observing data	Carry out the plan	Idea (to generate options)	Implement	Analyze the forces	Testing and evaluation of solutions	Generate answers

	Inducting	Evaluate the solution	Select (to choose among options)	Monitor results	Generate alternative strategies	Routinization: development of new routines	Forge the solution
	Planning	Consolidate the gains (learn)	Implement		Select the best strategy		Align resources
	Prechecking		Evaluate (to review and plan again)		Forecast potential problems		
	Activating plans				Test the strategy		
	Evaluating				Write a work plan		
					Implement and evaluate the plan		

TABLE 4.3

Review of Human Problem-Solving Frameworks: Automotive Industry Practices

Toyota	Honda	General Motors	Ford	Daimler Chrysler	DMAIC	AIAG Standard: THE-4
8-Step/A3 Problem-Solving Process	**5 Principles for Problem-Solving Sheet**	**5-Phase Action Plan**	**Global 8D Report**	**7-Step Corrective Action Form**	**Six Sigma Methodology**	**7-Step Problem-Solving Method**
Establish goal	Problem description, discovery, definition, and details	Identify issue (definition and history)	Symptom, emergency response action	Problem identification and description	Define	Identify the problem
Break down the problem	Identify root cause(s) and selection/ justification	Analyze (cause and root cause)	Problem description and statement	Interim action, containment	Measure	Determine and rank possible causes
Set target	Countermeasure(s) (initial and long-term schedule)	Plan (quick fix and corrective action)	Interim containment action(s)	Root cause analysis	Analyze	Take short-term action

Root cause	Countermeasure effectiveness (actual evidence of containment actions or countermeasure effectiveness)	Implement (current status of activities)	Root cause(s) and verification	Permanent corrective action	Improve	Gather data and design test
Develop countermeasure	Feedback/forward (vertical, horizontal, and to discovering associate)	Evaluate (verification and validation)	Chosen permanent corrective action and verification	Verification of corrective action plan	Control	Conduct test, analyze data, identify root cause, Select solution
Select and implement countermeasure			Implement permanent corrective actions	Controls and prevention		Plan and implement permanent solution
Check and monitor			Prevent actions and systemic prevention recommendations	Verify corrective action resolves issue and lessons learned		Measure and evaluate
Standardize			Team and individual recognition			Recognize the team

methodologies used in the automotive industry, including Toyota's eight-step process as well as the popularized Six Sigma define, measure, analyze, improve, control (DMAIC) process. The main difference in these techniques compared to previous models is an emphasis on solving problems that relate to defects.

4.6 LIMITATIONS WITH PROBLEM-SOLVING METHODOLOGY RESEARCH

Not all theorists share a positive view of structured thinking or methodologies in general that lead to productive thinking (Wertheimer, 1959). Theories that describe complex human behavior in general remain very much an approximation by hypothesizing extreme neatness of discrete behavior (Newell and Simon, 1972). There is also work to suggest that structured thinking does not follow natural cognitive patterns, which can prevent creativity, innovation, and insight (Nakamura, 1955). Guided thinking has the risk of preventing automatic control shifts between higher- and lower-level thinking that is natural in thinking processes (Scandura, 1977). Last, there is research that implies problem solving does not have to follow precise steps (Buswell, 1956) or that human problem solving is a linear process (Robertshaw et al., 1978).

Unfortunately, problem-solving research has not progressed as hoped. There have been problems reflecting comprehensive human behavior over a wide domain, between subjects and the environment (Scandura, 1977). There is also the issue that not all aspects of problem solving can be considered since the total human system varies in dimensions relating to task, performance learning, and individual differences (Newell and Simon, 1972). In addition, experiments that study the factors in problem solving are rarely useful because of the use of unanalyzed and nondimensionalized variables (Duncan, 1959; Sachse, 1981).

Surprisingly, with all the different types of problem-solving methodologies, it is still largely unknown which problem-solving methods lead to successful problem solving and which features and characteristics of these problem-solving methods lead to productive thinking (Wertheimer, 1959; Conner, 1999). Much of the research and literature is fixed on finding *one effective process for solving all problems rather than understanding their differences*, which could provide insight into their design.

4.7 PROBLEM-SOLVING THEORY AND CORE THINKING SKILLS

Core thinking skills make up the broader cognitive processes that provide a way to organize specific skills problem solvers must learn to become good thinkers. A list of core thinking skills, compiled from the literature, is shown in Table 4.4 for each major cognitive process. These skills can be further broken down into specific practices that indicate the manner in which the skill is to be performed. Table 4.4 lists the specific practices corresponding to each core skill. In most problem-solving frameworks, the discretion of applying the core thinking skills through the various

TABLE 4.4

Core Thinking Skills

#	Core Skill	Description (Examples)
1	Emergency skills	a. Interim action, quick fixes, containment
2	Acceptance skills	a. Team or individual involvement
3	Sensing skills	a. Problem awareness and impact (i.e., symptom)
4	Focusing skills	a. Goals, targets, objectives
5	Information-gathering skills	a. Data collection, etc.
6	Organizing skills	a. Defining and breaking down the problem
7	Analyzing skills (creative or logical)	a. Creative or logical skills
		b. Identify relationships
		c. Root cause analysis, problem analysis
8	Generating skills	a. Generate solution or plan
		b. Check the plan
		c. Choosing the best solution
9	Executing skills	a. Implement solution or countermeasure plan
10	Evaluating/decision skills	a. Verification
		b. Pick the best solution
		c. Validation (second verification step)
		d. Prevention, controls in countermeasure
11	Recording skills	a. Standardization
		b. Communication across groups
12	Feedback skills	a. Iterative process or cycle of problem-solving method
13	Team recognition skills	a. Team or individual recognition

practices is usually left to the user, which can have a drastic impact on the effectiveness of the problem-solving process. The list of thinking skills shown in Table 4.4 is not all inclusive; they have been extended to include other areas outside education theory proposed by earlier work (Marzano et al., 1988).

4.8 CRITERIA FOR EVALUATING EFFECTIVE COGNITIVE PROCESSES

Goal clarity is described as the interpretation of the gap between the ideal state and the actual condition or problem situation. Specific goals lead to better performance compared to vague or unclear goals (Locke, 1991). Goal clarity can be used to enhance quality and increase intensity (Bandura, 1993). However, well-defined goals can lead to some negative effects in which the user is prompt to use means analysis instead of spending adequate time learning and searching to gather information (Sweller and Levine, 1982). In this case, attention is spent on what appears to be goal-relevant information yet often involves nonrelevant aspects of the problem because of the heavy focus on the goal. A general goal achieves better results when more complex problems are presented. Goal clarity relates to two types of active problem solving: means–ends analysis and working forward.

Means–ends analysis (Figure 4.9) is a form of active problem solving associated with goal clarity. In this strategy, the solver compares the distance between the goal and the current state and then tries to find a way to reduce the distance. Well-structured problems can generally be specified in advance (Greeno, 1976), which allows means–ends analysis to be an effective approach. Well-structured problems imply clear goals and that information is available.

Working forward (Figure 4.10) involves taking what is given and using some heuristic to guide selection of operations without reference to a goal. For ill-structured problems, the solution state remains unknown until it has been achieved. Ill-structured problems require initial searches for information without knowing exactly how it will be used. Such a process is called working forward. Novices tend to use means–ends analysis, while experts tend to work forward, given the information that is available.

Productive thinking is the ability to distinguish relevancy of facts and information (Wertheimer, 1959). It is the ability to sense, gather, organize,

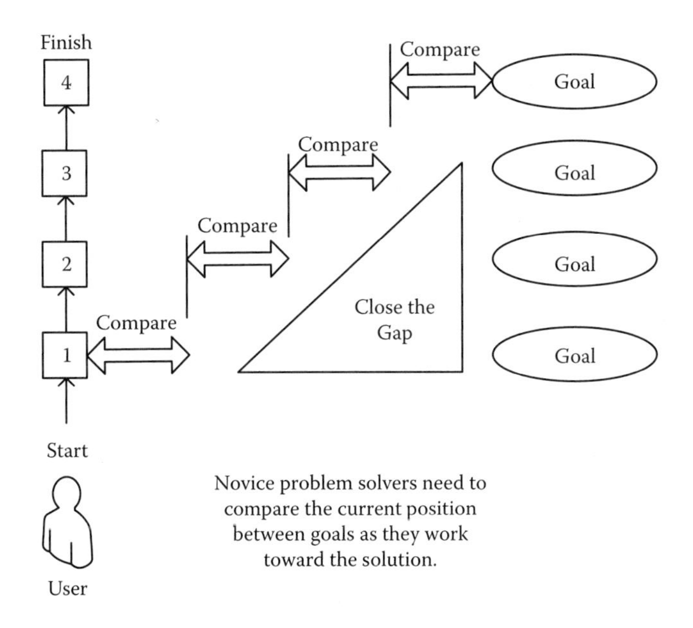

FIGURE 4.9
Means–ends analysis.

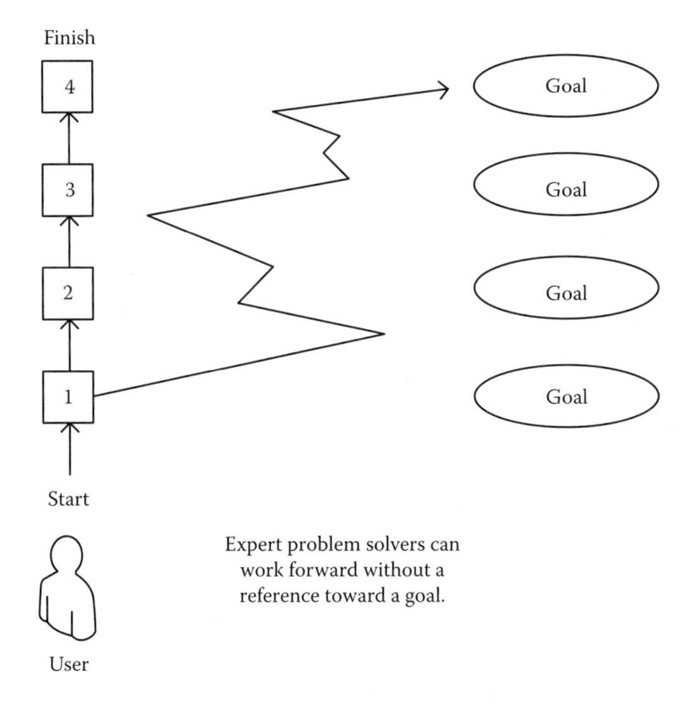

FIGURE 4.10
Working forward.

and evaluate information to make it usable and relevant in the later stages of problem solving. Productive thinking should not be confused with order and logic that accompanies the processing of information (Kepner and Tregoe, 1965). Traditional logic aspects in methodologies have the concern of not finding a solution but only focusing on the correctness of each step (Rogers, 1967). Productive thinking is not only about a logical process that solves an actual problem but also about discovering, envisaging, going into deeper questions. Novice and expert problem solvers are distinguished by their ability to engage in productive thinking.

Novice problem solvers tend to be distracted by surface structure and often exhibit difficulty in distinguishing relevant information in and outside the problem space. Although attitude of the problem solver is key in facing problems freely and openly, novice problem solvers have the tendency to go back to their old habits as a last resort when there is a lack of reasonable possibilities.

Expert problem solvers seem to comprehend the deeper structures and underlying information. They tend to grasp the inner relations between operations and their results and develop a level of understanding that considers the structural completeness of the question. Expert problem solvers can make cues distinctive and relevant and develop connections or insights of inner structural relationships (Wertheimer, 1959).

Problem analysis and decision making are two primary phases in problem solving that relate to problem finding and problem fixing. In the initial phase, subjects explore parameters of the problem space, whereas in the final phase subjects discover the solution (Kotovsky et al., 1985). Static or linear problem solving is when the user does not return to the problem space (Figure 4.11). Methodologies that are linear in nature create solutions with the information that is provided in the initial phase. In this context, no change occurs in understanding the problem space from beginning to the end. It is assumed that all relevant information has been obtained from the initial phase. This type of problem-solving methodology emphasizes pushing forward with the best information that is known at the time.

Some believe that problem analysis and decision making should be combined to make problem solving more effective since both aspects occur together dynamically and not in isolation (Cooper, 1961). This theory implies that subjects can constantly adjust the problem space by returning to earlier phases of problem analysis as new information is acquired and processed (Figure 4.12). Interestingly, the majority of problem-solving

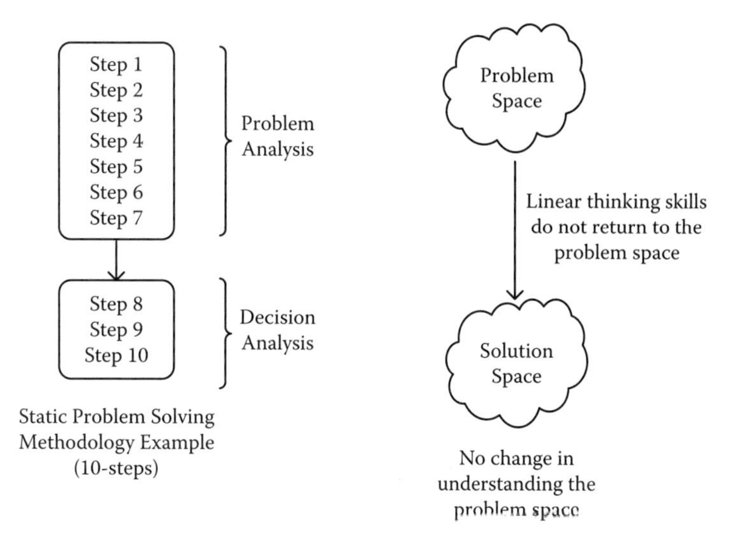

FIGURE 4.11
Static or linear problem-solving methodology.

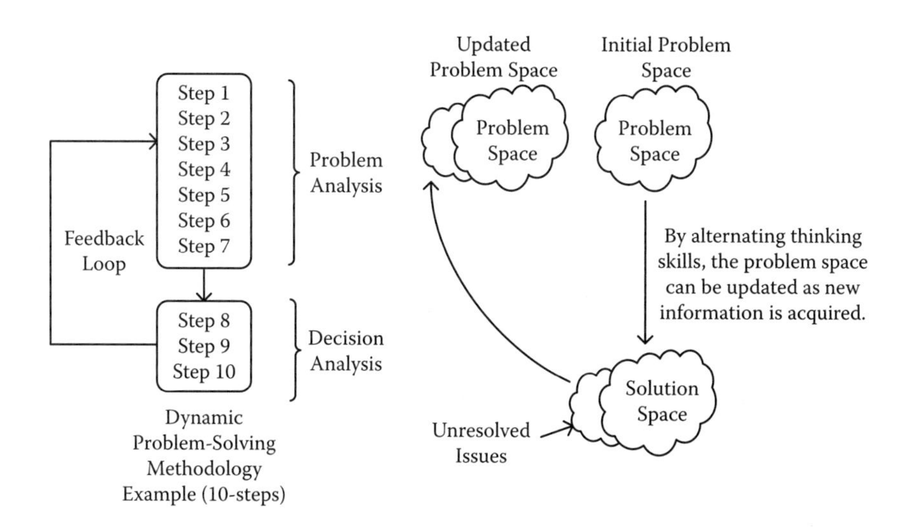

FIGURE 4.12
Dynamic problem-solving methodology.

frameworks listed in Tables 4.1, 4.2, and 4.3 organize these two phases separately into preparation and solution phases (Johnson, 1962).

4.9 APPLYING HUMAN PROBLEM-SOLVING THEORY TO TOYOTA'S EIGHT-STEP PROCESS

Figure 4.13 compares the types of thinking skills used in Toyota's eight-step process to various other well-known problem-solving methodologies. Each problem-solving methodology emphasizes dominant traits and characteristics that make it unique and distinguishable. While most practitioners view that all problem-solving methodologies are essentially the same (problem finding and problem fixing), they are not. Each methodology has a certain degree of narrowing to help users along the way to meet desired outcomes. As thinking skills are combined and prioritized, each methodology contains its own style for guiding users. In Toyota's case,

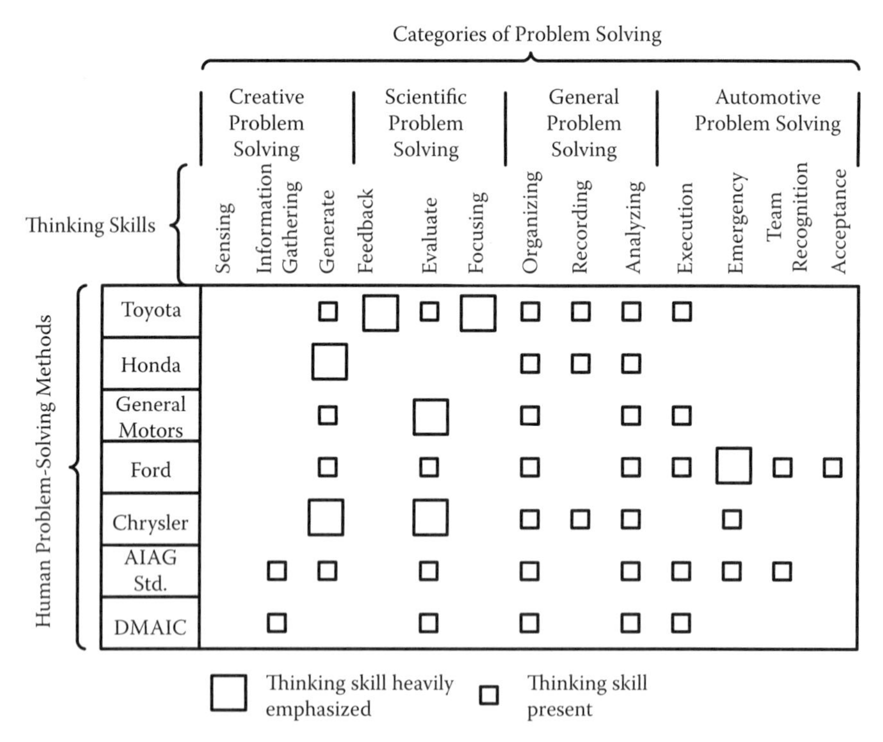

FIGURE 4.13
Summary of thinking skills and human problem-solving methods.

focusing and feedback skills are heavily emphasized; these relate to goal clarity and productive thinking, respectively. The goal in this section is to understand how these two thinking skills are significant in applying TPS.

4.9.1 Goal Clarity in A3 Thinking

One of the characteristics that make Toyota's problem-solving process unique is a strong emphasis on goal clarity. The first step in the A3 process is to define the ultimate goal, the standard, and the current situation. The ultimate goal is the overall reason why the department or business unit exists. Members of each group need to understand that the work they perform ultimately serves a greater purpose. Toyota utilizes the ultimate goal (like means–ends analysis) to help users keep their eyes on the big picture. It is often easy to lose focus throughout problem solving, which is why Toyota wants members to reference the overall objective of the group, department, or business unit.

Another focusing skill that is used in the A3 process is step 2, problem breakdown. In this step, the large and ambiguous problem space is defined into a well-structured problem space. If the problem space is left unstructured, even partially, there is a chance that the derived solution will not match the true solution space. When the problem space is not structured, there is a greater likelihood that the problem space and solution will not be congruent.

Step 2 is one of the most pivotal and demanding steps in the A3 process. The purpose of step 2 is to define the problem by answering the what, when, where, and who of the problem domain. Most problem solvers want to rush to the solution because they feel that they understand why the problem has occurred. When this happens, problems return because the problem space was never defined or structured properly. Toyota views that if employees can structure the problem space, the accuracy of the problem-solving process will increase.

A technique that Toyota uses to define the problem space into a well-structured problem is referred to as *problem breakdown*. In this step, the large and ambiguous problem is broken down into many smaller defined problems to be solved. This extreme structuring of the problem domain reduces the complexity of the problem-solving process by more easily matching each smaller problem to a solution. The idea is that accuracy is improved because solutions can more easily be confirmed, tested, and evaluated on a small scale rather than a large one.

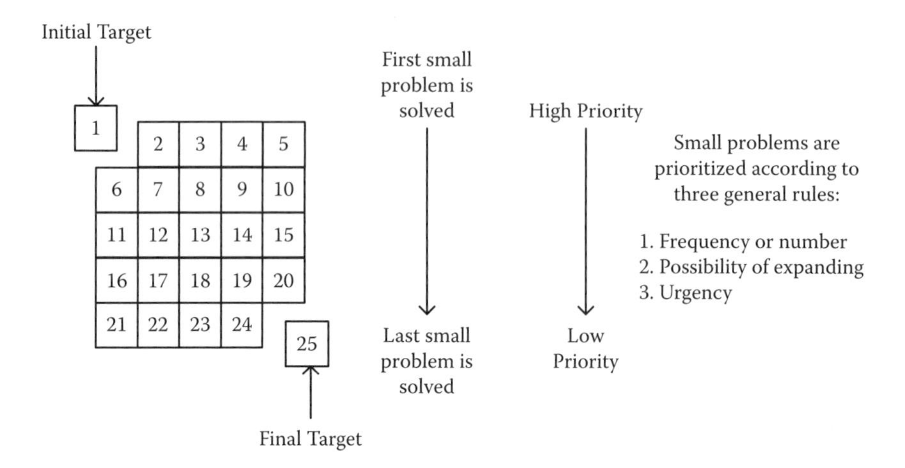

FIGURE 4.14
Step 3 target setting: prioritizing the problem domain.

Another way to understand how step 2 can improve the accuracy of problem solving is the degree to which the problem is broken down or structured. Ideally, when the solution is applied to the system, the response or effect of the solution should be noticeable and direct. When the response of the solution cannot be noticed directly, much like turning on a switch, one probable cause is that the problem domain has not been broken down. When problems are well structured, solutions can be more easily tested and verified because there is a direct response to the system.

The last focusing skill in the A3 process is step 3, target setting (Figure 4.14). Step 3 decides which small problem to work first. Multiple small problems are not allowed to be solved simultaneously because of matching issues between the problem space and solution explained previously. In step 3, the entire problem domain is prioritized among each of the small problems. Toyota prioritizes problems according to three general rules: the frequency or occurrence of the problem, the possibility that the problem can expand, and the urgency of the problem. Step 3 helps users to narrow in on a particular aspect of the problem domain that is the most important to overcome first.

4.9.2 Productive Thinking in A3 Thinking

One of the ways that the A3 process encourages productive thinking is how the problem space is updated during problem and decision analysis. The A3 process devotes equal attention to problem and decision analysis.

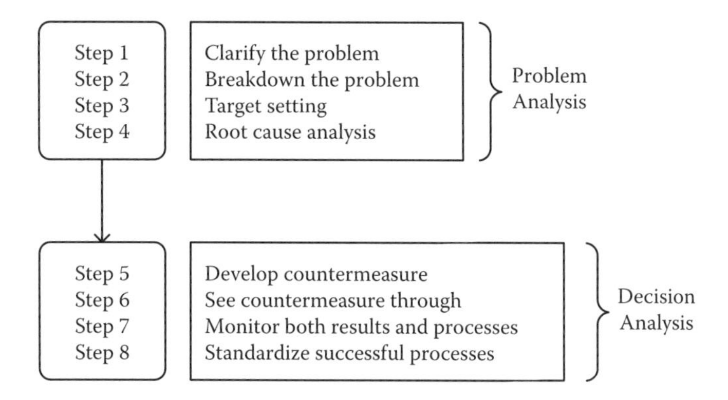

FIGURE 4.15
The two basic phases of the A3 process.

The first four steps are problem analysis and the last four are concerned with decision analysis. All problem-solving methodologies exhibit these two basic phases (Figure 4.15), yet how they interact greatly influences the manner of productive thinking.

The algorithm that is used in the A3 process to encourage productive thinking is shown in Figure 4.16. Once the problem has been clarified in step 1, the problem is broken down into many smaller problems in step 2. Next, problems are prioritized, and one is chosen in step 3 to solve. The first small problem selected will be named iteration 1 of the algorithm. The root cause is completed in step 4, followed by developing several possible countermeasures in step 5. Next, the most likely countermeasure is selected in step 6 and tested in step 7. If the countermeasure does not work, the algorithm directs the user back to step 5 to select another countermeasure to test. This loop continues until the problem space matches the solution space for the small problem. If the countermeasure is successful, the solution is standardized in step 8, and the next iteration begins at step 2, where the next prioritized problem is selected at step 3. These iterations continue until there are no prioritized problems left in the problem space to match to a solution space.

Toyota's A3 process encourages productive thinking because the problem space is revisited throughout the problem-solving process. As the problem space is updated and refined, it becomes easier to distinguish relevant and irrelevant information. This dynamic characteristic in the A3 process is crucial in productive thinking. If the A3 algorithm did not update the problem space or give the users time to rebalance known information, there would be much more uncertainty in developing solutions to

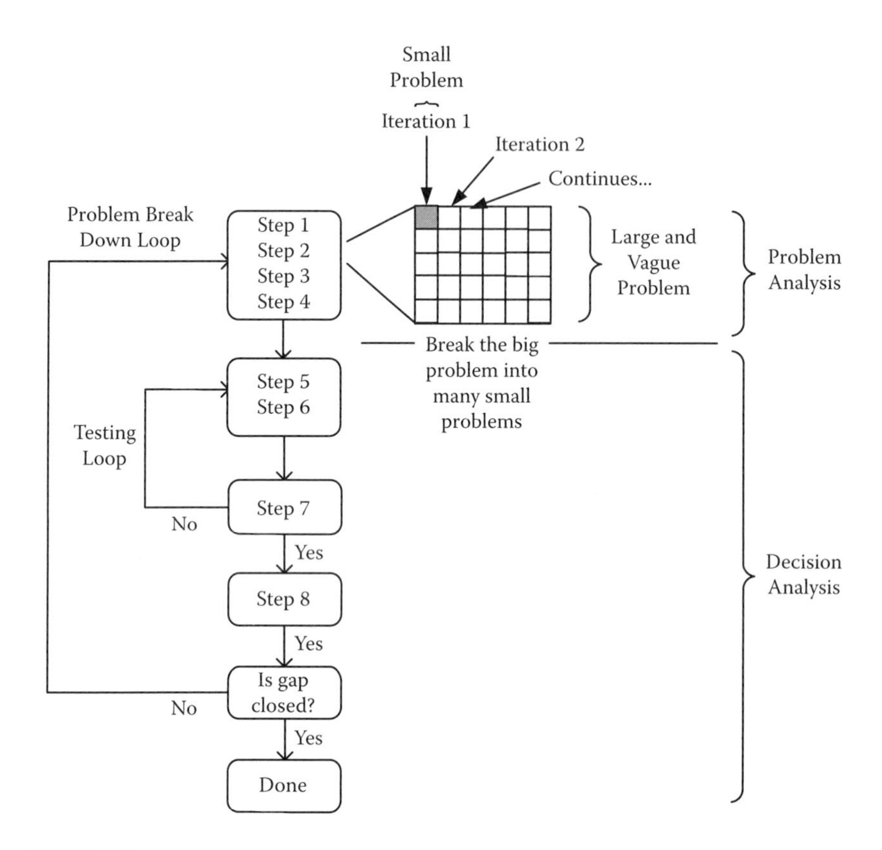

FIGURE 4.16
The A3 algorithm for productive thinking.

the problem space. Step 8 is critical in moving unknown information to known information because known variables are held constant so that the complexity of testing variables can be simplified.

Another characteristic of the A3 algorithm that helps users to engage in productive thinking is the testing loop between steps 5 and 7. Toyota employs the adaptive one factor at a time (OFAT) approach for testing countermeasures (Figures 4.17 and 4.18). For a user to distinguish relevant and irrelevant information, countermeasures need to be tested one at a time to understand the impact of each on the problem space. To obtain results as quickly as possible, each test is run beginning with where the last finished. OFAT seeks to optimize the response along the way, allowing investigators to find out more rapidly whether a factor has any effect (Friedman and Savage, 1947; Daniel, 1973). Compared to a design of experiments (DOE), in which the goal is to understand all interdependent relationships in the problem domain, OFAT is concerned with getting to improvement as quickly as possible.

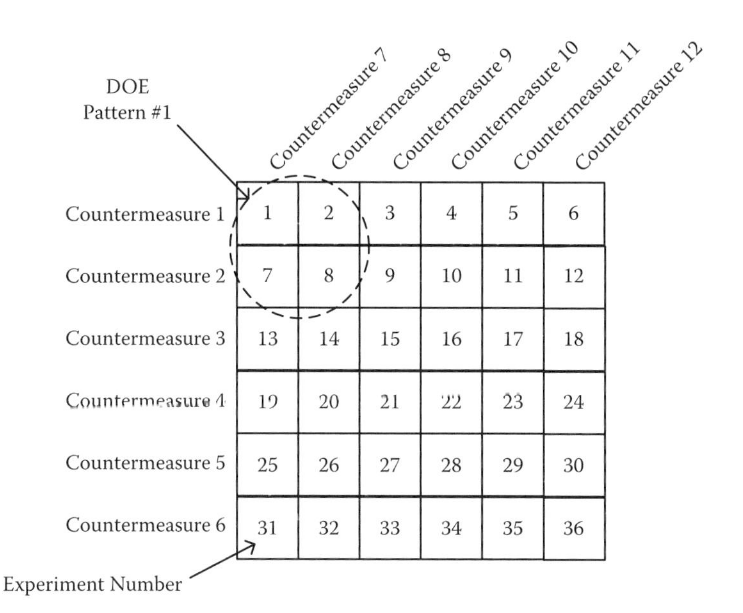

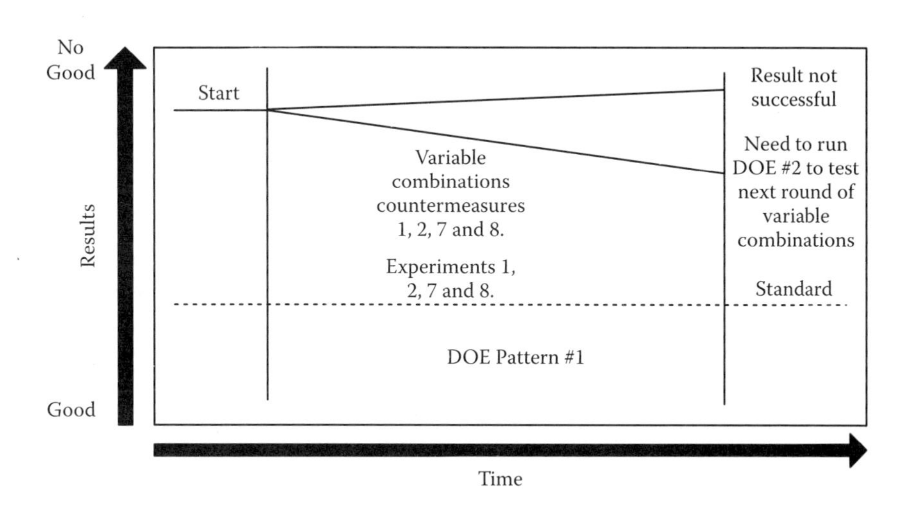

FIGURE 4.17
Design of experiment approach for testing countermeasures.

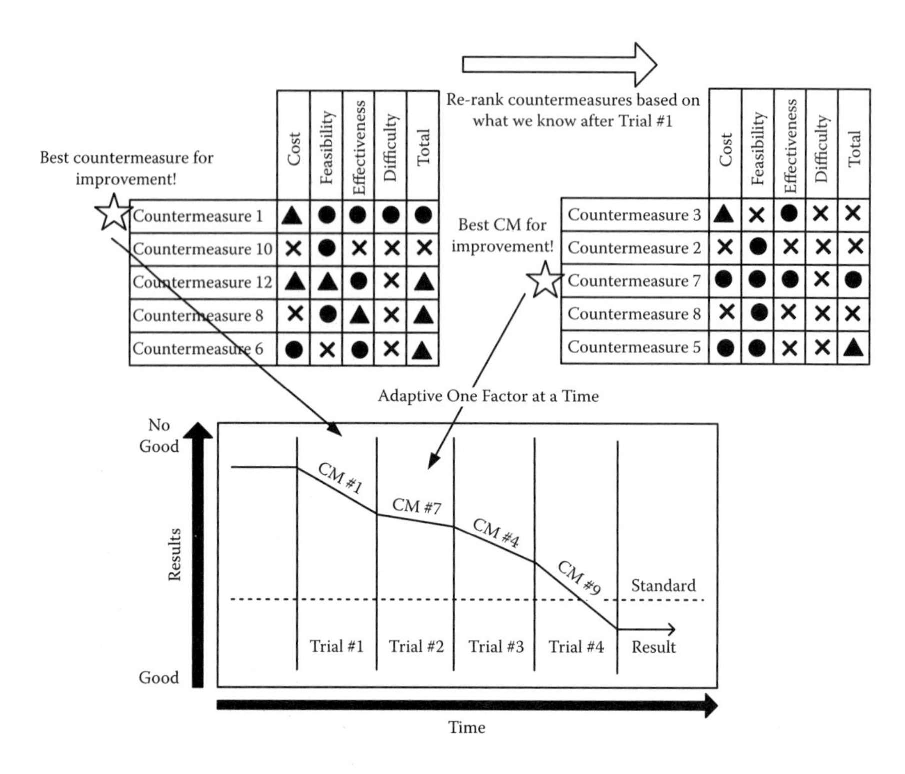

FIGURE 4.18
Toyota's OFAT approach for countermeasure testing. CM, countermeasure.

4.10 PUTTING IT ALL TOGETHER

The A3 process guides users into breaking down large and ambiguous problems into smaller problems while quickly generating and standardizing countermeasures one by one. This technique emphasizes the tendency to arrive quickly at testable solutions rather than perfect solutions. The method also encourages frequent decision making in evaluating "minisolutions" that contribute to the overall goal of the problem.

What makes the A3 process unique compared to other problem-solving methodologies are the types of core thinking skills used in problem and decision analysis. In problem analysis, the A3 method strongly emphasizes goal clarity using means–ends analysis. Users are expected to apply a series of focusing skills to limit and narrow the number of wrong choices made in the problem-solving process. In decision analysis, narrowing also occurs by employing the OFAT technique to help users to distinguish relevant and irrelevant information in the problem domain. Productive thinking is also encouraged in the A3 process because the problem space can be updated and refined as the user works toward the solution. By revisiting the problem space gradually, the users can learn as they go without having to balance all the unknown conditions of the problem at once.

The qualities of the A3 method suggest that Toyota's problem-solving process is suited for novice problem solvers. Focusing and iterative thinking skills are characteristics in problem-solving methodologies that help users to engage in productive thinking when their domain knowledge of the problem is low. When a user is trying to solve a problem that has never been attempted, more guidance is needed. Interestingly, Toyota's method does not resemble the qualities of an advance problem-solving method or one that is for expert problem solvers. This means that the A3 process tries to channel and guide the user rather than leaving it up to the user to more freely solve the problem the way the user wants. This does not imply that the A3 method cannot solve complicated problems; it only means that expert problem solvers who have domain knowledge of the problem space will not need as much guidance in obtaining the solution.

This discovery illustrates that TPS is again a system that is intended to be used by everyone in the organization. If Toyota's method exhibited expert problem-solving traits and characteristics, it would imply that only the elite would be able to apply the technique successfully. An example of a methodology that is suited for specialist or expert problem solvers is the DMAIC process. This methodology has no focusing skills, which means it relies on the user to work forward without a reference to a goal. This methodology cannot be used successfully by everyone in the organization because it leaves too much for interpretation in reaching the solution space. While a method like this may be useful for examining the problem domain, it does not provide enough guidance to get new users to the finish line in a shorter amount of time. The A3 process is highly structured, which means new learners can attempt problem solving and be successful. These findings show that Toyota expects everyone in the organization to be able to use the eight-step problem-solving process. While all methods are greatly dependent on numerous factors, in how they are completed, in their context and discretion (Charles et al., 1987), Toyota is at least trying to promote efficiency by everyone in the organization by solving problems the same way.

4.11 SUMMARY

This chapter proposed that another goal-seeking property of TPS is problem solving. Problem solving is used to drive organizational efficiency by standardizing the types of thinking skills when correcting and improving the workplace. The previous chapter illustrated that everyone in the company can benefit from learning and applying some basic industrial engineering practices to study and improve their jobs. Toyota adapted a form of scientific management to help everyone in the company to think like an engineer, which is effective at treating everything as a process that can be broken down, evaluated, and improved.

This chapter showed that employees in the workplace can be more effective if they are given a structured process to evaluate and test their ideas. Toyota's A3 process is a technique that can be used by anyone to explore the problem domain with success even if new to it. The A3 process is a way of thinking that helps users drive down to a level that makes it easier to identify solutions that work, specifically problems that do not return. Any problem-solving process can find and fix problems, but what makes Toyota's problem-solving process unique is that it can help novice problem solvers keep problems from returning after they have solved them.

REFERENCES

Anderson, J., Reder, L., and Simon, H. (1996) Situated learning and education. *Educational Researcher*, Vol. 25, No. 4, 5–11.

Bandura, A. (1993) Perceived self-efficacy in cognitive development and functioning. *Educational Psychologist*, Vol. 28, 117–148.

Bloom, B. and Broder, L. (1950) Problem solving processes of college students. *Supplementary Educational Monographs*, No. B, p. 238.

Bruner, J., Goodnow J., and Austin, A. (1956) *A Study of Thinking*, Wiley, New York, p. 117.

Buswell, G. (1956) Educational theory and the psychology of learning. *Journal of Educational Psychology*, Vol. 47, No. 3, 175–184.

Charles, R., Lester, F., and O'Daffer, P. (1987) *How to Evaluate Progress in Problem Solving*, National Council of Teachers of Mathematics (NCTM), Reston, VA.

Conner, J. (1999) Determining the construct validity of problem solving performance assessments through the use of verbal protocols, dissertation, Indiana University.

Cooper, J. (1961) *The Art of Decision Making*, Doubleday, New York.

Daniel, C. (1973) One at a time plans. *Journal of the American Statistical Association*, Vol. 68, 353–360.

Davis, J. (1969) Individual-group problem solving, subject preference, and problem type. *Journal of Personality and Social Psychology*, Vol. 13, No. 4, 362–374.

Deming, E. (1986) *Out of the Crisis*, MIT Center for Advanced Engineering Study, Cambridge, MA.

Dewey, J. (1910) *How We Think*, Heath, Lexington, MA.

Duncan, C. (1959) Recent research in human problem solving. *Psychological Bulletin*, Vol. 56, No. 6, 397–429.

Elias, D. and David, P. (1983) *A Guide to Problem Solving, The 1983 Annual for Facilitators, Trainers and Consultants*, University Associates.

Fattu, N., Mech, E., and Kapos, E. (1954) Some statistical relationships between selected response dimensions and problem solving proficiency. *Psychological Monographs*, Vol. 68, 1–23.

Friedman, M. and Savage, L. (1947) Planning experiments seeking maxima, in *Techniques of Statistical Analyses*, ed. Eisenhart, C., Hastay, M., and Wallis, W. McGraw-Hill, New York. pp. 365–372.

Gagne, R. and Brown, L. (1961) Some factors in the programming of conceptual learning. *Journal of Experimental Psychology*, Vol. 62, No. 4, 313–321.

Greeno, J. (1976) Cognitive objectives of instruction: Theory of knowledge for solving problems and answering questions, in *Cognition and Instruction*, ed. Klahr D. Hillsdale, NJ: Erlbaum.

Gregory, C. (1967) *Management of Intelligence: Scientific Problem Solving and Creativity*, McGraw-Hill, New York.

Guilford, J.P. (1967) *The Nature of Human Intelligence*, McGraw-Hill, New York.

Hayes, J. (1981) *The Complete Problem Solver*, Franklin Institute Press, Philadelphia.

Heppner, P., Witty, T., and Dixon, W. (2004) Problem solving appraisal and human adjustment: A review of 20 years of research using the problem solving inventory. *Counseling Psychologist*, Vol. 32, 344–428.

Hurson, T. (2007) *Think Better: An Innovator's Guide to Productive Thinking*, McGraw-Hill, New York.

Johnson, D. (1962) Serial analysis of verbal analogy problems, *Journal of Educational Psychology*, Vol. 53, 99–104.

Katona, G. (1940) *Organizing and Memorizing, Studies on the Psychology of Learning and Teaching*, Colombia University Press, New York.

Kepner, C. and Tregoe, B. (1965) *The Rational Manager*, McGraw-Hill, New York.

Koberg, D. and Bagnall, J. (1981) *The All New Universal Traveler: A Soft-Systems Guide to Creativity, Problem-Solving, and the Process of Reaching Goals*, Kaufmann, Los Altos, CA.

Kotovsky, K., Hayes, J., and Simon, H. (1985) Why are some problems hard? Evidence from Tower of Hanoi. *Cognitive Psychology*, Vol. 17, 248–294.

Locke, E. (1991) Problems with goal-setting research in sports—and their solution. *Journal of Sport and Exercise Psychology*, Vol. 13, 311–316.

Lyles, R. (1982) *Practical Management Problem Solving and Decision Making*, Van Nostrand Reinhold, New York.

Maier, N. (1930) Reasoning in humans. *Journal of Comparative Psychology*, Vol. 10, 115–143.

Marzano, R., Brandt, R., Hughes, C., Jones, B., Presseisen, B., Rankin, S., and Suthor, C. (1988) *Dimension of Thinking: A Framework for Curriculum and Instruction*. Association for Supervision and Curriculum Development, Alexandria, VA.

Moore, P. (1995) Information problem-solving: A wider view of library skills. *Journal of Contemporary Educational Psychology*, Vol. 20, 1–31.

Nakamura, C. (1955) *The Relation between Conformity and Problem Solving*, Office of Naval Research.

Newell, A. and Simon, H. (1972) *Human Problem Solving*, Prentice Hall, Englewood Cliffs, NJ.

Ohno, T. (1988) *Toyota Production System: Beyond Large-Scale Production*, Productivity Press, New York, NY.

Osborn, A.F. (1953) *Applied Imagination, Principles and Procedures of Creative Thinking*, Scribner, New York.

Polya, G. (1945) *How to Solve It*, Princeton University Press, Princeton, NJ.

Robertshaw, J., Mecca, S, and Rerick, M. (1978) *Problem Solving, a Systems Approach*, Petrocelli Books, Princeton, NJ.

Robinson, V. (1987) A problem analysis approach to decision making and reporting for complex cases. *Journal of the New Zealand Psychological Services Association*, Vol. 8, 35–48.

Rogers, C. (1967) *On Becoming a Person*, Houghton Mifflin, Boston.

Rossman, J. (1931) *The Psychology of the Inventor*, Inventors Publishing, Washington, DC.

Sachse, T. (1981) *The Role of Performance Assessment in Tests of Problem Solving Ability*, U.S. Department of Education, National Institute of Education.

Scandura, J. (1977) Structural approach to instructional problems. *American Psychologist*, Vol. 32, No. 1, 33–53.

Schank, R. (1973) *Identification and Conceptualization Underlining Natural Language, Computer Models of Thought and Language*, Freeman, San Francisco.

Shewhart, W. (1939) *Statistical Method from the Viewpoint of Quality Control*, Department of Agriculture, Washington, DC.

Simon, H. (1977) *The New Science of Management Decision,* 3rd rev. ed., Prentice Hall, Englewood Cliffs, NJ. (First edition 1960)

Sternberg, R. (1997) *Thinking Styles*, Cambridge University Press, New York.

Sweller, J. and Levine, J. (1982) Effects of goal specificity on means-ends analysis and learning. *Journal of Experimental Psychology: Learning, Memory, and Cognition*, Vol. 8, No. 5, 463–474.

Thompson, P. (1991) Goal specificity and understanding of a problem: A schema theory approach, dissertation, University of Kentucky, Department of Educational and Counseling Psychology, Lexington.

Tyre, M. (1989) Managing the introduction of new process technology: A study of organizational problem solving at the plant level, PhD dissertation, Harvard Business School, Cambridge, MA.

Wallas, G. (1926) *The Art of Thought*, Harcourt, Brace and Company, New York, NY.

Wertheimer, M. (1959) *Productive Thinking*, Harper, New York.

Wiener, N. (1948) *Cybernetics: Or Control and Communication in the Animal and the Machine*, Librairie Hermann & Cie, Paris, and MIT Press, Cambridge, MA.

5

The System Property of Regulation in TPS

5.1 SYSTEM PROPERTY: REGULATION

One of the most essential features of systems is the property of regulation. Before a system can be modified, changed, or improved, its regulatory features must first be taken into consideration to understand how behavior is currently maintained (Klir, 1969). The purpose of regulation in systems is to get back to a predetermined condition to maximize its chances of survival and growth. A system must be able to provide information about the degree to which near-steady-state functioning is being maintained and the means to alter its own behavior in reaching the goal of the system (Becvar and Becvar, 1998). Above all, regulation serves to keep systems alive.

A system is referred to as homeostatic if it can retain its state by internal adjustments (Ackoff, 1971). Homeostatic systems can be described as open, closed, or isolated systems that regulate their internal environments to maintain a stable and constant condition. Characteristics of open, closed, and isolated systems are shown in Table 5.1 and illustrated in Figure 5.1. The discipline of thermodynamics is a field in mechanical engineering that uses these classifications to simplify how a system works when it interacts with its environment. Thermodynamics is concerned with various forms of energy that can be described in terms of a system's change in volume or work. Boundaries are used to describe the amount of interaction or transfer in a system from the outside (Cengel and Boles, 1994). Boundaries are an imaginary surface with zero thickness used to simplify the mathematics of heat and mass transfer.

The state and equilibrium of a system are used to describe the driving forces and balance of a system. At any given time, all properties of a

TABLE 5.1

Characteristics of Open and Closed Systems

	Open System	Closed System	Isolated System
Mass transfer	Yes	No	No
Energy transfer	Yes	Yes	No
Boundary	Permeable and flexible	Impenetrable	Impenetrable and rigid
Actively interacts with its environment	Yes	Limited interaction	Not at all
Interplay of its subsystems	Dynamic capability	Static capability	Single function
Alternate functions	Yes	One way	Fixed
Development	Shows potential for individualization	Functions and operations are established and do not vary or adapt over time	None

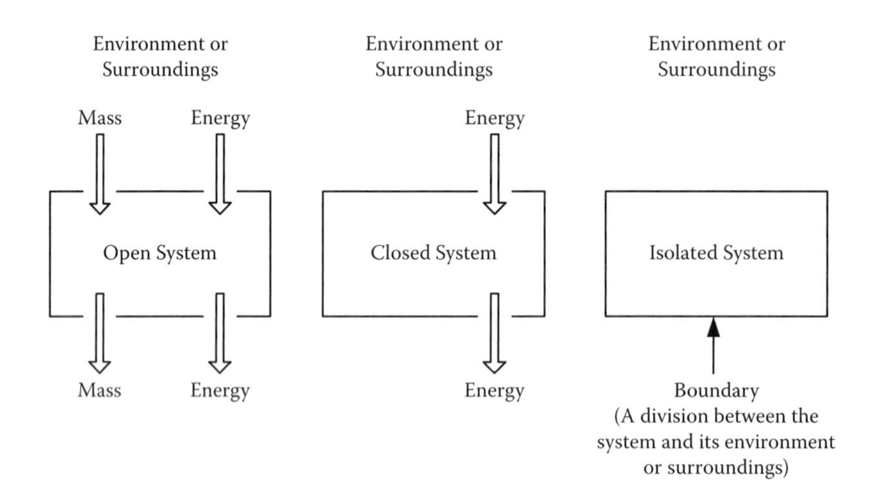

FIGURE 5.1

Representation of open, closed, and isolated systems.

system have a fixed state. If the value of one property inside the system changes, the state of the system will change. Different types of states and levels of equilibrium are used to describe how systems change over time. When there are no driving forces within or outside the system, the system has reached a state of equilibrium, meaning there are no unbalanced forces. Shown in Figure 5.2 are a few examples of states and the types of equilibriums.

Organizations are examples of both open and dynamic systems. Organizations are influenced by their environment and can change when nothing is acting on them. The type of regulation that is used in organizations dramatically influences their state and ability to reach equilibrium. Since organizations are not static, there is a greater need to understand how they regulate their internal functions.

Closed systems use error-controlled regulation. This type of control is after the fact. Closed systems have a greater degree of certainty because there are fewer unknowns and conditions that require balancing for the system to operate. Drifts, disturbances, and events in closed systems are better known, which justifies the use of reactive feedback controls.

Open systems are influenced by external conditions, which means the types of regulation used to keep them in dynamic equilibrium are generally more advanced. Open systems use anticipatory control. This form of regulation anticipates errors before they occur and takes corrective measures before the output is realized. Anticipatory regulation is often referred to as feed-forward control, in which variables in the system are managed before the output occurs.

Another aspect of regulation is how responsive the system is in returning to its predetermined condition. Delays can postpone the time when the effect of a change takes place and can cause systems to operate in an unstable state. The stability or resilience of a system can be described as the tendency of a system to return to its equilibrium when the system is disturbed (Putt, 1978). Lag and delays are often described in terms of first- and second-order feedback. First-order feedback is when there is only one variable to be controlled. Second-order feedback is when the system features a second variable in the delay (Gigch, 1978). Since organizations are complex, the types of regulation to reduce and prevent lag are higher-order feedback systems in which many variables exist to get the system back to the predetermined condition.

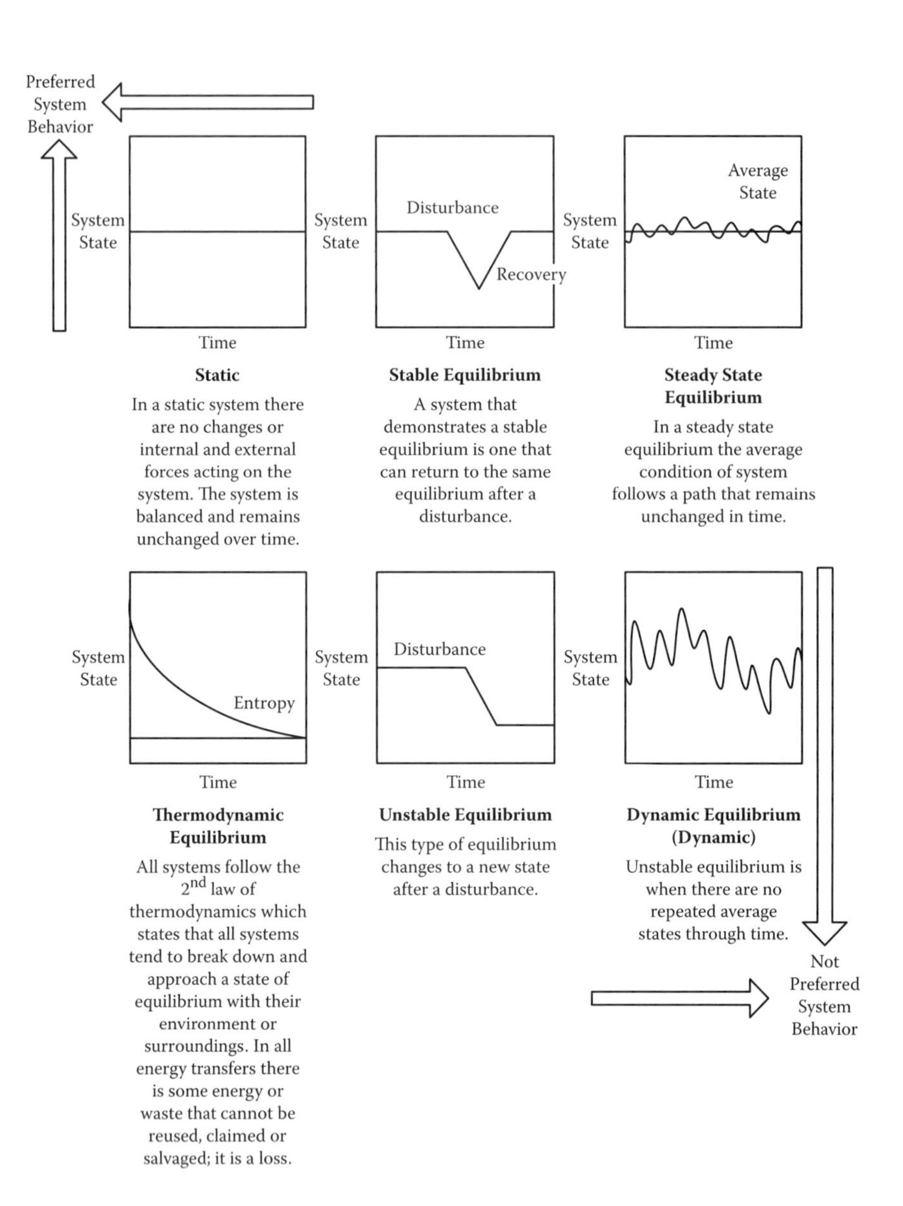

FIGURE 5.2
Representation of states and types of equilibrium. (Terms and definitions from Cengel, Y. and Boles, M., 1994, *Thermodynamics, an Engineering Approach*, McGraw-Hill, New York; and Incropera, F. and DeWitt, D., 1996, *Fundamentals of Heat and Mass Transfer*, Wiley, New York.)

5.2 THE SYSTEM PROPERTY OF REGULATION AND THE TOYOTA PRODUCTION SYSTEM

Before a system can be regulated, the purpose and the identity of the system must be defined. At Toyota, the identity of the Toyota Production System (TPS) is industrial engineering and problem solving. Toyota uses these techniques to guide workplace improvement. Systems theory would suggest that employees should have a way to regulate their thinking and understanding of workplace improvement. At Toyota, those mechanisms are known as jishukens and quality circles (QCs).

Jishukens and QCs are team-based problem-solving activities intended to drive workplace improvement (Figure 5.3). The difference between jishukens and QCs is determined by who leads the problem-solving activity. In jishukens, the problem-solving activity is led by management. In QCs, these activities are led by team members and team leaders (i.e., members of direct labor). These groups are also differentiated by the types of problems they solve (Figure 5.4). Management is traditionally charged with solving broad system-related problems, while direct labor are often assigned to work on small-scale process-related problems. In addition, QCs tend to emphasize activities that bring about consistency in the workplace, while jishukens focus on major improvement activities for the company. The commonality between jishukens and QCs is that they both use peer-based activities to solve workplace problems.

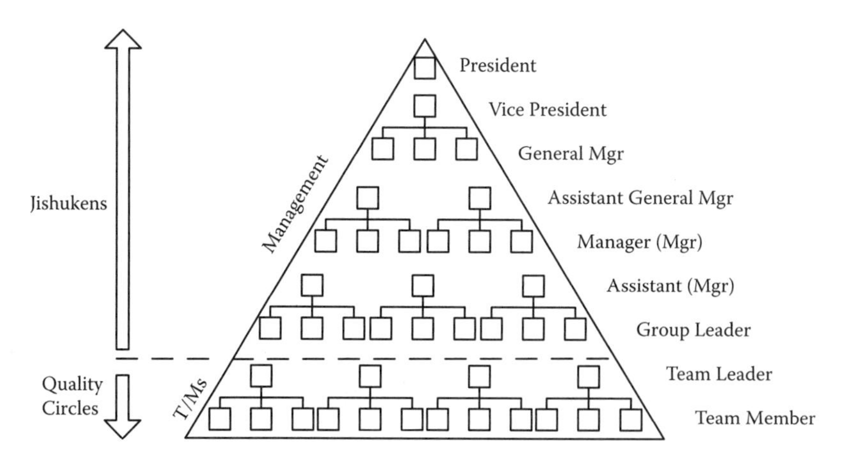

FIGURE 5.3
Representation of lead group in problem solving for jishukens and quality circles. T/M = Team Member.

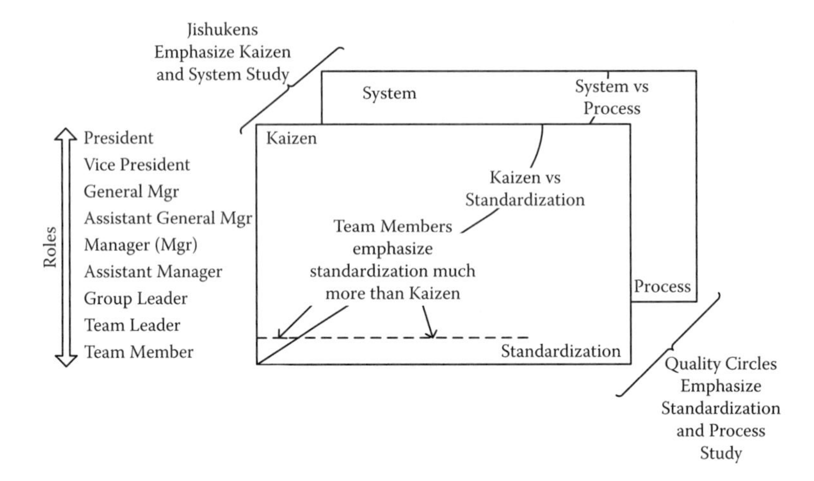

FIGURE 5.4
Major emphasis of quality circles and jishukens for kaizen, standardization, process, and system-related problems.

The effectiveness of TPS is strongly dependent on its ability to be regulated. Jishukens and QCs are one way that problem solving and industrial engineering are practiced consistently throughout the organization. Regardless of the work environment, situation, or context, jishukens and QCs act as a mechanism to normalize TPS thinking. Other reasons for regulating TPS include the following:

- Jishukens and QCs reduce resources. More resources are used to manage multiple systems rather than using a single system. While alternate systems are desirable for flexibility, there is a cost associated with managing and controlling multiple systems that accomplish the same thing.
- Jishukens and QCs prevent other systems from being established. Employees working to multiple systems in the organization will develop their own group identity that is separate from the organization. Members of the newly formed identity will seek to differentiate their group from the organization as it tries to emerge itself as the most dominant group.
- Jishukens and QCs bring certainty in using the company's system. Multiple operating systems that accomplish the same goals in organizations create more uncertainty. Members of the organization using methods different from what the organization has prescribed will not receive social recognition or vindication that those techniques are acceptable.

5.3 LITERATURE REVIEW OF JISHUKENS

Various attempts have been made to explain jishukens at Toyota. Jishukens have been described as supplier development activities (McNichols et al., 1998; Heckscher and Adler, 2006); kaizen events (Montabon, 1997; Worley, 2007; Heard, 1998); and kaikaku, a form of rapid improvement and change, to dramatically transform the workplace (Bicheno, 2000). There has also been some confusion on whether jishukens are management appointed or are voluntary activities (Smith, 1993; Liker and Hoseus, 2008). The work of Smith suggests that jishukens represent small groups of employees appointed by management, whereas the work of Liker and Hoseus proposes groups can decide on their own whether to initiate a jishuken (Smith, 1993; Liker and Hoseus, 2008).

Seen more clearly, jishukens, like many TPS activities, have both a learning goal and a productivity goal: As they harness manager teams for problem solving needed by the production process, jishukens help managers to improve their ability to coach and teach problem solving to others, specifically production workers (Badurdeen et al., 2009). Jishukens perform a vital role in TPS because they help managers to be better teachers by developing skills that can be passed on to future generations (Alloo, 2009; Hall, 2006). Jishukens regulate, reinforce, and maintain the company's values, beliefs, and behaviors. Participation in jishukens gives management a common language and a common approach to problem solving across the organization (Figure 5.5).

5.4 LITERATURE REVIEW OF QUALITY CIRCLES

Quality circles originated in Japan and have been adopted by American companies since the 1970s. They are generally more recognized than jishukens and have been credited for increasing productivity, enhanced employee morale, and cost reduction. Numerous authors have defined QCs as small groups of employees, usually from the same work area, who normally meet for an hour after work to identify problems, investigate causes, implement solutions, and report results (Erez, 2010; Pereira and Osburn, 2007; Olberding, 1998).

In the early 1960s, the Union of Japanese Scientists and Engineers (JUSE) initiated the "quality control revolution" and introduced the

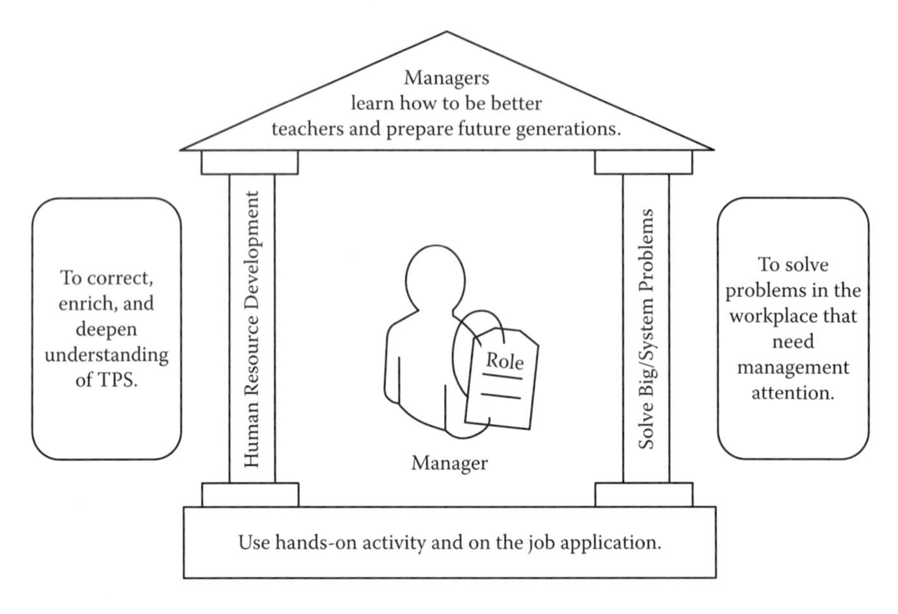

FIGURE 5.5
The jishuken house.

concept of "quality control" (which was then changed to the term *quality circles*) by establishing workplace groups for studying and applying quality methods (Ward, 2006). In the United States, the first QC was formally registered at the Lockheed Missile and Space Company in Sunnyvale, California, in 1974. By 1977, Lockheed had saved $3 million (Ward, 2006). Soon after, QCs became popular with companies such as Hughes Aircraft Company, Boeing, General Motors Corporation, Ford Motor Company, General Electric, Bank of America, and many others. At that time, QCs were considered one of the most promising total quality management (TQM) approaches toward improving productivity in the workplace. By the 1980s, 90% of the Fortune 500 companies had QC programs involved in a variety of workplace projects, including productivity, quality, cost reduction, employees' commitment, job satisfaction, and communication (Lawler and Mohrman, 1985; Tang et al., 1987).

By the end of the 1980s, QCs grew to be a management fad. Most QC practices in the United States ceased to exist mainly because companies did not focus on problem solving and employee training. Aravindan stated that QC programs failed because they could not evolve and challenge workers to solve more complex problems (Aravindan, 1996). Other studies showed QCs failed because of the lack of resources and management support (Walker, 1992). The main problem is that most organizations

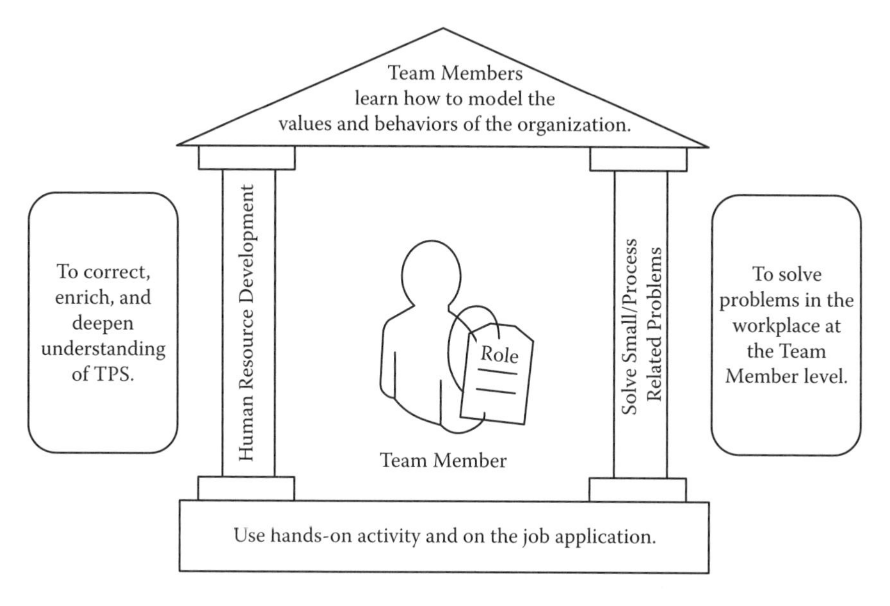

FIGURE 5.6
The quality circle house.

view QCs as a way to reduce cost and short-term financial impact instead of enhance employee development (Berman and Hellweg, 1989; Sillince, 1995; Kathawala, 1989). Several QC studies indicated that QCs were initially successful but later declined due to a lack of interest on the part of management (Gray, 1993).

At Toyota, QCs are viewed similarly to jishukens (Figure 5.6). QCs are shop floor activities that emphasize problem solving and team member development. The goal of QCs (like jishukens) is to learn how to model the company's values and behaviors during workplace improvement. QCs help to regulate TPS thinking at the team member level, while Jishukens regulate TPS at the management level (Figure 5.7). In many ways, Jishukens and QCs are nothing more than a scaled-down version of the plant performing an ideal version of TPS.

5.5 USING QUALITY CIRCLES AND JISHUKENS TO CREATE A SHARED VISION

A reoccurring theme in TPS is the importance of the organization working to one system. Jishukens and QCs regulate TPS so that everyone can

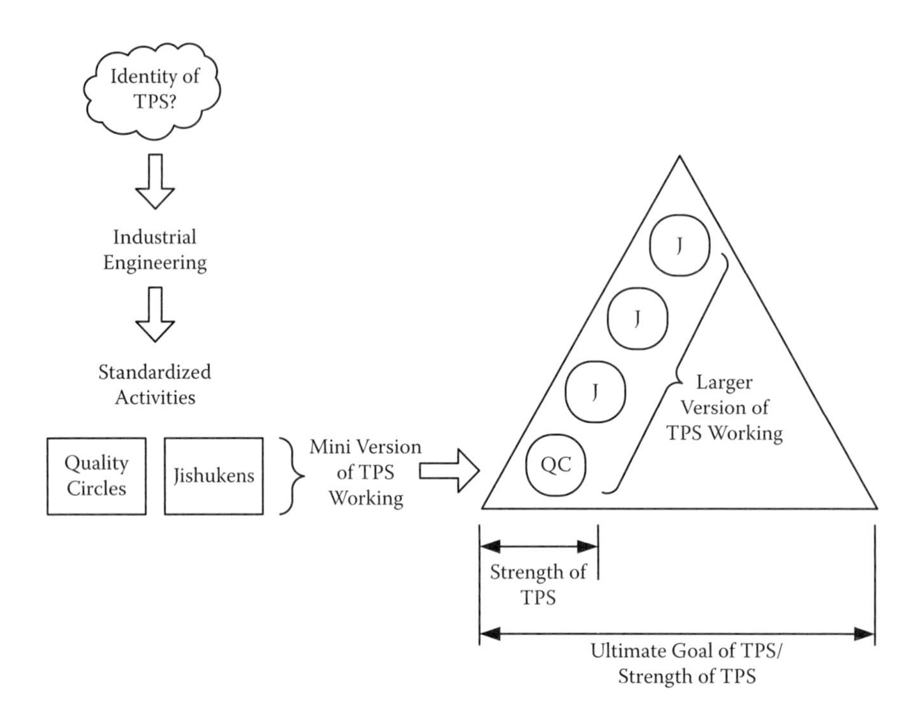

FIGURE 5.7
Conceptual image of TPS working at the QC and jishuken levels.

get back to a predetermined condition. In the context of organizational development, this is referred to as everyone in the company sharing the same idea or vision about the company. Senge, in his best-selling 1990 book, *The Fifth Discipline*, describes a shared vision as the practice of unearthing the shared "pictures of the future" to foster a genuine commitment within the organization. Shared vision is meant to seek enrollment of its members rather than compliance. When organizations have a shared vision (not to be confused with the all-too-familiar "vision statement"), people excel and learn not because they are told to, but because they want to (Senge, 1990). A shared vision has the power to be uplifting, encouraging, and motivational. It can promote experimentation and innovation and most importantly direct organizational members toward the future. A shared vision is the capacity of the organization to hold a shared picture of the future it seeks to create (Senge, 1990). Shared visions are about the sense of the long term.

Visions spread when people talk and group members want to pass on their group identity (Ledgerwood and Liviatan, 2010). Group members want their group to share a certain set of characteristics and attributes that is comprised of a desired group identity. Group identity serves to reduce uncertainty about how one should feel and behave (Hogg, 2007) and should be considered as a goal toward which group members can strive and seek to identify with others (Ledgerwood and Liviatan, 2010). A shared identity is important because it serves to make subjective experience seem factual (Sherif, 1935; Hardin and Conley, 2001). Shared visions or identities allow organizations to perceive the environment as predictable and controllable (meaning it fulfills the epistemic needs for understanding and certainty) and provides a foundation for social relationships (Hardin and Conley, 2001).

Shared visions are a lengthy process that takes much reinforcing, clarity, and enthusiasm. A shared vision cannot occur from a single person but from the interactions of people defining, sharing, and discussing their personal visions and the organization's goals and values. To make progress toward the completion of a socially shared group identity, it is necessary for the group to be capable of communicating aspects of the group identity to other people (Hardin and Conley, 2001; Hogg, 2007; Avery, 2004).

Once a shared identity is formed, the members of the group use the shared vision to guide their choices and actions (Kantabutra, 2008). Consequently, organizations that have a shared vision can more easily address the challenge of internal integration of resources through building capability, ownership, and responsibility. The involvement trait is one of the oldest and most researched cultural traits that have proven to increase organizational effectiveness (McGregor, 1960). High-involvement organizations can minimize transaction cost when members act from the same value consensus rather than bureaucratic rules and regulations (Ouchi, 1981).

For TPS to survive, the organization must have a way to communicate the aspects of the group identity to other people. The medium that Toyota utilizes to share the identity of TPS is hands-on activity and on-the-job application. In this setting, members can easily communicate, discuss, share, and reinforce the identity of TPS. Not like formal training, in which students are lectured in a classroom, members are expected to participate and be involved in problem solving. Toyota chooses this medium to communicate the identity of TPS because it values participative decision making by the ones who perform the work. QCs and jishukens allow members to share the future image of TPS. They allow a reference point for existing members to model their behavior and actions, and they are used to recruit new members who do not share the same identity. When the identity of TPS is shared at each level of the organization, TPS is strengthened (Figure 5.8).

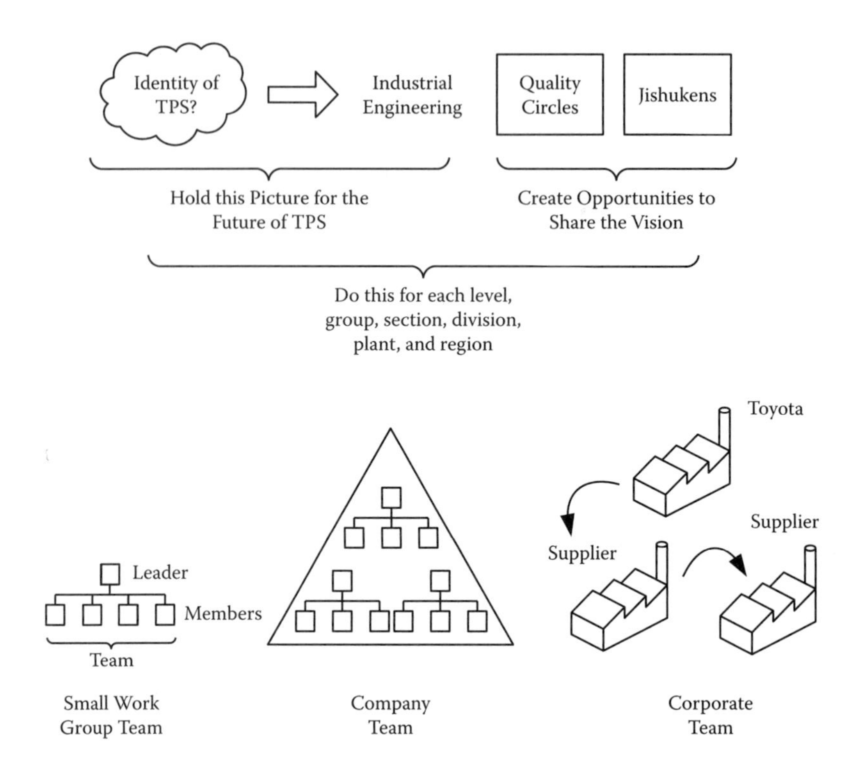

FIGURE 5.8
Using quality circles and jishukens to create a shared vision of TPS.

One of the benefits of establishing a shared vision or identity in TPS is that it can encourage the enrollment of the right behaviors among employees. Rather than using compliance to get employees to act accordingly, jishukens and QCs help define the identity of the group and seek to maximize how it is distinguished. Members of a group who share an identity will self-police, correcting and holding members to the same norms and values of the group. A few examples of those behaviors are shown in Figure 5.9.

5.6 THE WEAKENING AND STRENGTHENING OF TPS USING SHARED VISIONS

Organizations that do not create a shared vision run the risk of weakening organizational involvement. Just like a bad MapQuest encounter, if

Non-Lean Organization	Toyota
Asking for help is seen as weakness.	Asking for help is seen as a strength.
Admitting incompetence.	Models the right behavior for employees to see and feel comfortable exposing problems.
Problems are connected to the fault of someone performing badly.	Problems are inevitable because no system has been fully standardized.
Assigns blame for discovering or identifying problems.	Builds respect in the workplace by equipping employees with a system to stop the line when there are problems.
Conceals problems due to fear and negative consequences.	Encourages problems to be visible to avoid wasted problem solving due to fear and concealment.

FIGURE 5.9
Comparison of accepted behaviors that QCs and jishukens must demonstrate.

there is no future picture of Lean, employees within the organization will reach a different destination. This situation is further complicated when management does not understand the future picture of Lean. Because leadership and management are in a position to give direction, their interpretation of Lean will be given regardless how well it is understood. In this context, employees will do whatever makes their boss happy. "Boss thinking" is how subordinates find out how to perform their job, the accepted ways to obtain results, and most importantly, the discretion and behavior that is rewarded in doing the job. In this context, the boss is referred to as the norm gatekeeper who demonstrates the accepted ways to work in the organization. Boss thinking creates a hidden culture that is hard for each employee to think and act at work because the reference is in the boss's head and not formalized within the organization (Figure 5.10).

When managers have different thinking or understanding about Lean, workplace improvement varies. Each employee who reports to the manager will share the manager's views about Lean and create a unique identity that is in opposition of another view. In this context, Lean is weakened because the shared vision is unclear (Figure 5.11). Fewer employees are able to contribute to the vision and have no choice but to work to their immediate supervisor's system.

At Toyota, jishukens have the goal to get leadership and management on the same page when thinking and applying TPS. Toyota does not assume

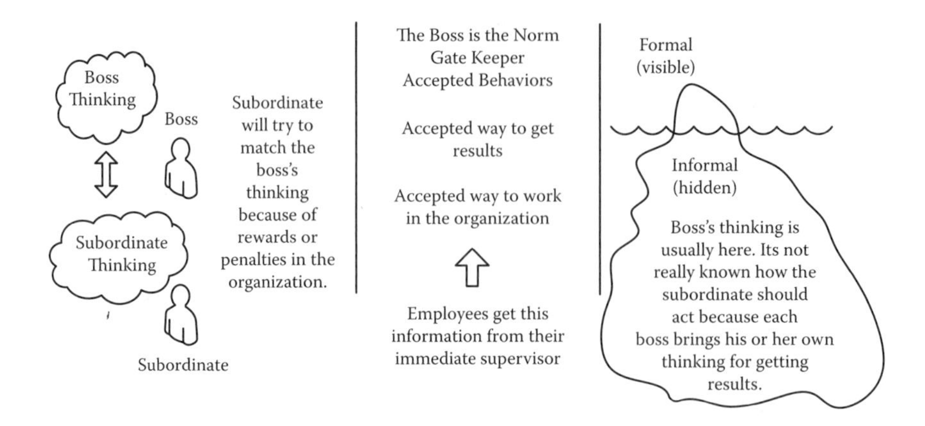

FIGURE 5.10
The hidden nature of boss thinking. (Iceberg model adapted from Hall E, 1976, *Beyond Culture*, Anchor Books, New York).

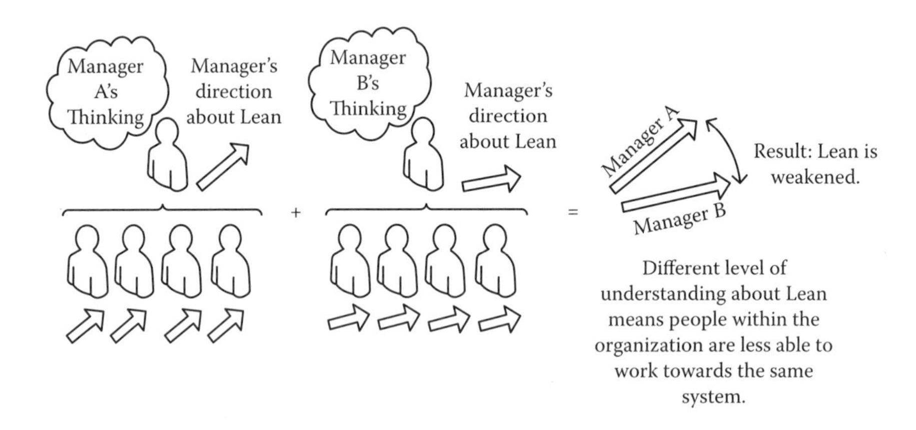

FIGURE 5.11
Representation of Lean weakening.

that its management shares a common understanding of problem solving or how TPS should be applied. Therefore, jishuken functions to stimulate discussion and communication among work groups as a means to reach a common understanding. When management shares the same vision about workplace improvement and problem solving, TPS is strengthened (Figure 5.12). Organizationally, this allows TPS to be more formalized and visible among its members. Instead of employees referencing their boss's thinking, employees will begin to reference the organization's thinking when making decisions about workplace improvement.

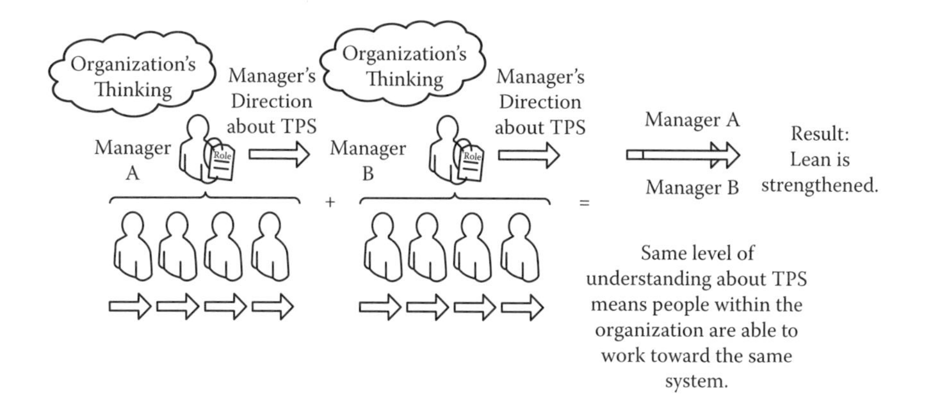

FIGURE 5.12
Representation of TPS when it is strengthened.

5.7 THE REGULATION CYCLE OF TPS

Can QCs exist without jishukens? Most companies start QCs without jishukens because they appear to fill a missing link in encouraging and promoting shop floor improvement. The problem occurs when QCs are not performed consistently with organizational practices. While most companies view any improvement or involvement by the shop floor as good, groups do not share the same kaizen mind or systematic approach to problem solving. In this situation, the best leadership can do is cheerlead. Leadership has not agreed how to perform problem solving and, most importantly, how to help and support team leaders and team members consistently when they need support. The outcome is a lot of perceived improvement without structure, consistency, or predictability.

Jishukens are a prelude to QCs because they prepare managers how to be teachers. When managers practice and develop skills in problem solving, TPS can be passed on more consistently throughout the organization. The best skills a manager can learn from jishukens are those that can be used by team members and team leaders. If jishukens trained managers to think like contemporary industrial engineers, TPS would be a system intended for the elite (i.e., kaizen specialists or master black belts). Jishukens do not emphasize the use of computers; instead, they emphasize basic charting and data collection techniques like those from the scientific management era. To become skilled at coaching problem solving, managers must use the same hands-on techniques at all levels of the organization

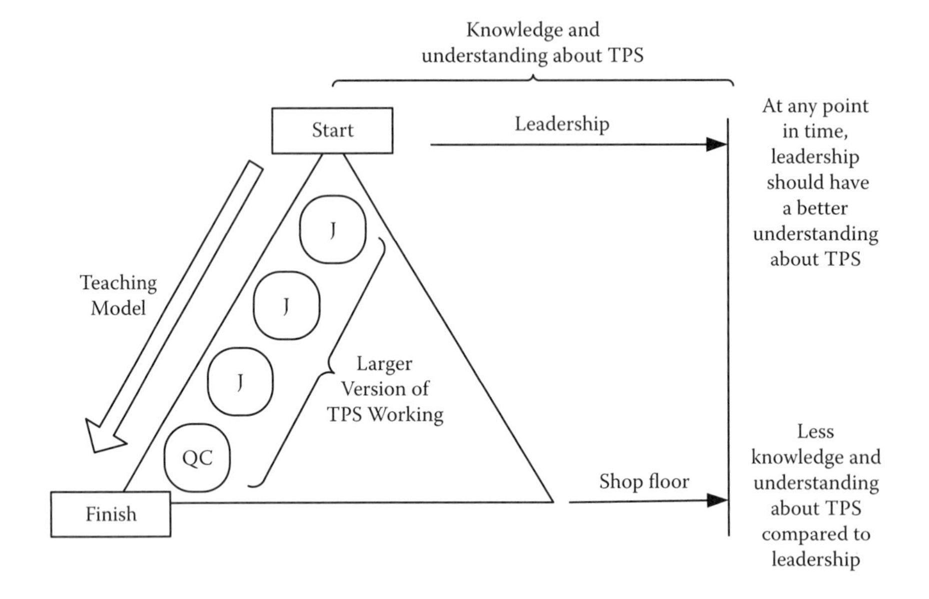

FIGURE 5.13
Correct regulation cycle of jishukens (J) and quality circles.

(Figures 5.13 and 5.14). Jishukens focus on refining a manager's skill to perform workplace study as if the manager was a team leader.

5.8 JISHUKEN AND QUALITY CIRCLE IMPLEMENTATION CONCEPTS

This section explains some of the commonalities between how TPS is regulated using jishukens and QCs. It is not the intent of this work to explain all the differences that exist but to highlight some of the more significant features. Jishukens are emphasized more in this section because they are less routine and are more complicated than QCs. QCs have a much different implementation pace than jishukens because of the problems that are solved. QCs typically last between six months and a year depending on the theme, while jishukens can be extremely focused, lasting several days or weeks without interruption. QCs are small process problems that can be completed by a group of team members, while jishukens are larger system problems that can affect many departments throughout the plant. Toyota does not intend for managers to solve small process problems or

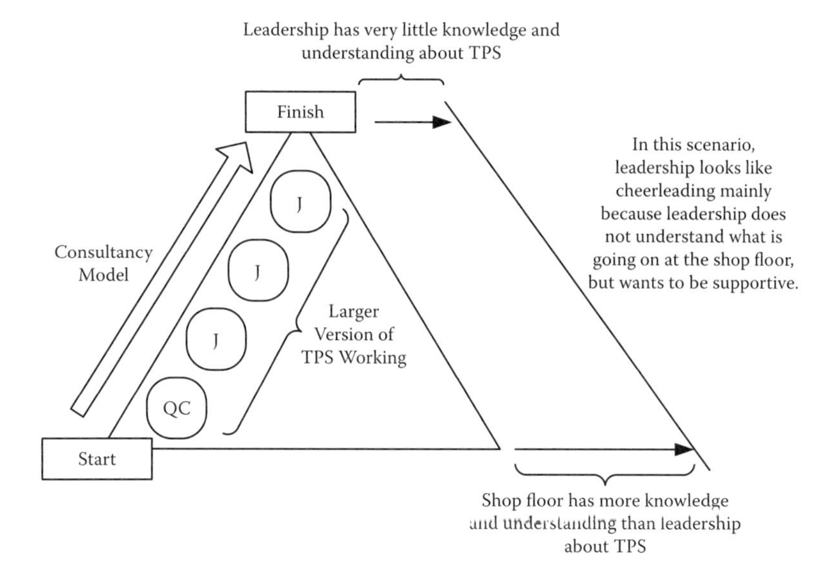

FIGURE 5.14
Incorrect regulation cycle of jishukens (J) and quality circles.

for team members to solve large-scale system problems. For example, if managers wrote standardized work, team leaders would wait until management saw the need to revise it. If this occurred, all involvement from the team member and team leader level would be lost. Jishukens and QCs are not combined to solve supermega problems but are examples of miniversions of TPS working at different levels (Figure 5.15). This section explains those TPS commonalities.

5.8.1 Asking for Help and Promoting Waste Elimination

Ideally, jishukens are voluntary. Jishukens are a way for managers to get help when efficiency in their department has begun to slow, something that happens periodically (Figure 5.16). It is relatively easy to eliminate waste at first, especially in a young system or a new process. However, over time, as more problems are solved and as workers become accustomed to a particular process, waste elimination can become much more difficult; problems are no longer as visible. The jishuken process becomes a way for managers to get help from other managers, who can bring a fresh eye, to stimulate waste elimination by helping waste become visible (Figure 5.17).

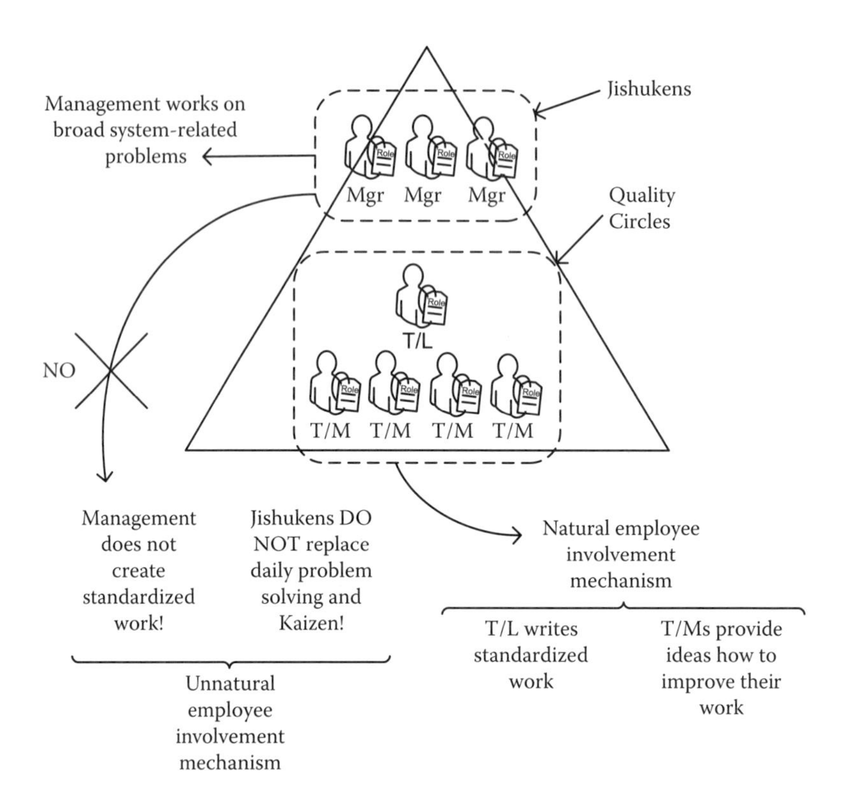

FIGURE 5.15
Jishukens and quality circles are not combined to solve supermega problems. TM = Team Member; TL = Team Leader.

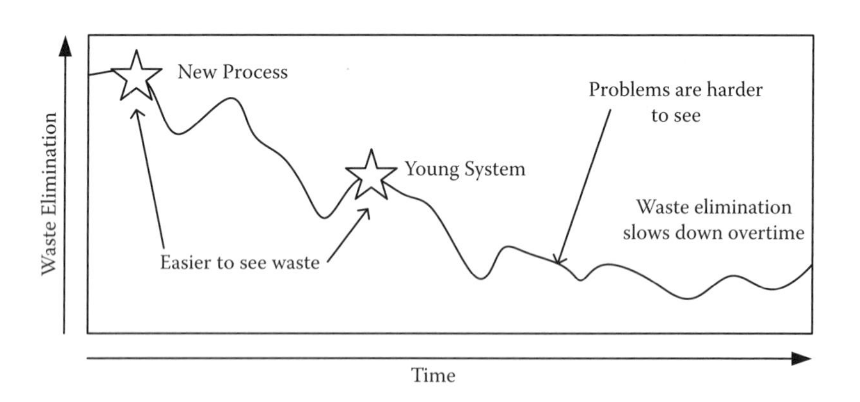

FIGURE 5.16
Waste elimination slows.

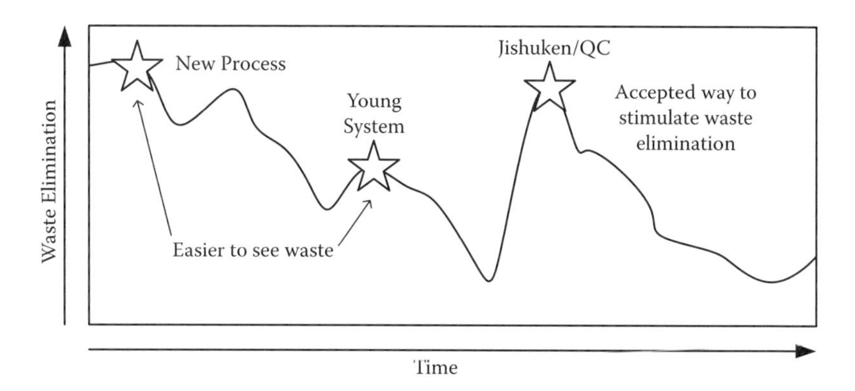

FIGURE 5.17
Accepted way to stimulate waste elimination.

It is not unusual for managers in industry or business to feel uncomfortable about asking for help. Larger cultural factors like gender roles can influence this; witness the standard joke about males in the United States—they will refuse to ask for directions when lost rather than admit to not knowing something. In U.S. culture generally and in some companies more than others, asking for help can be seen as an admission of weakness or incompetence. Another not-untypical kind of organizational culture assigns blame for problems; this cultural attitude can include blame for those who discover or identify problems, thus encouraging lower levels to conceal problems and mislead upper levels about any negative aspects of the actual situation. In such a culture, system effects will tend to be invisible. Problems will not be seen as more or less inevitable causes of aspects of a system that were not foreseen or have developed over time but as the fault of someone performing badly.

Thus, jishuken—because it is standard expected procedure—functions to lessen managerial reluctance about asking for help. Here is an example of why Jishuken cannot be used as a "tool" (like a hammer, for example, something than can be used effectively regardless of cultural factors) but only works when it is consistent with and supported by the larger culture, here with Toyota culture, which treats the identification of problems as a valued skill and a positive activity. This valuation in turn rests on the principle of continuous improvement, which, to continue, requires continuous improvement in the ability of team members and managers to see problems. It should be clear by now why a Lean implementation can fail if it is not consistent with the culture of the larger organization.

Kaizen and problem solving, for example, will not be successful if employees do not receive the right kind of coaching and support. Jishukens or some equivalent are needed. But, for Jishukens to be effective, managers at all levels need to feel comfortable and supported when highlighting and exposing problems. Thus, a company cannot outsource TPS to one group or level and ignore it otherwise. If leadership discourages the discovery of problems, jishuken will wither, no matter how many books on TPS are bought and read. If jishuken is not effective, kaizen will be ineffective, and Lean implementation will likely have failed in such an organization because the particular problem-solving approach to waste elimination it depends on will not take root. Consequently, if leadership is not wholeheartedly engaged in promoting this behavior, managers will conceal rather than identify problems (Alloo, 2009).

5.8.2 Starting Point for Workplace Improvement

An abnormal condition, a gap between the actual condition and the standard, triggers problem-solving activities at Toyota. While QCs are triggered more by themes scheduled throughout the year, jishukens are initiated when problems escalate to a level that needs management's immediate attention. Problem-solving activities are used to establish standards, to maintain the standards, and to improve the standards. The condition chosen depends on the actual condition of the system in relation to the standard. Figure 5.18 shows five examples where problem-solving activities could be initiated. Because problem-solving activities are team-based activities, Toyota discourages the use of QCs and jishukens when actual conditions are unknown. Simply, without some information regarding the scale and scope of the problem, teams cannot be properly established.

5.8.3 The Team Structure of Jishukens and Quality Circles

Once the need has been identified, a problem-solving team is assembled with help from the Operations Development Group (ODG; discussed in more detail further in the chapter). A typical team will have four to six members from various management levels, from group leader (GL) to the president. QC teams are typically formed with members from the same work area. The exact composition of the team is not specified in procedures because the emphasis is on what the particular problem demands.

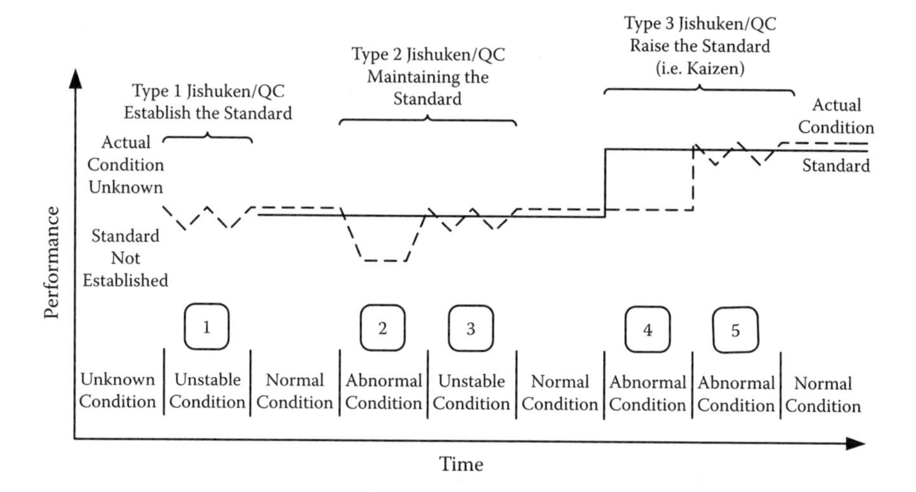

FIGURE 5.18
Trigger points for initiating jishukens/quality circles.

Figure 5.19 illustrates a typical organizational structure for a production environment in which material flows from Department A to D. Here, activity in A negatively impacts a downstream process at D. In this scenario, the jishuken leader would likely be chosen from A, ideally at the management level closest to the situation. The jishuken leader would then get assistance from the ODG in assembling the team. Here, the team would include members from Department A and Department D. Although other departments could also be tapped, depending on the nature of the problem, the jishuken team should be drawn from the department where the

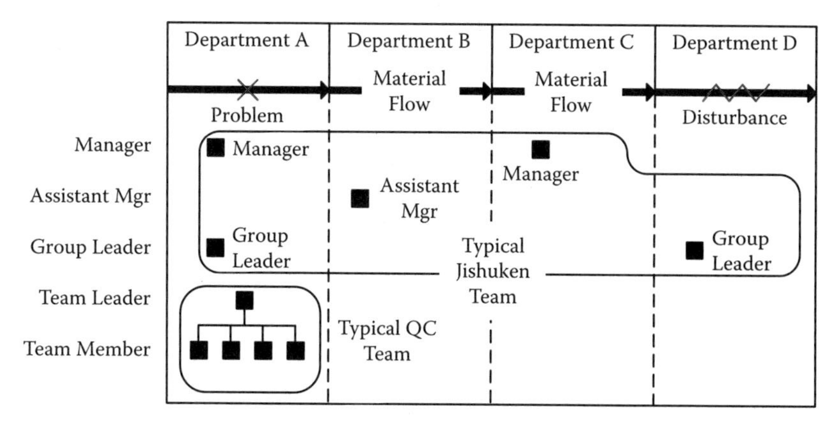

FIGURE 5.19
Example of jishuken and QC team selection.

problem appears to originate. The ODG facilitator will request that particular people join based on various factors, including jishuken experience and applicability of related process knowledge or with no prior knowledge to give a fresh eye. The facilitator will also try to include members from different shifts. In QCs, the teams are selected from members who work in the same department or work area.

5.8.4 Significance of Roles in Management-Directed Activities

Understanding the roles needed and playing these roles effectively in problem-solving teams are critical factors for success. Although the roles tend to be fixed, members are encouraged to develop and strengthen new skills by taking on different roles to learn and improve over time. Three functional roles within any problem-solving team are a leader to manage resources, a facilitator to teach new skills, and members who will contribute ideas but ultimately will support the team's chosen direction.

The QC or jishuken leader allocates team resources and sets the group's goals and targets. A good leader makes sure that everyone has what is needed, whether it is training, tools, or other resources. In jishukens, the leader should be a manager from the department where the problem originated so he or she is able to conduct follow-up after the jishuken. The leader has the final decision-making responsibility for the group but must take care not to block or inhibit the process. A good leader can encourage contributions and then help the team unite to support a decision that does not have full agreement of the group; although even the most successful teams rarely agree on everything, successful teams do support the group's decisions. Since most disagreements involve choosing the right approach to solve the problem, the leader can prevent most conflicts by following the problem-solving methodology with the help of a trained facilitator. Last, the leader keeps the team on track by setting goals.

A role common at Toyota, but rare elsewhere, is that of facilitator. Toyota's facilitator takes the role of a neutral person who guides the group through a structured process while encouraging team participation. The facilitator's main function is to ensure that the problem-solving methodology is applied consistently and team members are fulfilling their roles. A good facilitator not only will be effective at managing interpersonal dynamics within the team but also can step in when the team is stuck. In some cases, this role is confused with that of the tough sensei that refuses to provide solutions and only raises the hard questions. It is true that the facilitator should not provide answers but instead coach the group on how to complete each step in the problem-solving methodology. Last, the facilitator has to make sure that communication among all team members is good. For example, if team members have different understandings and do not communicate to make those explicit, progress can stop, and the team can fail.

Everyone plays the role of a supporting team member in problem solving. Members can also perform supporting roles such as coordinating, planning, tracking, and distributing information. With the exception of leader and facilitator, team roles can be rotated or shared within the group. There are no formal prerequisites for a member to be a part of a jishuken or QC; however, members should have working knowledge of the problem-solving process.

5.8.5 Support Functions in QCs and Jishukens

The ODG is an entity within Toyota that helps with the jishuken process as an aspect of its main role: to strengthen TPS (TMMK, 2009). This group provides assistance by offering training, facilitation, and expertise in various areas. In the case of jishukens, the ODG group helps assemble the team, teaches and facilitates the problem-solving process, tracks progress, and ensures that desired outcomes of the jishuken process are achieved. In the case of QCs, these ODG functions belong to the QC leader, the GL, and the manager. Figure 5.20 illustrates some of these differences in supporting QCs and jishukens.

5.8.5.1 Coach and Facilitate Problem Solving

Conducting problem-solving activities correctly requires much guidance and outside support for the team. With no facilitator, the problem-solving activity may have unexamined biases or otherwise be followed incorrectly. Even the most experienced teams have a tendency to rush to countermeasure selection by skipping and shortcutting necessary steps. This common desire to find and fix problems quickly can interfere with a major purpose of jishukens and QCs by reinforcing behaviors contrary to the principles of TPS and by communicating a mistaken idea of the problem-solving process. Problems prematurely "solved" in this way can cause problems to return.

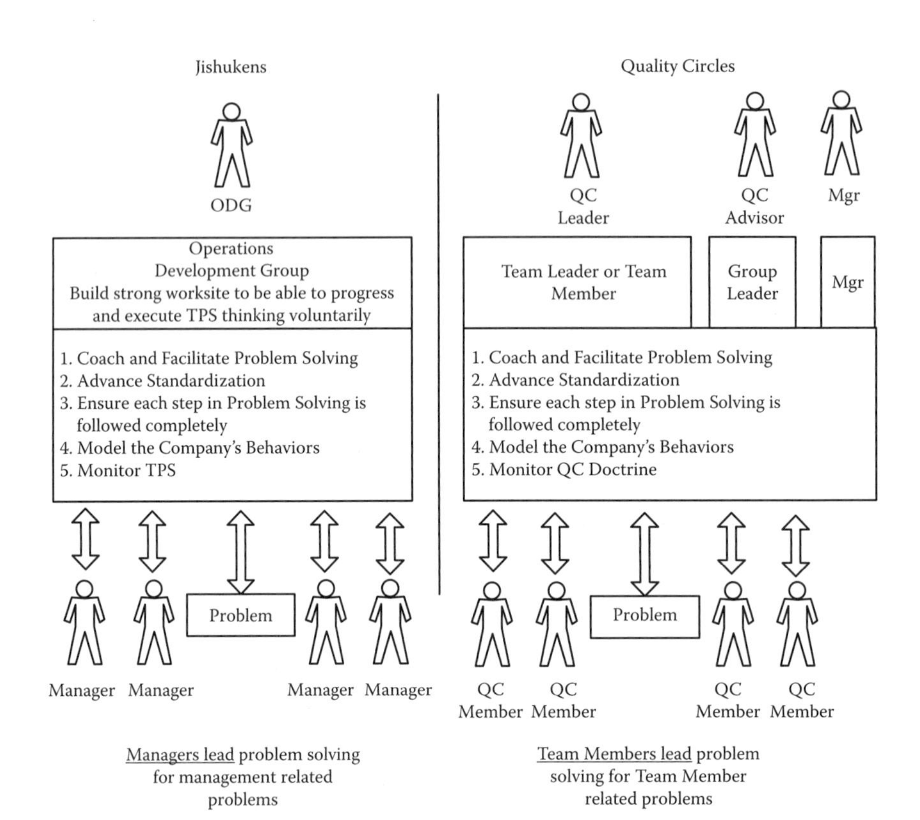

FIGURE 5.20
Support functions of quality circles and jishukens.

5.8.5.2 *Jishukens and QCs Must Advance Stability through Standardization*

A challenging aspect of workplace improvement is to control the expectation that instant improvement is going to occur from a process, whereas typically it can take time to understand the factors that make the process currently stable. Most organizations want to rush to a countermeasure or jump to kaizen. The Shewhart–Deming dynamic model of scientific inquiry (Plan-Do-Check-Act or PDCA) is one of the most widely accepted techniques used to make improvements in the workplace. Interestingly, most organizations follow the PDCA process literally; they start with P (Plan; often translated as a kaizen or idea) to test their improvements. When this happens, organizations rush to the solution without really understanding or knowing if their system is capable. Unstable systems cannot be improved unless they show characteristics of stability. Again,

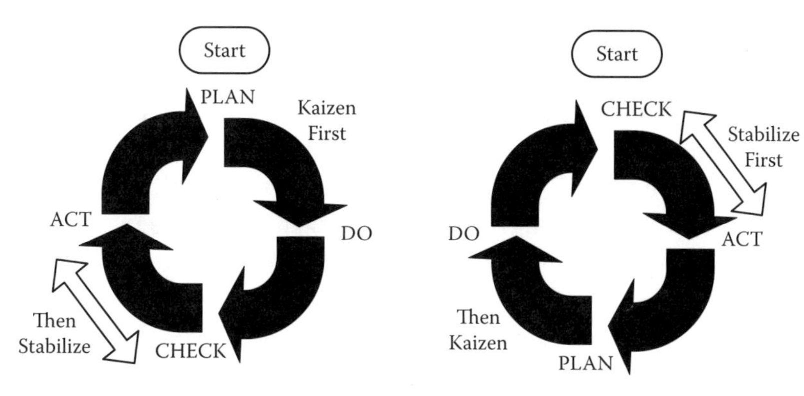

FIGURE 5.21
Representation of standardization and kaizen for the Shewhart–Deming cycle.

jumping from one point of instability to another is still moving from unknown to unknown.

Toyota follows the PDCA process, but it does not assume its systems are stable or even capable. The first step in PDCA is C—Check. Before any improvement is planned, Toyota would like to know the current situation. Is it stable, repeatable, or predictable? If not, current operations have to reach stability through standardization. In this context, Toyota spends much more time stabilizing and standardizing operations rather than improving them (Figure 5.21). TPS is about starting with standardization, then kaizen; rather than starting with kaizen and ending with standardization. In all cases, jishukens must invest time in studying and stabilizing the current condition before a new approach can be tried. This Check-Act cycle ensures that a new approach will not be applied to an unstable system, which is crucial because implementing a new approach in an unstable system has little chance of success.

5.8.5.3 Model the Company's Approach to Work

Jishukens and QCs need to be facilitated carefully to ensure that managers and team members model the company's values, beliefs, and behaviors while working through the problem-solving process. The facilitator can help raise awareness of non-TPS attitudes and work behaviors. If management does not learn the right ways to practice and implement TPS, damage to the organizational culture can outweigh any particular "fix."

Because of management's authority and power to influence, managers should mirror the company's standard for TPS rather than local variations. By the same token, if managers come away from a jishuken with a common understanding of TPS, TPS is strengthened.

Possibly the most difficult aspect of problem solving is managing the human element. Depending on how the jishuken or QC was initiated, several kinds of organizational dynamics could be introduced into the activity. For example, managers may feel personally responsible while their operations are under evaluation or defensive about their processes. The jishuken and QC process is intended to reinforce the TPS principles of trust, respect, and teamwork, in part because problem solving will not work well without them—analysis is difficult if the true behavior of the system is being concealed or falsified by a manager or team member who feels threatened by the process. But, jishukens and QCs are also intended to help employees communicate TPS to others so employees who do not model the right behaviors, attitudes, or demeanors pose the risk of distorting TPS for those with whom they work.

5.9 SUMMARY

The goal of this chapter was to illustrate how TPS is regulated. For TPS to reach its intended purpose—raise the efficiency of the workplace using industrial engineering and problem solving—a formal mechanism is used to monitor and control the system's behavior. One way that Toyota accomplishes this regulation is by conducting jishukens and QCs. These activities act as miniversions of TPS that aim to bring the identity of the system back to a predetermined condition. Jishukens and QCs are designed to work in abnormal and normal states, which means TPS can be regulated regardless of the state of the organization. If TPS can only be regulated in certain organizational states (i.e., when times are good or bad), TPS will become a fad that can start and stop. Consequently, for TPS to survive in the daily lives of organizational members, it must be applied during organizational activities and not after. Jishukens and QCs are a form of anticipatory control that tries to prevent the wrong application of TPS during activities rather than waiting until improvement activities are completed. This is why QCs and jishukens are integrated into normal and daily activities and are not treated as add-ons that do not relate to organizational goals and targets.

Because the strength of TPS is defined by the number of people who can work toward its achievement, the manner of regulation must also support the identity of TPS. For this reason, QCs and jishukens function as high-involvement activities that seek to pass on the identity of TPS through peer-based discussion and interaction. In this way, decision making is natural because managers lead problem solving for management problems, and team members lead problem solving for team member problems. Jishukens and QCs establish a culture among managers and team members that values identifying and solving problems among themselves and in the production workers they teach and guide; this is an essential step toward establishing a single, companywide system.

REFERENCES

Ackoff, R., (1971) Toward a system of systems concepts, *Management Science*, Vol. 17, No. 11, 661–671.

Alloo, R. (2009) Interview: Toyota Motor Engineering and Manufacturing North America (TEMA) and executive in residence, University of Kentucky, Center for Manufacturing.

Aravindan, P., Devadasan, S., Reddy, E., and Selladurai, V. (1996) An expert system for implementing successful quality circle programs in manufacturing firms, *International Journal of Quality & Reliability Management*, 13(7): 57–68.

Avery, G. (2004) *Understanding Leadership*, Sage, London.

Badurdeen, F., Marksberry, P., Hall, A., and Gregory, B. (2009) No instant prairie: Planting Lean to grow innovation. *International Journal of Collaborative Enterprise*, Vol. 1, No. 1, 22–38.

Becvar, D. and Becvar, R. (1998) *Systems Theory and Family Therapy*, University Press of America, Lanham, MD.

Berman, S. and Hellweg, S. (1989) Perceived supervisor communication competence and supervisor satisfaction as a function of quality circle participation. *Journal of Business Communication*, 26: 103–122.

Bicheno, J. (2000) *The Lean Toolbox*, 2nd ed., Picsie Books, Buckingham, England.

Cengel, Y. and Boles, M. (1994) *Thermodynamics, an Engineering Approach*, McGraw-Hill, New York.

Erez, M. (2010) Commentary culture and job design, *Journal of Organizational Behavior Special Issue: Putting Job Design in Context*, 31(2–3): 389-400.

Gigch, J. (1978) *Applied General Systems Theory*, Harper & Row, New York.

Gray, G. R. (1993) Quality circles: An update. *S.A.M. Advanced Management Journal*, 58(2): 41.

Hall, A. (2006). *Introduction to Lean Sustainable Quality Systems Design: An Integrated Approach From the Viewpoints of Dynamic Scientific Inquiry Learning & Toyota's Lean System Principles and Practices.* Published by Arlie Hall, Lexington, KY. ISBN 0-9768765-0-7.

Hall, E. (1976) *Beyond Culture*, Anchor Books, New York.

Hardin, C. and Conley, T. (2001) A relational approach to cognition: Shared experience and relationship affirmation in social cognition, in *Cognitive Social Psychology: The Princeton Symposium on the Legacy and Future of Social Cognition*, ed. Moskowitz, G. B. Erlbaum, Mahwah, NJ.

Heard, E. (1998) Rapid-fire improvement with short-cycle kaizen. *Annual International Conference Proceedings—American Production and Inventory Control Society*, May; 20(4): 15–23.

Heckscher, C. and Adler, P. (2006) *The Firm as a Collaborative Community: Reconstructing Trust in the Knowledge Economy*, Oxford University Press, New York.

Hogg, M. (2007) Uncertainty-identity theory, in *Advances in Experimental Social Psychology*, ed. Zanna, M. P. Academic, San Diego, CA. Vol. 29, pp. 66–126.

Incropera, F. and DeWitt, D. (1996) *Fundamentals of Heat and Mass Transfer*, Wiley, New York.

Kantabutra, S. (2008) Toward a behavioral theory of vision in organizational settings. *Leadership and Organization Development Journal*, Vol. 30, No. 4, 319–337.

Kathawala, Y. (1989) A comparative analysis of selected approaches to quality. *International Journal of Quality and Reliability Management*, 7:17.

Klir, G. (1969) *An Approach to General Systems Theory*, Van Nostrand Reinhold, New York.

Lawler, E. and Mohrman, A. (1985) Quality circles after the fad. *Harvard Business Review*, 63(1): 65–71.

Ledgerwood, A. and Liviatan, I. (2010) The price of a shared vision: Group identity goals and the social creation of value. *Social Cognition*, Vol. 28, No. 3, 401–421.

Liker, J. and Hoseus, M. (2008) *Toyota Culture—The Heart and Soul of the Toyota Way*, McGraw-Hill, New York.

McGregor, D. (1960) In *The Human Side of Enterprise*, McGraw-Hill, New York

McNichols, T., Hassinger, R., and Bapst, G. (1998) Quick and continuous improvement through Kaizen Blitz. *Annual International Conference Proceedings—American Production and Inventory Control Society*, May; 20(4): 1–7.

Montabon, F. (1997) Kaizen Blitz: Introducing a new manufacturing procedure based on the continuous pursuit of perfection. *Proceedings—Annual Meeting of the Decision Sciences Institute*, Vol. 3.

Olberding, S. R. (1998) Toyota on competition and quality circles, *Journal for Quality & Participation*, 21(2): 52–53.

Ouchi, W. (1981) *Theory Z: How American Business Can Meet the Japanese Challenge*, Addison-Wesley, Reading, MA.

Pereira, G. and Osburn, H. (2007) Effects of participation in decision making on performance and employee attitudes: A quality circles meta-analysis, *Journal of Business & Psychology*, 22(2): 145–153.

Putt, M. (1978) *General Systems Theory Applied to Nursing*, Little Brown, Boston.

Sherif, M. (1935) A study of some social factors in perception. *Archives of Psychology*, Vol. 27, 1–60.

Senge, P. (1990) *The Fifth Discipline: The Art and Practice of a Learning Organization*, Currency Doubleday, New York.

Sillince, J., Sykes, G., and Singh, P. (1995) Implementation, problems, success and longevity of quality circle programmes: a study of 95 UK organizations, *International Journal of Operations & Production Management*, 16(4): 88–111.

Smith, E. (1993) Japanese methods of "Lean production" make splash in U.S.—strategy: Business adopt the management techniques of "kaizen" and "jishuken." *Orange County Register*.

Tang, T., Tollison, P., and Whiteside, H. (1987) The effect of quality circle initiation on motivation to attend quality circle meetings and on task performance, *Personnel Psychology*, 40(4): 799–814.

Toushek, G. (2006) *Toyota Industrial Equipment Manufacturing Inc., Riding high, Manufacturing in Action*, Manufacturer US.

Toyota Motor Manufacturing Kentucky (TMMK). (2009) TPS Tools Training, May, Georgetown, KY.

Walker, T. (1992) Creating total quality improvement that lasts, *National Productivity Review*, 3: 473–478.

Ward, M. (2006) Omoshiroi: An appreciative narrative about quality circles in the Toyota Production System, PhD dissertation, University of Kentucky.

Worley, J. (2007) A comparative assessment of Kaizen events within an organization, IIE Annual Conference and Expo 2007, May 19–23, Nashville, TN.

6

The System Property of Differentiation in TPS

6.1 SYSTEM PROPERTY: DIFFERENTIATION

Differentiation is a transformation from a more general and homogeneous condition to a more specialized and heterogeneous condition made up of unlike components (von Bertalanffy, 1968). Differentiation occurs in systems so that certain components can perform specialized functions to better adapt to the open and dynamic nature of their environment and surroundings. Essentially, an organism has to divide its processes and functions to meet the complexity of its environment for it to survive. The degree to which certain units within a system specialize to perform different activities largely depends on the environmental factors and how the system adapts itself in creating modular building blocks that facilitate task performance (Lawrence and Lorsch, 1967). To this end, different parts in a system adapt in different ways, leading to a state of differentiation or segmentation (Figure 6.1).

Each specialized subsystem can help in dealing with fluctuation and variation from the environment, which ultimately leads to evolution (Figure 6.2). The evolution of differentiation means that each of the subsystems is guided primary by its own function. Subunits adapt best when they meet the demands of their immediate environment (Lawrence and Lorsch, 1967). For example, if the environment is unpredictable, the system will differentiate itself to be more organic, specifically less formal and nonlinear to adapt to continual changes. If the environment is predictable, the system will differentiate itself in a manner to be more mechanistic in nature. That is, the structure of the system will be highly organized, direct, and linear. This manner of differentiation follows the theory of environmental contingency, which states that different environments

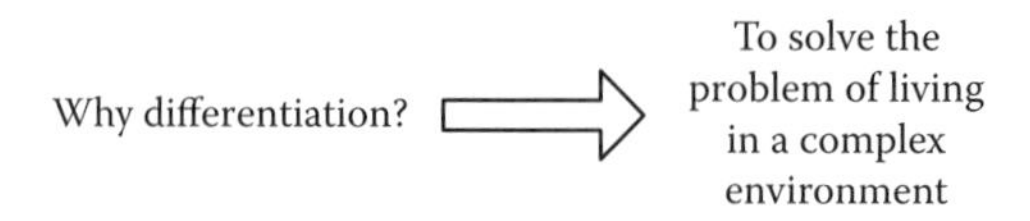

FIGURE 6.1
The purpose of the system property of differentiation.

create different pressures inside the system to differentiate (Lawrence and Lorsch, 1967; Burns and Stalker, 1961).

From the viewpoint of organizations, differentiation is a way of dealing with the complexity of environmental factors either inside or outside the company. Subdividing functions and tasks is a common way for organizations to concentrate and focus on specific outcomes (Weber, 1946). Differentiation in organizations is a process of increasing the complexity of systems to allow for more variation within the system to respond to variation in the environment. Like biological systems, increased variation facilitated by differentiation allows for not only better responses to the environment but also faster evolution, whether it is technological, sociological, or social–technical (Ritzer, 2007).

Differentiation can occur vertically and horizontally in organizations. Stratificatory differentiation is a form of vertical differentiation according to rank or status in a system conceived as a hierarchy. Every rank fulfills a

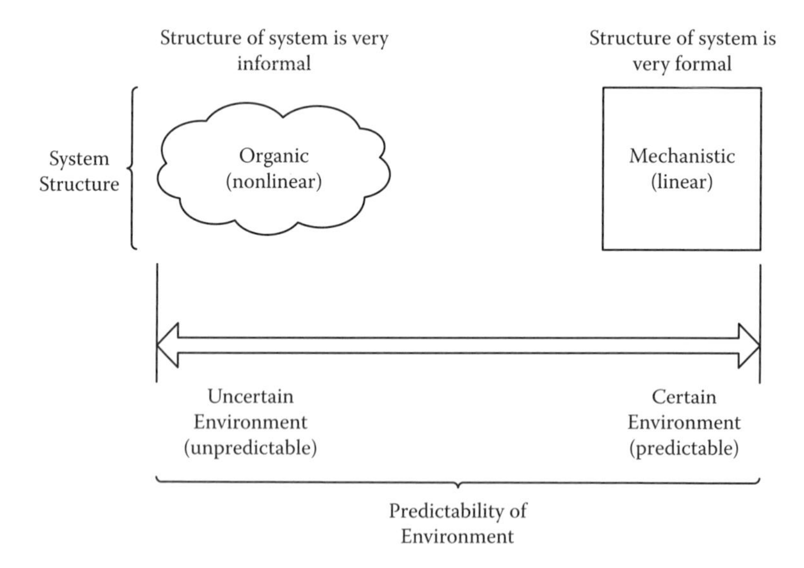

FIGURE 6.2
Comparison of a system's structure depending on the predictability of the environment.

particular and distinct function in the system. Subsystem differentiation in organizations is dependent on the differences in each subunit's environment. Organizations cope with large-scale problems by subdividing responsibilities horizontally in numerous ways (Blau, 1970). Integration is a driving characteristic in organizational differentiation because collaboration among subunits is needed to achieve common goals. The more differentiation there is, the more elaborate the integration devices that will be needed within a system. Turner proposed that an organization will integrate its subsystems depending on a variety of factors (Turner, 1981). The manner of integration will be influenced by the following:

- Degree of intra- and intergroup regulation and coordination
- Degree of subgroup formation around diverse productive activities
- Degree of coordination vested in centralized authority
- Degree of organized opposition to centralized authority
- Degree to which coordination is sanctioned

6.2 TOYOTA PRODUCTION SYSTEM AND SYSTEM DIFFERENTIATION

Differentiation has the goal of fulfilling system functions that cannot be substituted by any other part of the system (von Bertalanffy, 1975). An example of a differentiation in the Toyota Production System (TPS) is the role of leadership. Leadership is a complex process; its role is to establish the direction of the organization and to influence organizational members to work toward its achievement. Leadership is essential because it provides the destination, the path, and the justification for the existence of TPS. Leadership defines the identity of TPS and how it is to shape and evolve to meet the changing needs of the organization. Leadership can anticipate how the organization's management systems will need to be refined, organized, and differentiated to meet the complexities of a changing environment. The leadership function within TPS is not a subsystem that can be replaced by a kaizen promotion group, the business excellence office, or most importantly, through a dispersed leadership approach that encourages small pockets of the company to practice one system while its neighbors work to another. Leadership's role is to integrate and unify the business systems of the organization to accomplish companywide efforts

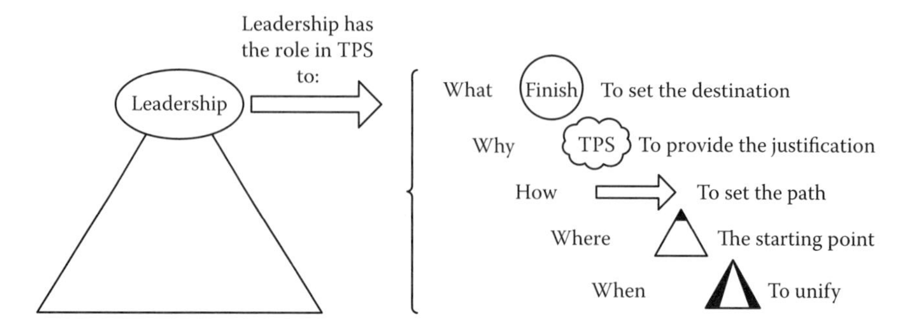

FIGURE 6.3
The role of leadership in TPS.

(Figure 6.3). Only leadership can promote cooperation and collaboration among units in achieving organizational goals that cut across sections, departments, plants, divisions, and regions.

6.3 LITERATURE REVIEW OF LEADERSHIP

Over the last century, many descriptions and terms have been given to define the meaning of leadership. Leadership is a broad and multidisciplinary topic that ranges over several areas, from the social sciences to psychology and from economics to politics. The development of leadership has traditionally been described as sequential, meaning scientists generally like to organize the development of leadership as categories of ideas built over many years.

In the early 1900s, leadership was viewed as a process of control that directed the behavior of others (Cooley, 1902; Blackmar, 1911). Leadership was believed to be successful, largely due to the traits and characteristics of the individual. Successful leaders were those who demonstrated certain behaviors, skills, and traits. A few decades later, leadership broadened to include areas outside a leader's traits or particular style. In the 1940s, leadership studies focused on followers and the aspects of group cooperation and socialization. Successful leaders were defined as those who could influence organizational members to follow them from a social perspective (Pigors, 1935).

In the 1950s, leadership shifted attention toward the functions of leadership rather than how they behaved as individuals (Beense and Sheats, 1948). Leadership processes include strategy development, financial systems, planning, and communication. In the 1970s, leadership emphasized the context and situations that leaders must work in and adapt to be successful. There is evidence to suggest that persons who are leaders in one situation may not necessarily be leaders in another (Stogdill, 1974). In this context, situational leadership and contingent theories were used to describe the types of factors that could make a leader effective in certain situations.

In the 1980s, motivational theories and concepts about leadership became extremely popularized. Leadership was viewed as successful when it could create changes in a follower's behavior by attending to their motivations. Various researchers studied leadership based on extending Maslow's hierarchy of needs (Maslow, 1946). In this context, the leader and the follower could be characterized by a series of exchanges or transactions. Transaction leadership has been the traditional model for most organizations focused primarily on the bottom line (Bolden et al., 2003).

In the era of globalization, leadership focused on its effectiveness to lead in turbulent times (Drucker, 1980). Transformational and change management leadership approaches became widespread in the attempt to evaluate the needs of the organization over the long term compared to the immediate (Bass, 1990). In this same era, the economic boom of Japan created much interest in Japanese management techniques. Theory Z proposed that effective leadership is the building of cooperation and shared meaning throughout the organization (Ouchi, 1981). Japanese management techniques tend to display participative forms of leadership, including social activities outside work in an attempt to promote homogeneous culture (Fukushige and Spicer, 2007).

6.3.1 The Distinction between Management and Leadership

A common comparison in literature is the distinction between management and leadership. Management refers to what the organization does to control the organization's processes and activities to meet organizational targets and goals. Management systems are concerned with satisfying customer requirements by establishing processes that monitor, control, and maintain current operating conditions. Management and management systems are designed to bring stability to the organization by managing activities that bring predictability (Barker, 1995). Management is a rational activity that exists formally within the organization, which means it is regulated by the organization's structures, rules, and operating practices.

Leadership has the role of articulating the vision, giving it legitimacy, and making it a priority for organizational members to follow (Tichy and DeVanna, 1986). Good leaders work at the vision continuously and make it the core of every conversation (Clutterbuck and Hirst, 2002). Leadership is vitally important to the success of the organization because it promotes change when there is a difference between what exists and what is desired (Hopper and Potter, 1997). In this context, leadership is founded on responding to environmental influences. It is changes in the environment that acts as a catalyst for the leadership process (Barker, 1995). Crisis gives people incentive to consider actions, trade-offs, and sacrifices that they would not have considered otherwise. Leadership systems refer to how leadership is exercised, formally and informally, throughout the organization and the basis for the way decisions are made, communicated, and carried out. Leadership systems are those that enable mechanisms for decision making across all locations, at all levels, by all leaders, all the time. While the function of management is to create stability by anticipating change, the primary function of leadership is to create change (Barker, 1995) (Figures 6.4 and 6.5).

Leadership and leadership systems take place outside the organization because they create it. Once a structure is created, it is maintained and adapted by the management function (Barker, 1995). Leadership is disruptive to the management process because leadership challenges the organization to change. Deming (1986) referred to change as something essential for the continued survival of American management: Requiring a whole new structure, leaders cannot be restricted by structures that already exist in the organization.

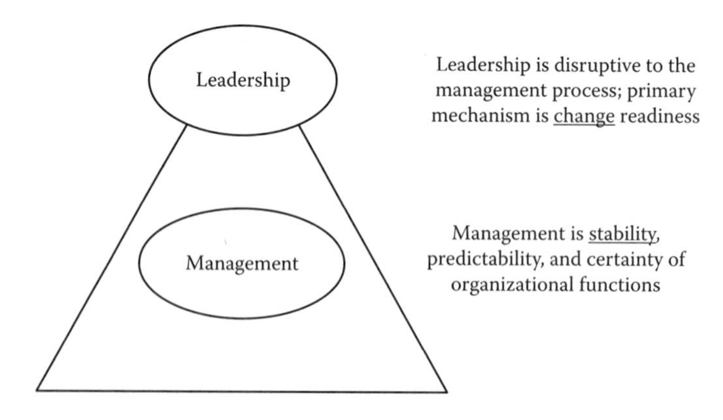

FIGURE 6.4
The role of management and leadership.

FIGURE 6.5
The position of management and leadership in the organization.

6.4 LITERATURE REVIEW OF LEADERSHIP THEORY

All leadership theories indicate that some form of hierarchical leadership is always important in influencing subordinate satisfaction or performance (Kerr, 1977). Even if leadership is redundant or relevant, all theories about leadership assume that leadership is significant to organization performance (Rowe et al., 2005). There is a romantic notion that leadership is always a determinant of organizational outcomes (Meindl, 1993). Despite the occasional success and failures in organizations, leadership theory has been shown to have significant effects on subordinate behaviors in some situations and no effects in others (Padsakoff et al., 1993). Leadership theories vary in their prescription of how to accomplish behavior change among subordinates given the situation, context, and conditions. Leadership models and frameworks that emphasize behavior change have been suggested to be universal regardless of cultural differences (Fukushige and Spicer, 2007). The common denominator in leadership theories is that organizations are open systems that depend on human direction to succeed and call for a more person-centered investigation (Armenakis and Bedeian, 1999).

Leadership theory has been used to answer the five key questions about leadership effectiveness: What? Why? How? Where? and When? The *what* refers to the goals that leadership tries to achieve. The *why* gives justification for pursuing the new goals (Mostovicz et al., 2009). The *how* describes the approach that leadership will use to guide the organization in meeting its new goals. The *where* illustrates the starting point for achieving these new goals, which should redirect to leadership's own behavior. The *when* describes the pace of change and the adjustments needed along the way and the manner of involving and unifying the organization as a whole.

6.5 LEADERSHIP THEORY FOR TPS

A popular leadership theory that has been used to describe Japanese management as it is applied to American companies is known as theory Z (Ouchi, 1981). Theory Z is based on many of the same qualities of Japanese management, such as lifetime employment, slow promotion, nonspecialized career paths, participative decision making, collective values, holistic concern for people, and succession planning. Ouchi made the case that each type of company has its own Z distinctiveness, and that organizations are rarely purely of the form A or Z. While theory Z was not written to describe TPS, it does share many of the same national cultural tendencies of Japanese management.

Currently, no single American or Japanese theory completely describes the distinctiveness of Toyota's leadership approach. Each theory emphasizes a specific approach used by leadership to accomplish a certain goal or organizational outcome. Depending on the organizational state, this would leave practitioners to decide which theory or leadership style to select based on their immediate needs. Since no theory can universally meet the needs of each business condition and climate, several leadership theories exist to describe different leadership behaviors and functions that are most effective in different organizational circumstances. The end result is a leadership style and behavior that changes for each environment and for each leadership function.

It is important to note that the leadership style should not be confused with the leadership direction. The direction of the organization will change and differentiate itself according to system theory. However, there is a notion that the leadership style and behavior of TPS is not dependent on organizational circumstances. This is illustrated in Toyota's strong dependence on procedure-based systems, succession planning, and the use of highly defined roles. In Toyota's case, a new leader does not imply a new leadership behavior or style or new way of carrying out leadership functions. The end result is a consistent way of behaving and carrying out functions regardless of the organizational state. Again, this does not imply that leadership does not set a new direction when the organization is influenced by environmental factors; it only suggests that approach to leadership is consistent and carried out in a predictable way for each leadership function. The best way to describe the theory of TPS as it relates to leadership is to match the particular framework with each leadership function.

6.6 THE *WHAT* FUNCTION OF TPS LEADERSHIP

Leadership has the role to set the direction of the organization and the means to get there. In TPS, the role of leadership is no different. Leadership has the role to define the destination and the future image of TPS. Simply, leadership must create a positive and self-rewarding vision of TPS to the extent that members will embrace it in their lives voluntarily. According to Senge (1990), there are two types of visions leadership can use to influence organizational members: negative and positive visions. Leadership can use a negative message, for example, "change or die," to pull the organization together by creating a state of fear. Deming (1986) would argue that management who uses fear to motivate employees is creating unnecessary anxiety and stress in the workplace. Employees who have fear lose creativity and in turn substitute it with anger, which cannot be expressed at work (Hall, 2008). Fear is often associated with defensive visions, such as wanting to beat a competitor. Once the goal is accomplished, the vision is no longer applicable. In this context, fear-based visions are short term, extrinsic, and transitory.

One of the most popular visions of Lean is the elimination of workers. Organizations that emphasize head-count reduction give the impression that "less employees are needed" (LEAN). TPS is a management system that seeks to raise involvement of the workplace; firing employees after cost reduction efforts is counterintuitive. Why would an employee contribute an idea if there is the fear of losing one's job? Toyota wants to eliminate waste, not its people.

The leadership role in TPS is to create a positive vision that is compelling, inspiring, imaginative, and exhilarating. Womack and Jones (1990) described Lean as the continuous pursuit for perfection. While it is impossible to ever reach perfection, organizations are encouraged to strive for it. This view of Lean is congruent with the quality philosophy of continuous improvement. Various quality initiatives have been proposed over the years. However, is the idea of continuous improvement enough to influence organizational members to follow TPS? Perfection is positive, yet is it motivating?

Another perspective is to view what lies at the end of the finish line for employees who follow TPS. Does Toyota expect employees to buy into the idea of continuous improvement from an ideological perspective? According to Senge (1990), the best visions are intrinsic motivators that coincide with the personal vision of individuals. Intrinsic motivators offer the greatest source of human energy because it is driven by interest, enjoyment, and aspiration, which is enduring and inspirational. Intrinsic motivation is what an employee engages in without reward (Reiss, 2004). Extrinsic motivation is driven by external pressure to succeed using factors outside the individual such as money or threat of punishment. Some work suggested that several factors promote intrinsic motivation, such as challenge, curiosity, fantasy, control, competition, cooperation, and recognition.

One of the most powerful forms of intrinsic motivation is control. Human beings have the basic tendency to want to control what happens to them. TPS uses control to enhance intrinsic motivation because Toyota wants workers to have a voice in how the work is performed (Figure 6.6). A work environment that would allow workers to display their full capabilities themselves through decision making would be a foundation of human respect of the highest order (Sugimori et al., 1977). Toyota believes that it is human nature for workers to want to see their ideas realized. Workers will not be motivated to eliminate waste or take part in improvements if they are not involved in decision making. What is better than perfection, and possibly more motivating, is having the ability to control one's environment and be a part of the continuous improvement effort personally. Only leadership can define TPS as something that is internally motivating for employees and long lasting. If TPS appeals to the extrinsic side of the business, TPS will only live when the system is paying (i.e., using money) or someone is stimulating (i.e., using fear) to achieve TPS. Research has shown that the more extrinsic motivation is used in the workplace, the greater intrinsic forms of motivation are diminished. TPS is not based on extrinsic rewards or tangible things that exist outside the self of the individual. The leadership function is to define TPS as something that is shared between the company and the individual that is a source of positive energy.

Various leadership theories exist with vision-communicating components that relate to intrinsic motivation. The best explanation of Toyota's

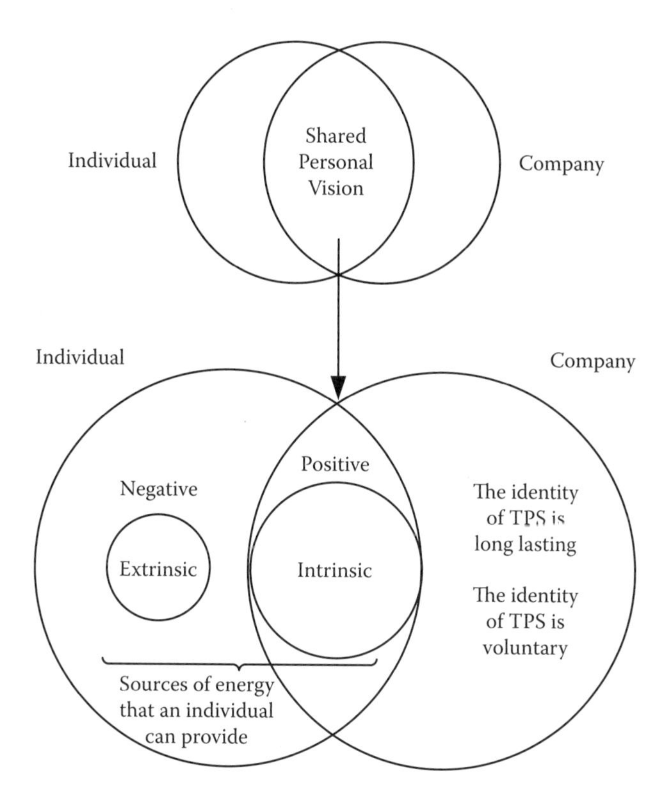

FIGURE 6.6
Motivating factor for the vision of TPS.

leadership function as it relates to defining TPS is a higher order of Maslow's needs theory. Maslow proposed that individuals abide by a hierarchy of needs in which lower-level needs such as psychological needs, safety, and belonging must be fulfilled before an individual can be motivated to achieve higher-order needs such as esteem or self-actualization (Maslow, 1943). Maslow's hierarchy of needs is the basis of various motivational theories in leadership that separate external (lower-level) and internal (higher-level) needs.

TPS favors the higher-order needs of esteem and self-actualization because the needs of competence, mastery, and control rest more on the inner self. Familiar to the theme of continuous improvement, Toyota has defined TPS as a way for employees to reach their full potential while having personal control in improving their jobs in the workplace. TPS is positive and motivational because it inspires employees to become everything that one is capable of becoming by continuously and voluntarily exercising control to make their jobs better.

Another critical function of leadership, as it relates to defining TPS, is making sure it is something that is clear. The vision of TPS should be a clear destination that everyone in the company can look to when measuring their progress toward its achievement. Leadership has the role to provide a clear vision of TPS so that the right things may be done in the organization. *Clarity* is defined as the extent to which a vision can be made clear in approximately five minutes (Kotter, 1999). When the vision is clear, an organization can accomplish its business priorities in the most direct and effective fashion. When the vision is abstract, organizational members can misinterpret organizational goals and think of the vision as all inclusive rather than a way to establish organizational interest. Unfortunately, Lean has been described similarly.

Lean is often defined as an endless journey of waste elimination, improved quality, and the reduction of lead time (Figure 6.7). The Lean journey is often an endless one. A common definition of Lean is often philosophical: waste elimination, shorter lead times, and improved quality. Often, companies focus on the philosophy of continuous improvement rather than building the management system that achieves continuous improvement. Consider the example of total quality management (TQM). The reason why TQM failed for so many organizations is that it was not defined as a management system. It was defined as a philosophy, which cannot bring action to move people in the organization. A management system is one that defines standards, assigns responsibilities, and

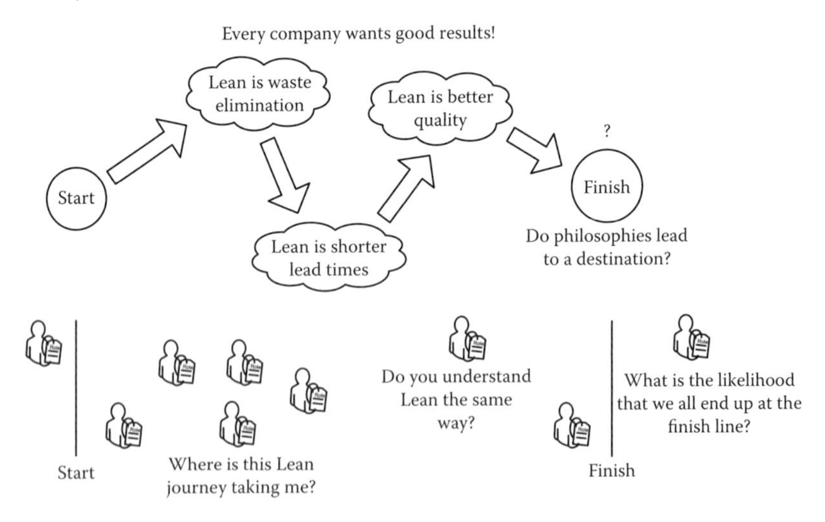

FIGURE 6.7
Traditional approach to defining Lean (unclear vision or destination).

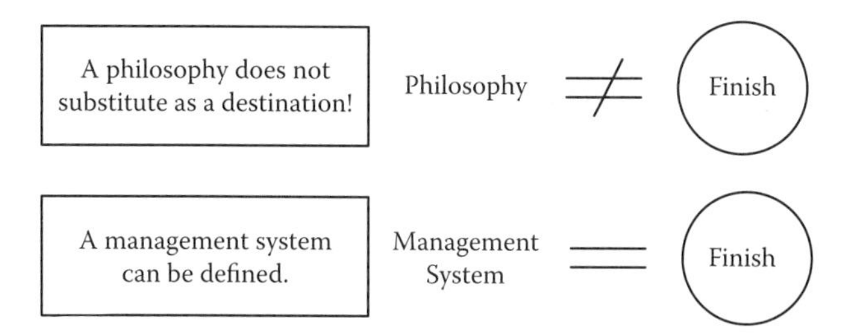

FIGURE 6.8
Better to think about TPS as a management system rather than a philosophy (management systems can be installed, philosophies cannot).

documents and controls organizational functions. Management systems can be installed in organizations much quicker and more effectively than philosophies. While management systems do not guarantee the immediate sanction of behaviors, they do represent a more concrete way to think about TPS (Figure 6.8).

A leadership theory that best describes Toyota's approach in defining a clear vision of TPS is participative leadership theory. Participative leadership theory is a style of leadership that encourages input from those persons responsible for carrying out the work. The idea is that the quality of decision making can be improved by those who have knowledge of the process. Participative leadership raises the commitment of members because they have a sense of control and ownership to see their ideas realized. Participative leadership theory is particularly effective in group settings because collaboration and cooperation among work groups are enforced rather than competition.

According to the identity of TPS and participative leadership theory, the destination of TPS is simply a group of employees performing industrial engineering and problem solving in achievement of the company's goals and targets (University of Kentucky, 2011). Toyota's management system is nothing more than a systematic way for its members to achieve results while working toward organizational goals. Organizations wanting to emulate Toyota only have to install a management system that structures workplace study that appeals to the worker's primary intrinsic motivation (i.e., decision making). When workplace study and improvement occur voluntarily by a group of workers, TPS is working.

The ultimate goal in TPS is to get all levels of the organization and groups to perform workplace study voluntarily. An organization that is able to achieve participation from all connected groups is effective and ideal. In

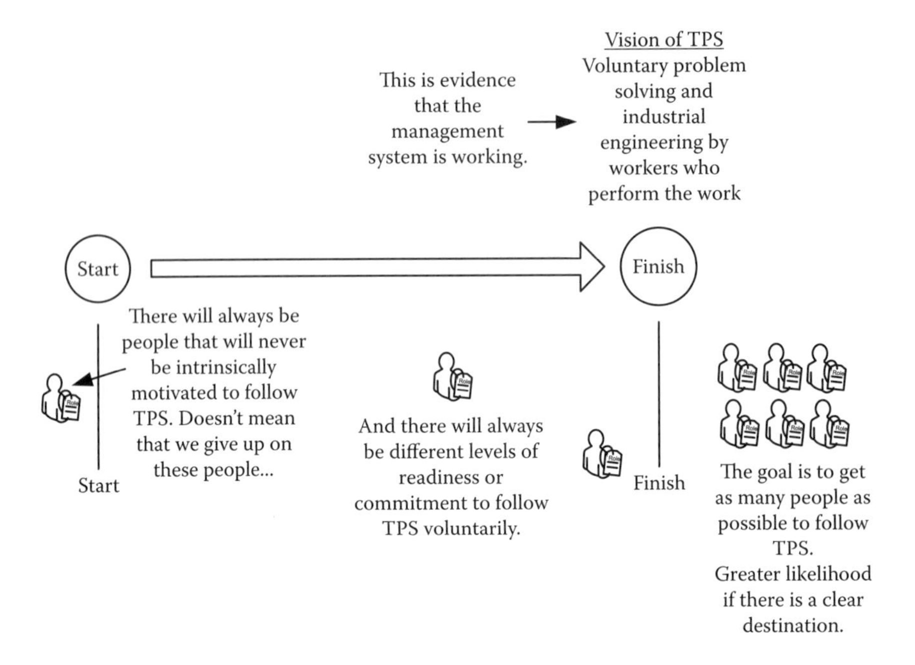

FIGURE 6.9
Improved approach to defining Lean (clear vision or destination).

addition, the more levels and groups that are able to perform workplace study strengthen TPS. A clear destination and vision of TPS ensures that more members are able to work in alignment with TPS without misinterpreting or taking on other interests that are not in line with the organization's targets and goals (Figure 6.9).

6.7 THE *WHY* FUNCTION OF TPS LEADERSHIP

One of the major roles of leadership is to establish a management system to improve the way the organization operates. A management system is defined as a way to control and allocate organizational resources in the achievement of targets and goals. TPS is one example of a management system used to make organizational functions more efficient. In this sense, TPS plays a critical role in Toyota's success.

What makes TPS interesting is the type of management system it represents. There are a variety of different operating and management systems in existence. The common denominator in these systems is that they emphasize a single dimension, such as quality, cost, or productivity, as important for achieving organizational goals. TPS differs from most management systems because it encourages improvement of organizational functions based on need rather than maximizing a certain performance characteristic. While most management systems would improve productivity for the sake of improving productivity, TPS would improve organizational practices based on need.

In 1949, Toyota suffered the worst financial crisis and labor disputes of all time. Toyota did not have any money to emulate Ford's large-scale inventory production system or the market to support the capital investment of several dedicated lines. Toyota had to learn to be resourceful, mainly because the only two real resources the company had after the war were its people and the effort of its people (Kreafle, 2010). TPS was not born to solve the problem of a modified Ford production system; TPS was created as a way to solve three basic needs:

- The need to move a group of individuals with a common purpose to a higher level of achievement than they would achieve individually
- The need to reestablish a relationship with employees without coercion.
- The need to challenge all employees beyond their comfort zone

The leadership theory that describes the approach that Toyota applied in establishing the existence of TPS is transformational leadership. Transformational leadership occurs when a leader transforms, or changes, his or her followers, which results in followers trusting their leader, performing behaviors that contribute to the achievement of organizational goals, and being motivated to perform at a high level (Bass, 1985). Transformational leadership is defined as a leadership approach that causes change in individuals and social systems. Transformational leadership is a model that encourages meeting organizational goals in changing times.

A leadership theory that is often contrasted with transformational leadership is transactional leadership. Transactional leadership theory is primarily concerned with the bottom line. It is concerned with the short term and relies on a series of exchanges between the leader and the follower to encourage subordinate behavior (Bolden et al., 2003). Toyota could have very well established TPS as a form of transaction leadership considering the condition of the company. If this were true, TPS would be known as a management system based on rewards and penalty. In other words, if a

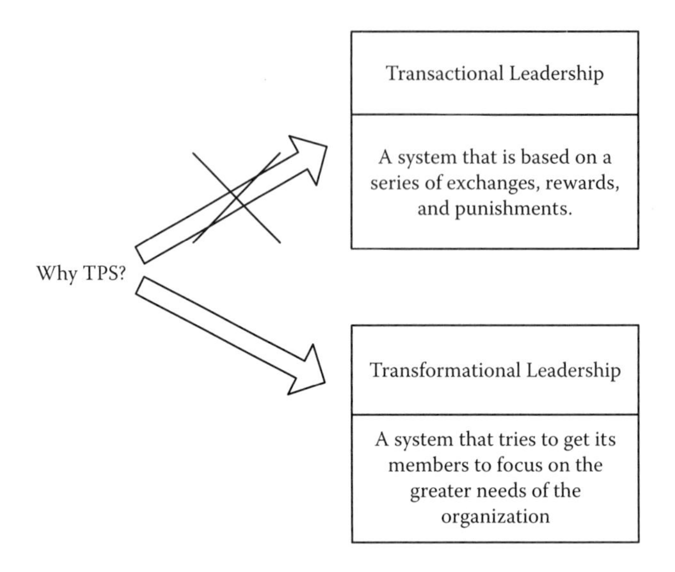

FIGURE 6.10
The reason for TPS.

team member did something good, the team member would be rewarded; if the person did something wrong, he or she would be punished.

Transformational leadership encourages behavior change by increasing members' awareness of what is important beyond their own self-interest (Bass, 1997). *The reason for TPS is to get followers to accomplish more than what is expected of them by instilling in them a greater sense of purpose that goes beyond their self-interest* (Figure 6.10).

6.8 THE *HOW* FUNCTION OF TPS LEADERSHIP

Leadership must also be responsible for implementing TPS. Toyota views that TPS is not easy and cannot be achieved without guidance. Leadership has the role of providing assistance when the path to TPS is uncertain or nonroutine. Leadership has the role of removing roadblocks that are stopping the achievement of TPS and to increase the confidence of members along the way. A model that is often used to describe leadership's basic role in implementation is the path–goal theory of leadership.

Leadership is justified when the path and goal are not clear in organizations (Figure 6.11). In this situation, it is assumed that followers are at a lower level of readiness or capability and need help in reaching

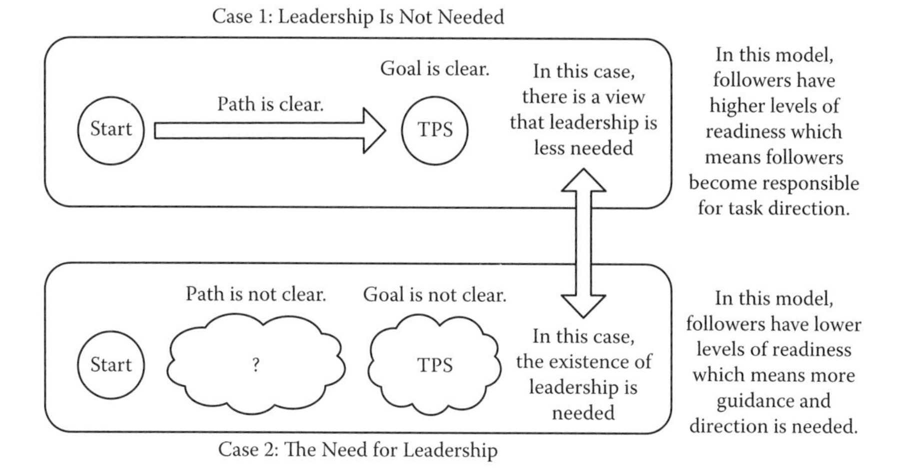

FIGURE 6.11
The need of leadership defined under the path–goal theory context.

organizational goals. On the other hand, when the path and goal are clear, followers are less dependent on the leadership function. Followers become responsible for task direction, which decreases the dependency of the leadership function (House, 1971).

Surprisingly, most organizations do not consider leadership as a key role in defining and guiding their Lean transformation. Organizations that view that the destination and path to Lean is clear simply do not consider leadership as an important function. In this context, Lean is delegated to a lower level at which tasks are allowed to be self-managed.

A leadership theory that supports a nonacting role of leadership is known as substitute leadership theory. Substitute leadership theory suggests that the characteristics of the organization, the task, and the subordinates may substitute or negate the effects of leadership (Kerr and Jermier, 1978). Substitute leadership theory was proposed in the late 1970s primarily because there was a lack of data to substantiate that leadership is relevant. Despite many attempts to confirm the importance of hierarchical leadership, in many situations leadership behaviors were determined to be irrelevant, leading to claims that the concept of leadership has outlived its usefulness (Miner, 1975). A leadership substitute is any person or thing that is acting or used in place of another. Kerr and Jermier (1978) identified three major variables that can be used as potential substitutes for leadership; namely, characteristics of subordinates, task characteristics, and organizational characteristics.

Characteristics of the organization that can act as a leadership substitute include

- How well an organization uses formalized/inflexible rules
- How close-knit the group/group cohesiveness is
- The use of organizational rewards not under control of the leader
- The spatial distance between the superior and subordinates

Characteristics of subordinates that act as a leadership substitute include

- The degree of ability, experience, training, and knowledge
- The need for independence
- The indifference toward organizational rewards

Characteristics of tasks that act as a leadership substitute include

- The clarity and routineness of the job
- The intrinsically satisfying degree of the job

The concept behind substitute leadership theory is to move the organization from its dependency on leadership. The theory proposes that, in time, direct leadership will become less and less important, and self-management will take hold as a substitute for leadership.

Self-management, also known as self-control, is defined as the ability of a person to display self-control when in the relative absence of immediate external constraints, so engaging in behavior by choice when more than one alternative is present (Thoresen and Mahoney, 1974). The idea is that the more professionalism is emphasized in the organization, the less leadership is needed. Kerr and Slocum (1981) have noted that professional employees in organizations often resist efforts by their hierarchical superiors to guide and control the employee's work behavior because, to them, the path and goal are clear. The theory goes on to propose that the more leadership is emphasized, the less likely professionalism is to develop.

At Toyota, leadership has the function to define and implement TPS. Toyota does not consider that the less it is involved, the more employees will develop an understanding of TPS on their own. In fact, leadership considers TPS to be difficult to understand and implement without help. A theory that is in direct opposition to substitute leadership theory, and more closely resembles Toyota's view on the implementation aspect of TPS, is servant leadership theory. Servant leadership theory emphasizes a style of leadership that gives a leader the duty to serve his or her followers rather than a desire to lead (Greenleaf, 1969). The idea is that organizational goals will be achieved by fulfilling the needs of followers (Bolden et al., 2003). Servant-style leadership encourages collaboration, trust, foresight, listening, and most importantly, teaching. Servant-style leaders are not bound by status and look toward building loyalty of their followers through the process of serving others. Servant-style leadership is based on the principle of reciprocation: When you do something for another person, that person is psychologically obliged to return the favor. Servant-style leadership shows concern for followers' welfare and aims at creating a friendly work environment by building self-esteem in followers.

Servant-style leadership best describes Toyota's approach to implementing TPS mainly due to the emphasis on teaching, coaching, and supporting team members when learning or practicing new skills. Leadership has the role to understand TPS to the point they can demonstrate the skill firsthand and support team members when help is needed. The best help a supervisor or manager can provide to his or her subordinates is to show the workers how to do the job directly. Toyota believes that the student shall never fail. Every Toyota manager should behave as a teacher who takes full responsibility for student learning. Consequently, the best teachers are those who practice and stay active in their profession. Servant-style leadership is essential in TPS because it keeps leaders and managers grounded in understanding and applying TPS. Industrial engineering is a challenging discipline that takes careful study and examination of workplace problems. Working on the right problem or solving the problem in the right way requires help. Toyota views that leadership is in the best position to help employees because they have solved similar problems in the past, on a larger scale when applying TPS.

6.8.1 What If Leadership Is Not on Board with Defining, Coaching, or Implementing Lean?

This chapter introduced a theory that describes how an organization can adjust various internal conditions of the company to replace or substitute leadership. It is not the goal of this book or chapter to lead organizations ideally into this direction. This book provides ideal conditions that practitioners should consider when designing and implementing Lean systems. Substitute leadership theory shows why some leadership behaviors are effective in some situations and not in others (Kerr and Jermier, 1978). The theory can help practitioners identify when organizational factors may have a greater effect on subordinates than leadership. The theory can also identify conditions and design in conditions for which it might be desirable to build in substitutes into the organization if leadership is traditionally uninvolved in Lean-type initiatives.

True substitutes are those that only duplicate the leadership function and do not negate, neutralize, or enhance leadership's ability to influence. For example, neutralizers are attributes of subordinates, tasks, and organizational aspects that interfere with the leader's aim to influence workers. Enhancers are variables that have the real possibility to influence and strengthen the leadership function; examples include supportive and cohesive work group norms. Substitutes are those that make leadership redundant and not needed.

6.9 THE *WHERE* FUNCTION OF TPS LEADERSHIP

Where should TPS start? Most organizations start Lean where value is created, where a process transforms a raw material into a finished product, or where a service is provided for a fee. By eliminating processes that contain waste or anything that the customer is not willing to pay for, organizations are justified in starting Lean. Following this logic, anything that drives cost reduction is justified by most organizations. TPS argues that the way organizations go about achieving cost reduction is fundamentally different. All management systems encourage cost reduction, but how leadership defines the system to achieve cost reduction is important. This implies that not all systems that achieve cost reduction are allowed or encouraged at Toyota. Unfortunately, most Lean practitioners misinterpreted Ohno's cost reduction principle because it gave the impression that any and all cost reduction measures are good for the company. From Ohno's point of view, sales prices are fixed in a global economy, which meant to raise profits the cost to produce had to be lowered. However, the ways to achieve cost reduction are just as important as the cost reduction principle itself.

The starting point of TPS is not the shop floor or even where value is created. The starting point of TPS has a prerequisite: It is leadership that must first establish the way the organization will achieve its organizational goals. If not, everyone will approach cost reduction in his or her own way or, even worse, for personal interest. Managers could establish cost reduction activities for their departments and jeopardize organizational functions in other areas. While cost reduction is essential, the right type of cost reduction is a leadership issue.

Leadership cannot be pressured by outside influences or ideas that all cost reduction activities are good for the organization. There are moral and ethical obligations associated with cost reductions. Leadership must exercise a balance between information processing and the protection of organizational values (Lester, 1993); specifically, there needs to be an unbiased collection and interpretation of negative or positive information when achieving cost reduction. Leadership has to set examples of desired behavior and implement reward systems that are consistent with desired outcomes (Barker, 1995).

The leadership theory that best emulates the starting point for TPS is authentic leadership theory. Authentic leadership theory is based on the idea that values, standards, and integrity play a direct role in doing the right thing for the organization in the long run (Harter, 2002). Authentic leaders lead from personal conviction rather than short-term rewards. Leadership has the role to mentor and role model the organization's systems, values, and practices (Lester, 1995). Authentic leadership theory emphasizes self-awareness that gives an organization a sense of self that provides an anchor for decisions and actions (Walumbwa et al., 2008). Toyota's leaders are authentic because they lead TPS through their own behaviors and actions. TPS is not something that is delegated or left up to another group to perform. TPS starts when leadership changes personally.

6.10 THE *WHEN* FUNCTION OF TPS LEADERSHIP

Organizations that continuously show concern for production rather than people run the risk of creating an authoritative and directive leadership style. By continuously telling employees what to do at specific times, with no orientation to people, gives the impression that employees are inexperienced and need constant control to meet organizational objectives. A leadership theory that encapsulates this Western behavior style is the Blake and Mouton managerial grid (Figure 6.12) (Blake and Mouton, 1964). This grid describes five behaviors when leadership is applied, which

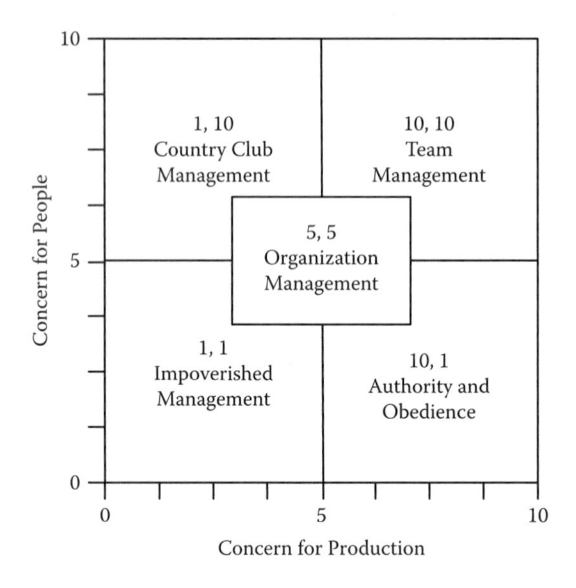

FIGURE 6.12
The Blake–Mouton managerial grid. (Adapted from Blake, R. and Mouton, J., 1964, *The Managerial Grid*, Gulf, Houston, TX.)

is composed of two axes: concern for people and concern for production. Blake and Mouton proposed that "team management," a balanced composition of people and production is the most effective form of leadership behavior (Bolden et al., 2003).

McGregor's theory X (Figure 6.13) implies that it is human nature to be lazy, which means people have to be coerced, controlled, directed, or threatened with punishment to get them to put forth adequate effort to achieve organizational goals (McGregor, 1960). Theory X proposes that leadership is always exercising control without any other forms of influence or intrinsic motivation. This approach would suggest that leaders treat the average employee as a person who tries to avoid responsibility, has little-to-no ambition, and can best be described as an isolated wage earner. The average employee dislikes work and will make every attempt possible to avoid it.

McGregor also suggested a more positive view of human nature that he referred to as theory Y (Figure 6.14). A theory Y leader would view employees as self-directed, committed to achieve objectives, and with a high capacity to learn and be creative. Under this view, the average employee would seek responsibility, exercise self-control, and look to be involved in organizational objectives.

FIGURE 6.13
McGregor's theory X approach to leadership.

FIGURE 6.14
McGregor's theory Y approach to leadership.

While theories X and Y illustrate two extreme views of leadership behavior, most organizations fall somewhere in between. A model that describes how leaders adjust these X and Y behaviors is the Hersey–Blanchard model of leadership (Figure 6.15). The Hersey–Blanchard model is a situational leadership model that considers the readiness and maturity of followers in achieving organizational outcomes. In this context, followers who have less experience or who are not ready to follow a leader may need more of a directing leadership behavior. And, followers who have a high maturity and willingness to meet organizational objectives may perform better with a leadership style that is more delegating.

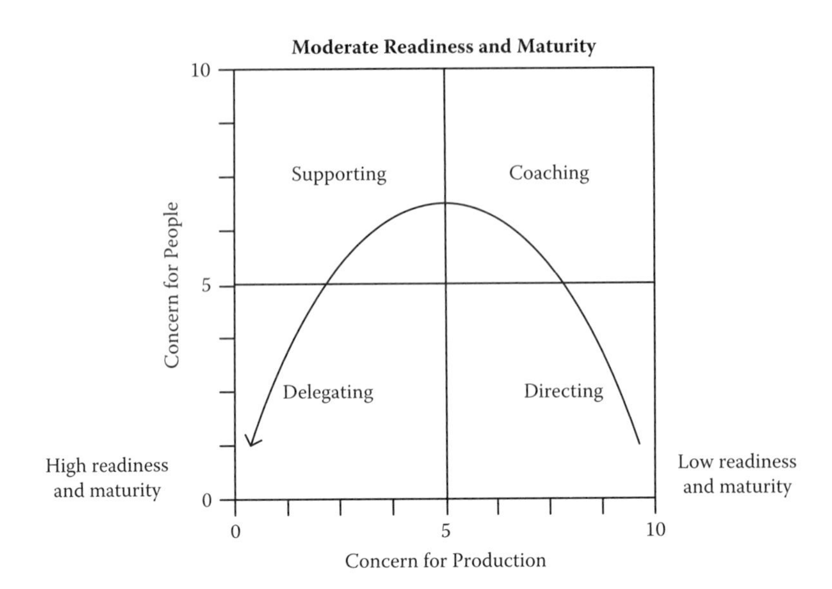

FIGURE 6.15
Blake and Mouton's grid and the Hersey–Blanchard readiness model combined.

The leadership theories described previously in this section give the impression that effective leadership is situational. Toyota does not wait for a certain member, a task, or organizational characteristics to apply TPS, which is why most theoretical models do not accurately describe the differentiated function of leadership (Figure 6.16). Leadership constantly has to reconcile and coordinate the needs and goals of the work group members (Scarpello and Vandenberg, 1987). TPS has to be continuously cascaded through the organization, layer by layer, to ensure complete and thorough adoption (Lester, 1993). If TPS waited for certain organizational conditions, it would be viewed as a secondary system inferior to the organization's primary management system (Figure 6.17).

A leadership theory that best describes how TPS is applied in varying organization conditions consistently is the performance–maintenance (PM) theory of leadership (Misumi and Peterson, 1985). PM leadership theory, similar to the Blake and Mouton theory, proposes that the performance of tasks and the maintenance of relationships are equally essential for meeting organizational goals (Figure 6.18). The primary difference between the Blake and Mouton model is that the PM model emphasizes the group function compared to the individual (Misumi and Peterson, 1985; Numata, 2002). The "performance function" concerns the forming and accomplishing of group goals, while the "maintenance function"

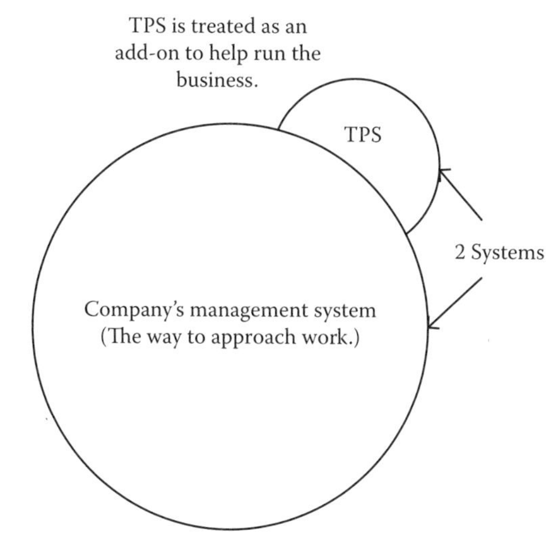

FIGURE 6.16
TPS is treated as an add-on: Organizations are looking for when to apply TPS.

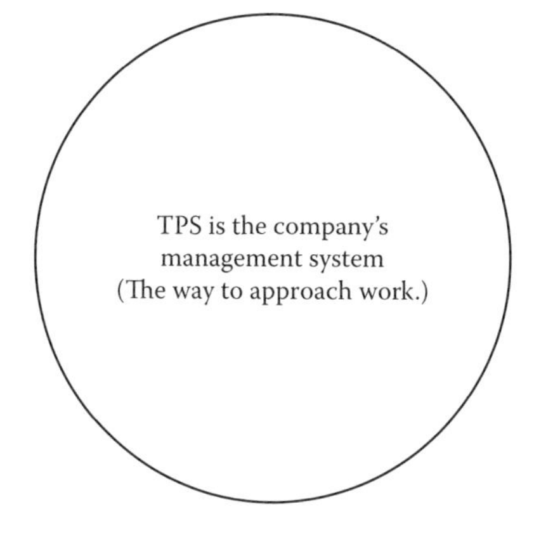

FIGURE 6.17
TPS is treated as the company's management system: TPS is always applied.

focuses on preserving the social stability or harmonious relations of the group. PM theory further divides leadership behaviors into four categories based on the relative emphasis placed on the respective functions: PM type (both performance and maintenance are high); Pm type (performance high, maintenance low); pM type (maintenance high, performance low); and pm type (both performance and maintenance are low). Misumi and Peterson's research (1985) in Japan indicated that the PM type is the most effective type of leadership behavior in terms of results for group performance and satisfaction.

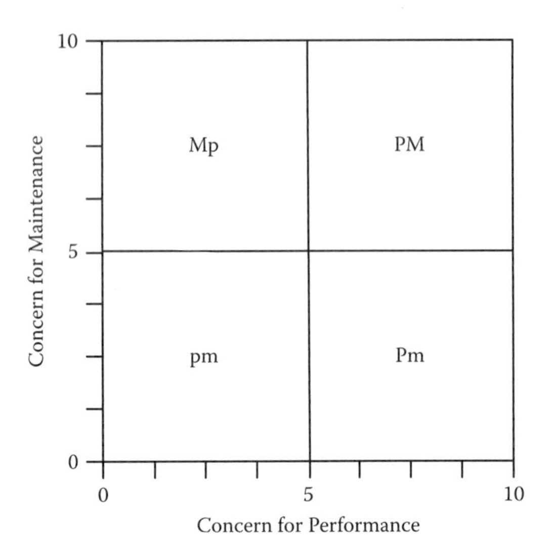

FIGURE 6.18
Performance-maintenance. (Adapted from Misumi, J. and Peterson, M., 1985, The performance-maintenance (PM) theory of leadership: Review of a Japanese research program, *Administrative Science Quarterly*, Vol. 30, No. 2, 198–223.)

Misumi and Peterson's PM model illustrates that the best way for leaders to accomplish organizational goals is through teamwork. Leaders who have concern for the performance and maintenance of groups have the best chance of meeting organizational goals consistently and continuously. The PM model best describes Toyota's application of TPS because group goals emphasize the tendency to work toward one system. Leadership theories that emphasize the performance and maintenance of the group imply a certain level of consistency across the organization. In contrast to most organizations, which look for a trigger to apply TPS, the PM leadership theory proposes the best way to preserve group goals and functions is through interdependent activity. In this context, leadership is constantly trying to unify organizational members to work toward one system.

6.11 SUMMARY

Systems must differentiate themselves to perform certain functions based on environmental conditions. Differentiation occurs differently at each subsystem level, and the functions of each subsystem also perform unique operations. Differentiation is a property of systems adjusting to their environments, where new operations and features are created, not duplicated or redundant, but unique subsystems that achieve specific functions that cannot be substituted by other subsystems. A unique and differentiated property of TPS is leadership.

This chapter illustrated that leadership cannot be substituted or replaced by other entities inside the organization for TPS to survive. Various leadership theories have been used to illustrate Toyota's specific approach for each function of leadership. Leadership has the role to define TPS as a management system that is positive and motivating. TPS must be clear to attract organizational members and most importantly articulate the reason why TPS must exist. Toyota's approach to helping organizational members can best be described as servant-style leadership. Leaders are expected to serve the needs of their employees and provide a larger sense of the organization's goals than their own. Last, leadership must be able to be genuine in how they apply and model the behaviors and actions of TPS. Leadership should work toward unifying TPS using teams and make an effort to integrate the organization through the use of interdependent activities and functions.

REFERENCES

Armenakis, A. and Bedeian, A. (1999) Organizational change: A review of theory and research in the 1990s. *Journal of Management*, Vol. 25, No. 3, 293–315.

Barker, R. (1995) The philosophical links between leadership and total quality. *Journal of Leadership and Organizational Studies*, Vol. 2, No. 27, 27–41.

Bass, B. (1985). *Leadership and Performance beyond Expectations*. Free Press, New York, NY.

Bass, B. (1990) From transitional to transformational leadership: Learning to share the vision. *Organizational Dynamics*, Winter, 140–148.

Bass, B. M. (1997) From transactional to transformational leadership: Learning to share the vision. *Organizational Dynamics*, Winter, 19–31.

Benne, K. and Sheats, P. (1948) Functional roles of group members, *Journal of Social Issues*, Vol. 4, 41–49.

Blackmar, F. (1911) Leadership in reform, *American Journal of Sociology*, Vol 16, 27–39.

Blake, R. and Mouton, J. (1964) *The Managerial Grid*, Gulf, Houston, TX.

Blau, P. (1970) A formal theory of differentiation in organizations, *American Sociological Review*, Vol. 35, No. 2, 201–218.

Bolden, R., Gosling, J., Maturano, A., and Dennison P (2003) *A Review of Leadership Theory and Competency Frameworks*, Centre for Leadership Studies, University of Exeter, United Kingdom, June.

Burns, T. and Stalker, G. (1961) *The Management of Innovation*, Tavistock, London.

Clutterbuck, D. and Hirst, S. (2002) Leadership communication: A status report. *Journal of Communication Management*, Vol. 6, No. 4, 351–354.

Cooley, C. (1902) *Human Nature and the Social Order*. Scribners, New York, NY.

Deming, E. (1986). *Out of the Crisis*, MIT Press, Cambridge, MA.

Drucker, P. (1980) *Managing in Turbulent Times*, Harper & Row, New York, NY.

Fukushige, A. and Spicer, D.P. (2007) Leadership preferences in Japan: an exploratory study, *Leadership & Organization Development Journal*, Vol. 28, No. 6, 508–530.

Greenleaf, R. (1969) Leadership and the individual: The Dartmouth lectures, in *On Becoming a Servant Leader*, ed. Frick, D. and Spears, L., Jossey-Bass, San Francisco, CA, pp. 284–338.

Hall, A. (2006) *Introduction to Lean: Sustainable Quality Systems Design—Integrated Leadership Competencies from the Viewpoints of Dynamic Scientific Inquiry Learning and Toyota's Lean System Principles*, Published by Arlie Hall, Ed.D., Lexington, KY.

Harter, S. (2002) Authenticity, in *Handbook of Positive Psychology*, ed. Snyder, C. R. and Lopez, S. J. London: Oxford University Press, pp. 382–394.

Hopper, A. and Potter, J. (1997) *The Business of Leadership*, Ashgate, Aldershot, UK.

House, R. (1971) A path goal theory of leader effectiveness. *Administrative Science Quarterly*, Vol. 16, 321–338.

Howell, J. and Dorfman, P. (1981). Substitutes for leadership: Test of a construct. *Academy of Management Journal*, vol. 24, 714–728.

Kerr, S. (1977) Substitutes for leadership: Their meaning and measurement. *Organizational Behavior and Human Performance*, Vol. 22, 375–403.

Kerr, S. and Jermier, J. (1978) Substitutes for leadership: Their meaning and measurement. *Organizational Behavior and Human Performance*, Vol. 22, 375–403.

Kerr, S. and Slocum, J. (1981) *Controlling the performances of people in organizations*, In P.C. Nystrom and W.H. Starbucks (Eds.), *Handbook of Organizational Design*, Vol. 2. Oxford University Press, New York, NY, pp 116–134.

Kotter, J. (1999) *What Leaders Really Do*, Harvard Business School Press, Boston.

Kreafle, K. (2010) Lean Expectative Leadership Institute, University of Kentucky, Lean Systems Group, Seminar, April, Univeristy of Kentucky.

Lawrence, P. and Lorsch, J. (1967) Differentiation and integration in complex organizations. *Administrative Science Quarterly*, Vol. 12, 1–30.

Lester, R. (1993) Creative leadership for total quality. *Journal of Leadership and Organizational Studies*, Vol. 1, 141–145.

Lester, R. (1995) Leadership for a quality organization. *Journal of Leadership and Organizational Studies*, Vol. 2, 3–8.

Maslow, A. (1943) A theory of human motivation. *Psychological Review*, Vol. 50, No. 4, 370–396.

Maslow, A. (1946) A theory of human motivation. In P. Harriman (Ed.), *Twentieth-Century Psychology: Recent Developments in Psychology*, Philosophical Library, New York, NY, pp. 22–48.

McGregor, D. (1960) *The Human Side of Enterprise*, McGraw-Hill, New York.

Meindl, J. (1993) Reinventing leadership: A radical, social psychological approach, in *Social Psychology in Organizations: Advances in Theory and Research*, ed. Murnighan, J. K. Prentice Hall, Englewood Cliffs, NJ, pp. 89–118.

Miner, J. (1975). The uncertain future of the leadership concept: An overview. In J. G. Hunt and L. L. Larson (Eds.), *Leadership Frontiers*. Kent State University Press, Kent, OH, pp. 197–208.

Misumi, J. and Peterson, M. (1985) The performance-maintenance (PM) theory of leadership: Review of a Japanese research program. *Administrative Science Quarterly*, Vol. 30, No. 2, 198–223.

Mostovicz, E., Kakabadse, N., and Kakabadse, A. (2009) A dynamic theory of leadership development. *Leadership and Organization Development Journal*, Vol. 30, No. 6, 563–576.

Numata, A., (2002) *A Comparison of Leadership Preference by Culture*, Ball State University.

Ouchi, W. (1981) *Theory Z: How American Business Can Meet the Japanese Challenge*, Addison-Wesley, Reading, MA.

Padsakoff, P., Niehoff, B., MacKenzie, S., and Williams, M. (1993) Do substitutes for leadership really substitute for leadership? An empirical examination of Kerr and Jermier's situational leadership mode. *Organizational Behavior and Human Decision Processes*, Vol. 54, 1–44.

Pigors, P. (1935) *Leadership or domination?* Houghton Mifflin, Boston, MA.

Reiss, S. (2004). Multifaceted nature of intrinsic motivation: The theory of 16 basic desires. *Review of General Psychology*, vol. 8, no. 3, 179–193.

Ritzer, G (2007) *Contemporary Sociological Theory and Its Classical Roots: The Basics*, Second Edition. McGraw-Hill, New York, NY.

Rousseau, J. (1960) *The Social Contract and Other Discourses*, Modern Library, Random House, New York.

Scarpello, V. and Vandenberg, R. (1987) The satisfaction with my supervisor scale: Its utility for research and practical applications. *Journal of Management*, Vol. 13, No. 3, 447–466.

Senge, P. (1990) *The Fifth Discipline: The Art and Practice of the Learning Organization*, Random House, London.

Stogdill, R .(1974) *Handbook of Leadership*, Free Press, New York.

Sugimori, Y., Kusunoki, K.,Cho, F., & Uchikawa, S. (1977) Toyota production system and Kanban system: Materialization of just-in-time and respect-for-human system. *International Journal of Production Research*, Vol. 15, No. 6, 553–564.

Tichy, N. and DeVanna, M. (1986) *The Transformational Leader*, Wiley, 1986.

Thoresen, E. and Mahoney, M. (1974) *Behavioral self-control*, Holt, Rinehart and Winston, New York.

Turner, J. (1981) Emile Durkheim's theory of integration in differentiated social systems. *Pacific Sociological Review*, Vol. 24, No. 4, 379–391.

University of Kentucky, Lean Systems Certification (2011) Certification Series, Course 1, January, Lexington, KY.

von Bertalanffy, L. (1968) *General System Theory: Foundations, Development, Applications*. Braziller, New York, NY.

Von Ludwig, B. (1972) *Perspectives on general system theory: Scientific-philosophical studies*, Braziller, New York, NY.

Walumbwa, F., Avolio, B., Gardner W, Wernsing, T., and Peterson, S. (2008) Authentic leadership: Development and validation of a theory-based measure. *Journal of Management*, Vol. 34, No. 1, 89–126.

Weber, M. (1946) *Essays in Sociology*, Oxford University Press, New York.

Womack, J. P., Jones, D. T., & Roos, D. (1990). *The Machine That Changed the World: The Story of Lean Production*. HarperCollins, New York, NY.

7

The System Property of Hierarchies in TPS

7.1 SYSTEM PROPERTY: HIERARCHY

Hierarchies serve as the structural basis of all living systems. A hierarchy is a collection of parts with asymmetric relationships inside a whole. Hierarchies are used in systems to describe the complexities of structures. Specialized subsystems serve as modular building blocks that facilitate task performance along and throughout a hierarchy. Hierarchies in systems imply both separateness and connectedness. Hierarchies are used to describe the possibility of energy transfer (Becvar and Becvar, 1999).

The better an organism can adapt to a turbulent environment, the more complicated the system will be (Overton, 1975). When a system differentiates itself, functional specialization occurs, which enables a system to adapt to its environment. Hierarchization is a property of mature systems that shows differentiation over time.

The hierarchical arrangement of systems gives insight in how systems are organized (Peery, 1972). Hierarchies can be arranged by the level of complexity, level of mechanization, decision-making choices, level of communications, rank, layer, and the degree of the whole. Hierarchies tend to behave as wholes when facing downward while acting as parts when facing upward (Gigch, 1978). Higher-order subsystems take the leading role, while lower-order subsystems generally operate with more constraints. The higher the hierarchy, the more diverse a system's functions and properties will be.

The advantages of hierarchical systems is that they can specialize to form nested hierarchies. Nested hierarchies involve levels that consist of and contain lower levels. Nested hierarchies allow specialized functions to occur within subsystems. The entire basis of hierarchies allows systems to operate as a more complex process. Hierarchies simplify decision making within a system because ranking allows higher levels to have control over lower levels.

7.2 THE HIERARCHICAL PROPERTY OF THE TOYOTA PRODUCTION SYSTEM

An aspect of Toyota's management system that relates to the system property of hierarchies is hoshin kanri. Hoshin kanri or annual business planning is a process that coordinates the organization's decision-making processes on a global level. Toyota originated the hoshin kanri process in 1961, when the vice president at the time, Eiji Toyoda, wanted to modernize the company's management operations (Figure 7.1).

In the automotive industry, there are various factors and environmental conditions that make production challenging and difficult to predict. Automotive is broadly based and is integrated with various manufacturing processes, which involves many complex operations. The actual product contains more than 15,000 components, with the majority produced by outside suppliers. Customer tastes are constantly changing, which means the ability to design and produce a car has to be predicted several years in advance. Customers want choices and various

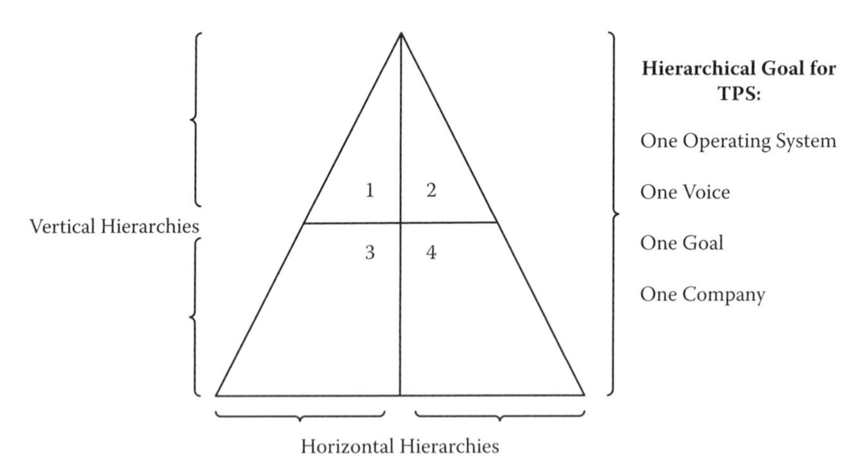

FIGURE 7.1
Purpose of hoshin kanri: everyone working to one system.

alternatives that require design and manufacturing to work together in producing various models and options. Last, the investment to produce one model is costly. Costs have to be managed through the design stage to avoid unreliable production processes and after production to ensure operations maintain target cost. Toyota wanted management to have a way to clarify company goals and encourage companywide cooperation. Ideally, everyone would share the company's vision and work to one operating system, with one voice.

7.3 DECISION MAKING IN HIERARCHIES

According to systems theory, the more turbulent the environment is, the more complex the organism will have to arrange its hierarchies to process information to make decisions.

The complexity of a system is dependent on the number of interactions it has with other systems. Simply increasing an organization's hierarchy does not lead to more effective interactions. No division of tasks can do away with all the interdependencies of a system so that subunits do not need to coordinate activities with one another. Complex organisms exhibit extremely decomposable and permeable boundaries, which encourages information exchange. Complex systems have the characteristics of open systems, which encourage more interaction across hierarchical boundaries. As a system decreases in complexity, the need for the system to interact with other internal subsystems become less dependent. Consequently, the hierarchical boundary of the system also becomes more distinct and less permeable and reliant on other subsystems for information. The permeability of hierarchical structures becomes increasingly important the more an organism is expected to survive in a turbulent environment.

7.3.1 Vertical Decision Making

One of the most distinguishing properties of hoshin kanri is how decisions are made across hierarchies. Hoshin encourages top-down direction setting and bottom-up flow means to accomplish organizational targets and goals. When units across different hierarchies work together in the achievement of common goals, decision making is improved. Hoshin

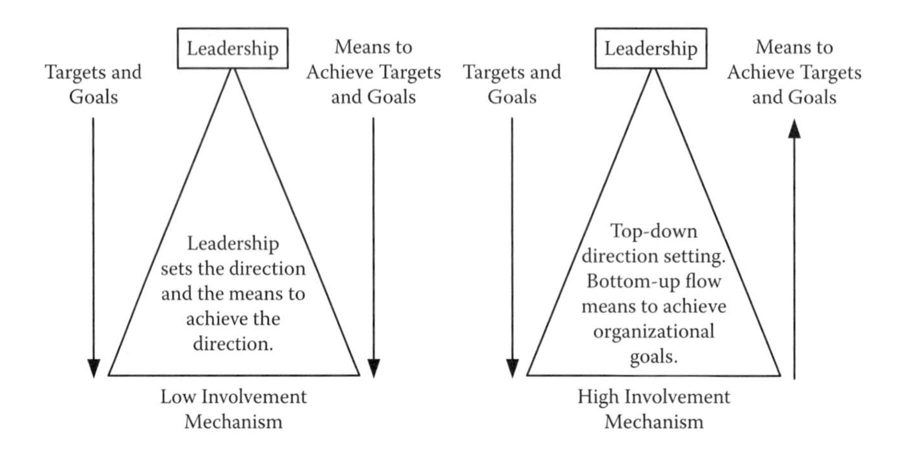

FIGURE 7.2
Comparison of decision-making systems: low employee involvement versus high employee involvement.

kanri shows characteristics of a system with high employee involvement because leadership does not establish the means for achieving targets. In a high-involvement system, decision making is delegated to the most appropriate level in the organization that has the best knowledge and expertise to make the decision. A system that demonstrates a low level of participation in decision making would rely on leadership to establish targets and tell subordinates how to achieve them. In a low-involvement system, leadership does the thinking and the telling (Figure 7.2).

7.3.2 Horizontal Decision Making

Cooperation among units is another area of hoshin that is critical for achieving companywide goals and targets. Toyota experienced weak horizontal cooperation among business units in the early 1960s. In that time, managers only cooperated if it avoided problems for them or brought them gain. The reason why managers acted on their own self-interest was because they were measured in how they performed in isolation compared to their horizontal counterparts. What is revealing is that the politics of each unit was more dominant than the company's culture to be homogeneous or willingness to work in groups. Toyota established hoshin to encourage communication across hierarchical boundaries and the setting of joint agreements to accomplish companywide goals. Toyota's approach to changing the company's culture was led by first changing the company's management system (Figure 7.3).

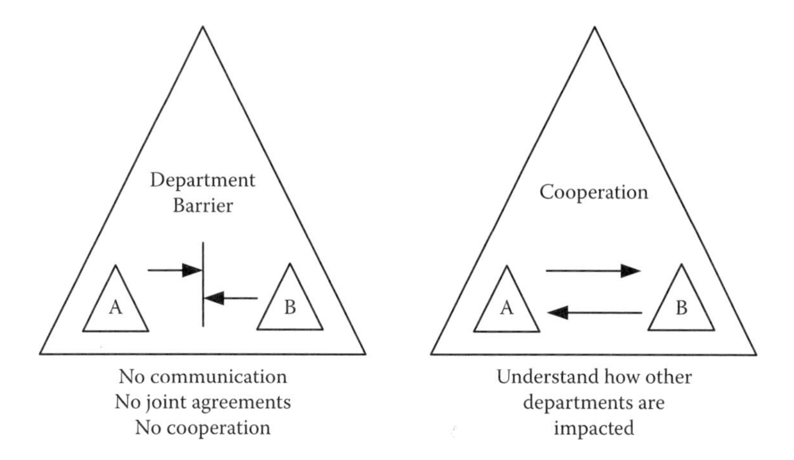

FIGURE 7.3
Comparison of departmental cooperation: traditional versus Toyota.

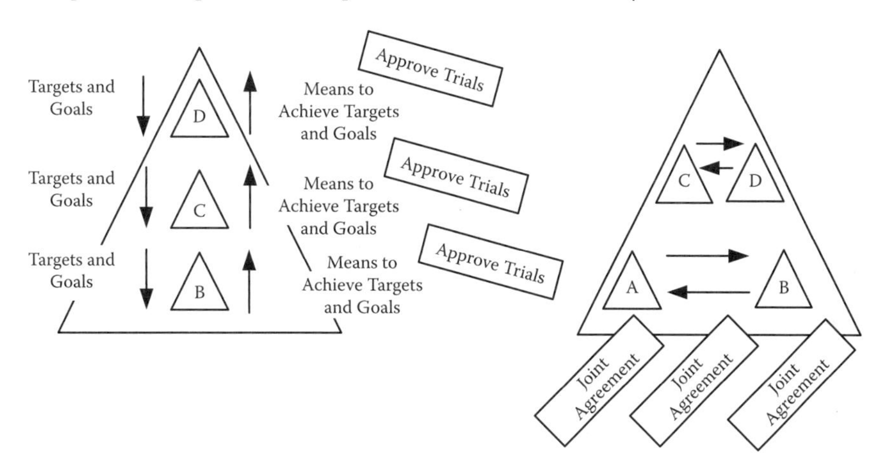

FIGURE 7.4
Vertical and horizontal hierarchical boundaries.

Decision making across boundaries in hoshin is driven by two factors: the degree of approval for testing ideas and the degree of cooperation among departments to establish joint agreements. Established properly, hoshin would engage team members at every level and every location (Figure 7.4).

7.3.3 Interdependent Decision Making

Employees need to know how their role fits into the role of the unit and how the unit fits into the needs of the organization. Hoshin reinforces the

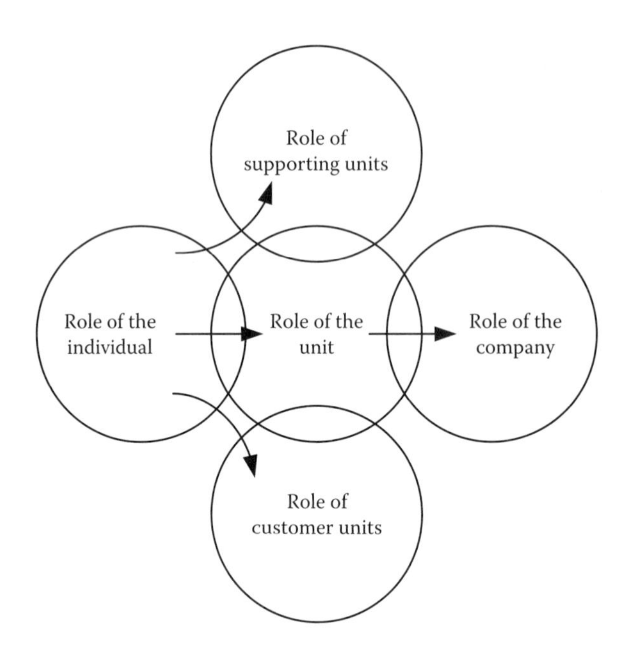

FIGURE 7.5
Concept of hoshin: interdependency of units.

idea of teamwork within and outside departments to increase the understanding and effectiveness of interdependent functions. Hoshin requires departmental managers to pass on their plans to units that require support and to consider how they can support other units asking for assistance. Hoshin represents an annual mechanism that formalizes how departments should work together by providing a system of checks and balances between all involved units. Hoshin raises questions in the planning stages how units should interact with other units to prevent problems and to enhance the effectiveness of interdependent tasks (Figure 7.5).

7.3.4 Interdependent Performance

Hoshin kanri is a tool for managing resources for overall performance. One of the ways that Toyota maximizes performance outcomes is through linking the needs of the company to the needs of each employee (Figure 7.6). An example is the human appraisal system. The performance appraisal is linked to hoshin so that the structure of the organization can be reinforced. Hoshin links company performance measures to appraisal factors so employees know what is expected from them, their unit, and the manner they will cooperate in meeting company objectives. On a broader

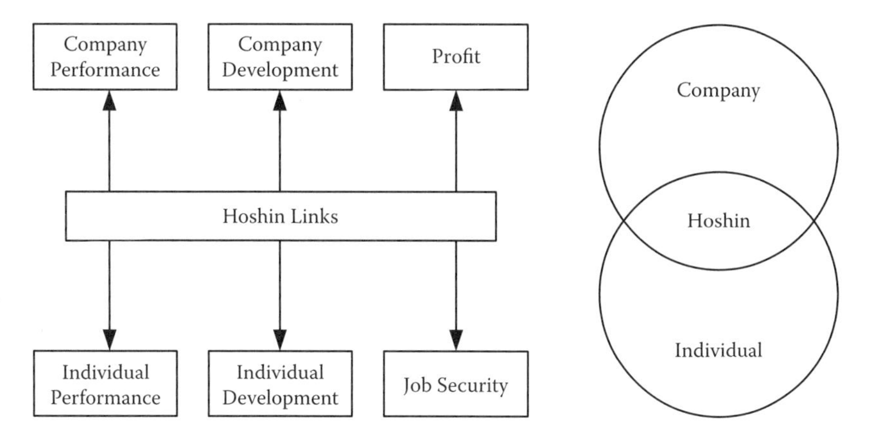

FIGURE 7.6
Hoshin links the needs of the company to the needs of the individual.

scale, hoshin links the development of the company to the development of each employee. If employees are not developed, the company cannot expect to develop. Consequently, employees cannot expect job security if the company does not earn a profit.

7.3.5 Monitoring and Progress

Hoshin is a visual management system that tracks results and processes. Hoshin is a way to monitor progress so that management can make high-quality decisions in a timely manner. Hoshin is not a one-shot attempt at meeting organization goals or a way to provide feedback at the end of the business plan. Instead, hoshin is a process that allows managers to make informed decisions about the business plan and provide feedback in any situation. The speed and quality of management are poor when the status of the business plan is not known. Hoshin provides managers information about the status of the organization so that decisions can be based on facts rather than intuition. Hoshin can help managers to prioritize, improve decision making, and allocate resources in a timely manner.

7.3.6 Routine Work and Nonroutine Work in Hoshin

Hoshin kanri is used to coordinate a wide variety of projects at Toyota. For most companies, annual business planning is something that is used to plan large-scale breakthroughs and innovation. At Toyota, hoshin is used similarly but is also a way to strengthen day-to-day operations, such

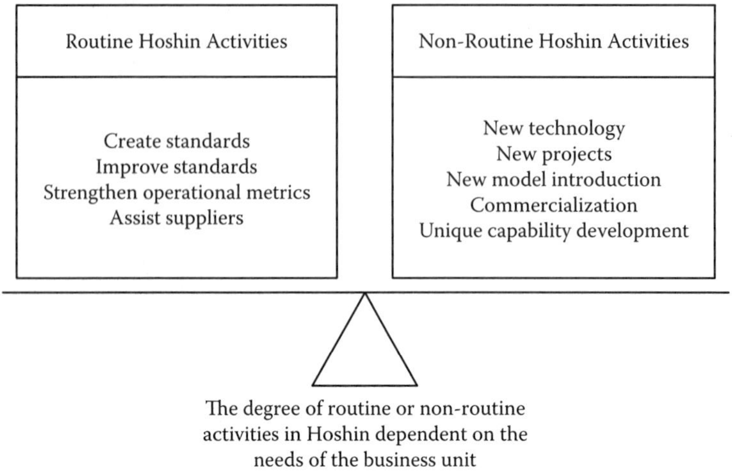

FIGURE 7.7
Examples of routine and non-routine hoshin activities.

as workplace standardization. Hoshin can be used to improve organizational functions or as a way to focus on unique capabilities of the organization. Regardless, the degree that hoshin is routine or nonroutine is largely dependent on the needs of the business (Figure 7.7).

7.4 LITERATURE REVIEW OF STRATEGIC PLANNING

Strategic planning (sometimes referred to as policy deployment or strategic management) is a discipline for helping organizations identify priorities by examining external and internal conditions (McLean, 2006). Strategic planning involves the implementation of actions and controls that allow resources to be aligned and decisions to be evaluated based on long-term objectives. The overall role of strategic planning is to identify an organization's core competencies and use them to provide a competitive advantage, using innovation, reputation, and organizational structure (Kay et al., 2003; Abro et al., 2009).

Strategic planning has evolved over the years to include corporate planning (Chandler, 1962), product diversification, portfolio planning, development of core business functions (Porter, 1980), and recently a focus on people-oriented processes (Tischler, 2002). The overall approach in strategic planning has remained basically the same over the last 50 years: examine the company, its competitors, and the environment; set goals; determine a strategic approach; allocate resources; and evaluate performance. Strategic planning is said to be effective when the organization adopts courses of action that compound the competitive advantage of the company.

Researchers have proposed various perspectives in strategic planning largely due to environmental factors acting on organizations. For example, there is work that shows that strategic planning is dependent on the alignment of resources (Wright et al., 2007); internal capability (Doving and Gooderham, 2008); innovation (Andrew et al., 2009); market orientation; global opportunities (Wood et al., 2009); and the ability to focus on customer needs (Morgan et al., 2007). Other strategic planning techniques take a softer approach by emphasizing direct and honest communication, shared learning (Smith et al., 2009), and collaboration (Holman et al., 2007).

A common approach in the implementation of strategic planning is the generalization of success factors (Abro et al., 2009). There is work that showed that organizations should be outward focused by analyzing customer requirements, market penetration strategies (Smith et al., 2009; Subramanian and Gopalakrishna, 2009; Wood et al., 2009), and changes in the industry (McLean, 2006). Other views stressed the ability to align resources and establish consistency within the organization (Faludi, 1986). This view proposes that strategic planning is more about understanding internal capabilities, both historical and present, rather than scanning the environment. Research also showed that strategic planning should be employee centered and motivational (Kim and Mauborgne, 2005), involving larger parts of the organization (Singh, 2009), while other techniques indicated a more direct and top-down approach exercising authority and power (Rodgers et al., 1988).

Various tools and techniques have been developed to assist strategic planning. One of the most popular strategic planning tools is Humphrey's SWOT (strength, weakness, opportunities, threats) analysis. Derivatives have also been proposed, such as SOAR (strengths, opportunities, aspirations, results) (Stavros et al., 2003) and PEST (political, economic, social, ethnological) (Cooper, 2000). Other tools in strategic planning include the X-matrix (Jackson, 2006), Porter's matrix, decision trees (Scholtes, 1988), stakeholder analysis (Freeman, 1984), scenario planning (Chermack, 2004), future search, and process management (Hammer and Champy, 1993). The use of tools in strategic planning still appears to be a major source of understanding and improving performance in organizations.

7.4.1 Problems and Challenges in Strategic Planning

Unfortunately, with all techniques and success factors known today there still exist many problems when implementing strategic planning (Figure 7.8). Possibly, the most challenging aspect of strategic planning is the balance between strategy formation and implementation (Kay et al., 2003). There is perhaps no process in organizations that is more demanding of human cognition than trying to sense what is going on inside and outside the organization (Mintzberg, 2007). All strategy makers face an impossible overload of information, and as a result there is no optimal process to follow (Mintzberg, 2007). Most strategies are based on a misconception that they can be implemented without adjustment, learning, or feedback. This is especially true when behavior changes are required to convert decisions into actions. This thinking raises many other important questions, such as the extent businesses become adaptable in their planning. Thus, researchers continue to debate that an optimum balance must exist between strategy formation and implementation (Chaffe, 1985).

Issues relating to the implementation of strategic planning vary but mostly deal with the ability to motivate and involve employees in achieving organizational goals and the ability to integrate them into daily work practices (Tennant and Roberts, 2001). Consequently, most companies have difficulty in coordinating, obtaining management commitment, and failing to follow the plan. Other problems in the implementation of strategic planning are that actions are often pushed forward, which causes plans to become obsolete quickly (Holman et al., 2007). When this happens, plans often are not reviewed or evaluated, which causes relevant information to be lost rather than updated (Donohue, 2005). Consequently, most decision makers have the view that strategy formation and implementation

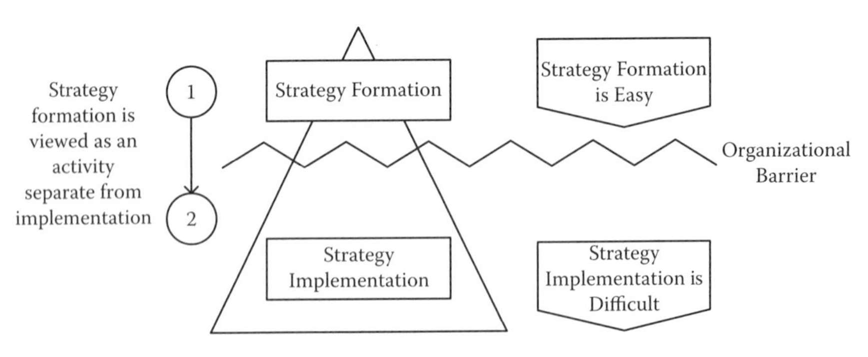

FIGURE 7.8
Difficulties in traditional strategic planning and implementation.

are parts of a regular, nicely sequenced process running on a standard cycle or schedule (Mintzberg, 2007). Although regular reviews are a positive success factor in strategic planning, an organization must be able to address and react to external changes by environment.

7.5 STRATEGIC PLANNING THEORIES

Strategic planning theories have drawn on a variety of disciplines, including marketing, finance, economics, and organizational behavior. Some of the earlier theories relating to the field include chaos and complexity theory. These theories describe the difficulties and uncertainties associated with future planning, specifically the influences of variation and control. Recent theories take a more holistic approach by considering various aspects of both strategy formation and implementation. A common distinction among strategic planning theories is their preference between industrial and sociological approaches. Industrial organization theories include resource-based theory (Crook et al., 2008), dynamic capability theory (Doving and Gooderham, 2008), and contingency theory (Lavie and Rosenkopf, 2006). The common theme in industrial organization theory is the mechanical aspects or steps of the strategic planning process. Sociological theories take a softer approach by describing the human aspect in strategic planning. These theories include knowledge-based theory (Anand et al., 2007); agency theory (Arthurs et al., 2008); social judgment theory (Smith, 2004); upper-echelon theory (Wright et al., 2007); institutional theory (Suddaby and Greenwood, 2005); and emergent/deliberate theory (Mintzberg, 2007). For a more detailed explanation of some of these theories, please see Table 7.1.

A common debate is the degree an organization should emphasize industrial organizational approaches over sociological approaches in strategic planning (Figure 7.9). Consequently, the order that organizations place on these approaches can also have a drastic impact on the overall effect of strategic planning. It has been shown that organizations that emphasize the "soft side" in strategic planning often run the risk of creating more politics in decision making compared to more linear and mechanistic approaches. On the other hand, when industrial organizational techniques are applied in isolation, employee involvement and participation can decrease in strategic planning. In all cases, the degree and

TABLE 7.1

Brief Literature Review of Strategic Planning Theory

Industrial Organizational Approach			**Sociological Approach**		
Resource-Based Theory (Crook et al., 2008)	**Dynamic Capability Theory (Doving and Gooderham, 2008)**	**Emergent/Deliberate Theory (Mintzberg, 2007**	**Knowledge-Based Theory (Anand et al., 2007)**	**Agency Theory (Arthurs et al., 2008)**	**Upper-Echelon Theory (Wright et al., 2007)**
Economic theory relates to the use of resource exploitation to develop a competitive advantage. Resources include a company's assets, capabilities, and organizational processes. Resource-based theory is often criticized as focusing on internal attributes of the company compared to analyzing external forces in the industry.	Dynamic capability theory relates to innovation and inward ability of a company to respond rapidly to changing environments. This theory is based on an organization's ability to understand outside market conditions by sensing and interpreting external information. Consequently, a company must also be able to apply new learning internally to adapt to market conditions.	Two types of approaches are believed to exist in strategy formation and implementation, namely, deliberate and emergent theory. Deliberate strategic planning defines planning activities as directed, disciplined, and controlling. Emergent strategic planning views that implementation is dynamic and unpredictable, which means that certain aspects of planning should emerge from circumstances.	Knowledge-based theory has the view that human capital and learning are primary drivers in strategic planning. This approach supports the view that training is a necessary aspect in organizational development and performance.	Agency theory is based on the view that groups in isolation have different interests, which means they are less likely to understand system influences and interdependencies. This particular theory is applicable to strategic planning since strategy formation and implementation are often created and achieved in isolation.	Upper echelon theory describes the ability of leadership to influence and control decisions within the organization. This view suggests that decision making is left up to higher levels of the organization, while lower levels perform as they are told. This theory also suggests that both strategy content and strategy process are more tightly controlled the higher in the organization.

Industrial Organizational Approach
1. Dynamic capability theory 2. Resource-based theory 3. Deliberate/emergent theory

Sociological Approach
1. Upper-echelon theory 2. Agency theory 3. Knowledge-based theory

FIGURE 7.9
Comparing industrial organizational and sociological approaches.

manner these theories are applied are important in how they shape the strategic planning process.

7.6 A SYSTEM OF HIERARCHIES: THE HOSHIN PROCESS

According to systems theory, a lot can be understood about a system by studying the way hierarchies are arranged within a structure and their degree of interaction. An illustration of that structure can be seen by outlining the hoshin kanri process. As shown in Figures 7.10 and 7.11, the hoshin process is made up five phases: planning, preparation, implementation, evaluation, and reflection. The interdependency of those phases relate to how each unit or nested hierarchy can function internally and externally.

7.6.1 Planning Phase of Hoshin

The planning phase of hoshin can best be described as communicating and studying what needs to be done for the upcoming year. Most annual business-planning processes want to identify immediately what can be done to contribute to the organization's objectives. Realistically, this places a unit in jeopardy of over- or undercommitting because it is assumed that the capabilities of the unit have not changed from the previous year. Toyota understands that each unit is constantly growing in its capabilities and has changed to some degree over the past year due to extraneous factors. Without proper planning, the business unit can miss factors that impeded the unit's ability to accomplish its basic functions. One of the most significant factors is not knowing how other units can contribute or impact the unit's capabilities. The planning phase in hoshin encourages the unit to understand the needs of the organization and the capabilities of the unit before jumping to conclusions (Figure 7.12).

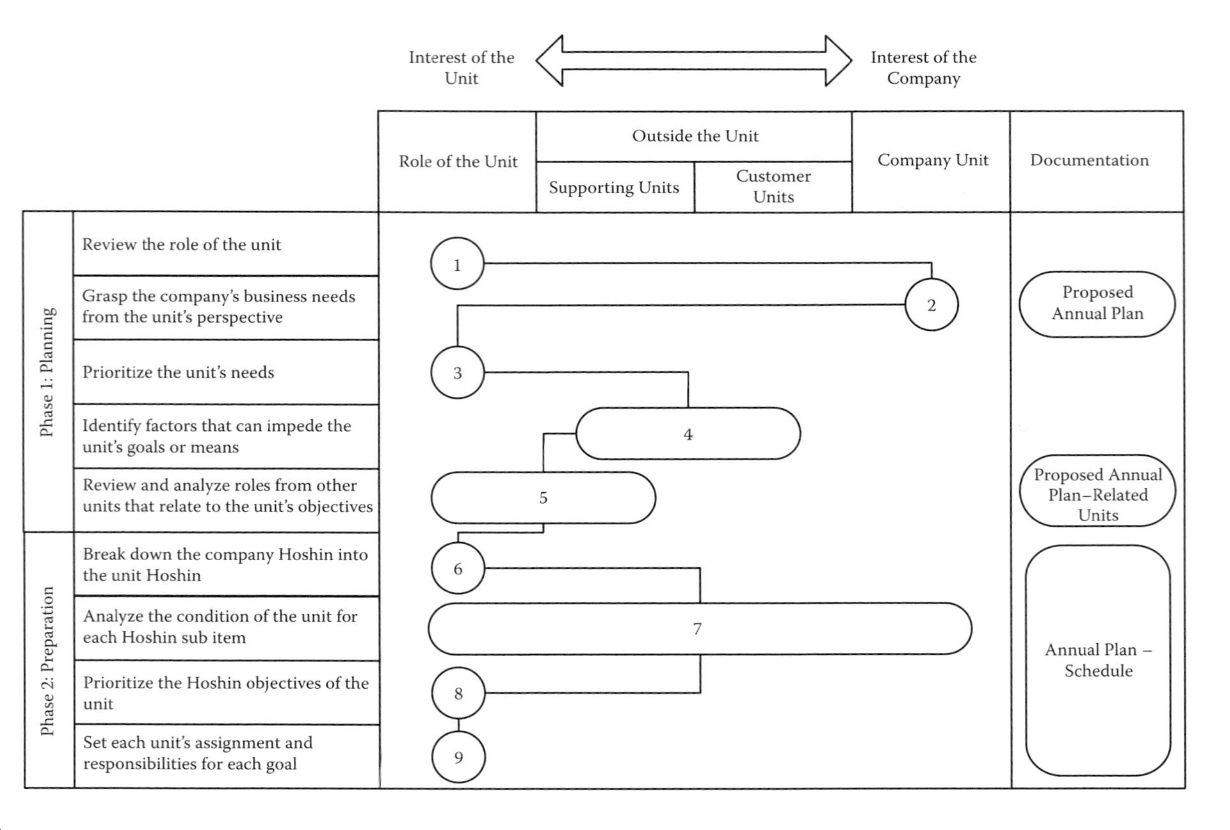

FIGURE 7.10
Hoshin algorithm: steps 1 through 9.

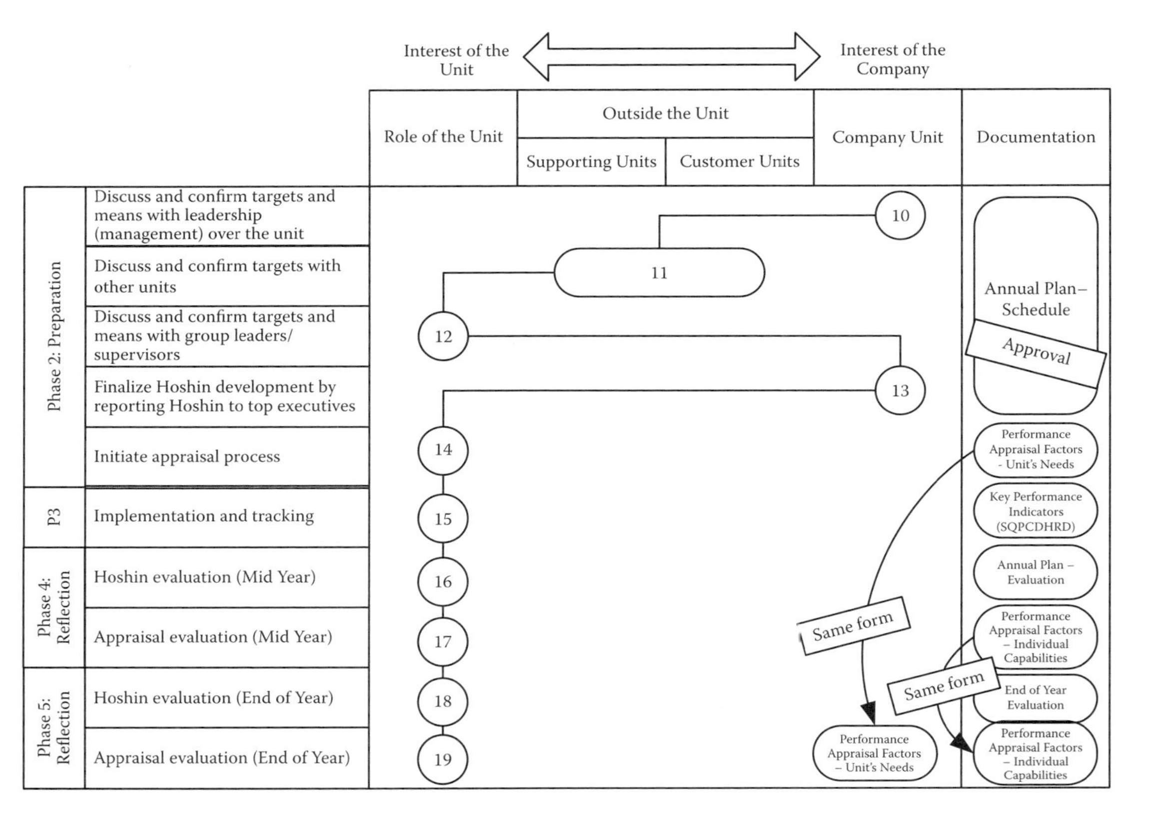

FIGURE 7.11
Hoshin algorithm: steps 10 through 19.

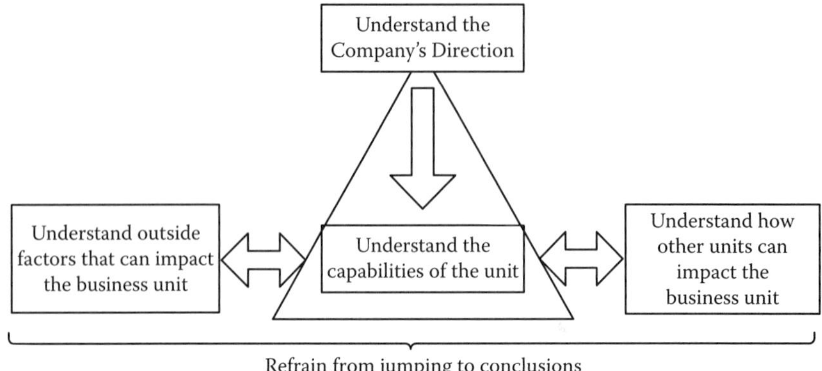

FIGURE 7.12
Planning phase for hoshin.

One of the goals of hoshin in the planning phase is getting the unit to look at the company goals from a big picture. The planning phase requires each unit to do some preliminary homework about the problem domain before starting. Units are required to benchmark, compare to other departments, and most importantly, understand what the unit can do and cannot control. Data are not enough for a unit to anticipate the forces acting within the unit and outside the unit. Department managers need to understand the unit's essential functions, priorities, and critical areas. Through study, each the unit can identify strong and weak points to be prepared for future challenges. The best way the unit can be prepared is to analyze the organization's current situation and pay attention to factors external to the unit. Examples of Toyota's forms for the planning phase are shown in Figures 7.13 and 7.14.

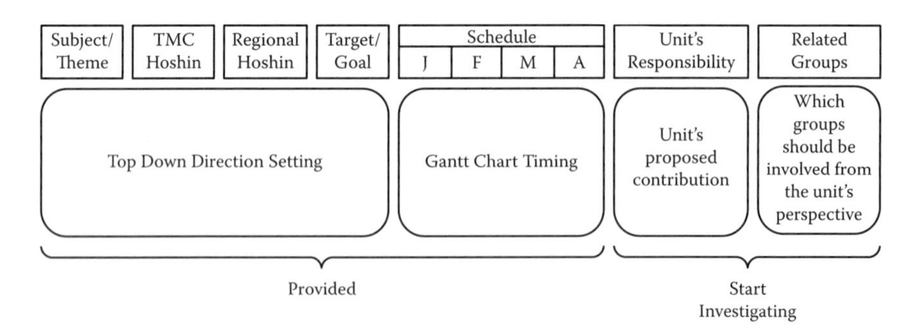

FIGURE 7.13
Proposed annual plan (adapted example): hoshin form.

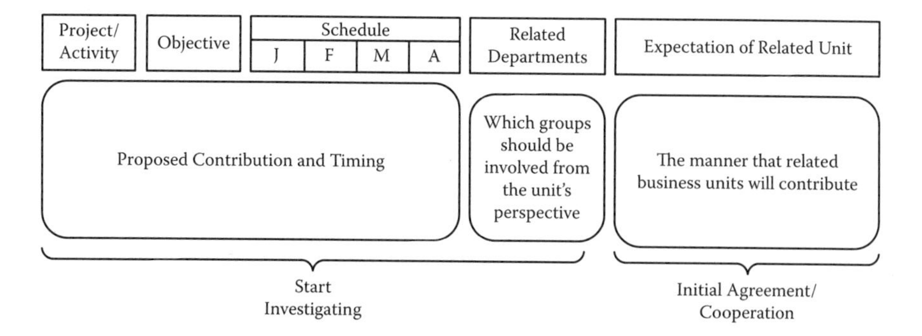

FIGURE 7.14
Proposed annual plan, related units (adapted example): hoshin form.

7.6.2 Preparation Phase of Hoshin

The main goal of the preparation phase is to understand how each unit will contribute to the company hoshin. A successful hoshin will encourage trust building and listening at every level of the organization for ideas. The preparation phase is one of the most critical aspects of hoshin because it captures how the organization intends to go forward, importantly, how everyone in the organization intends to participate.

In this phase, each unit is expected to break down the company hoshin into its unit hoshin (Figure 7.15). Larger companies will have a unit hoshin for the global, region, plant, department, and section levels. The ideal is to

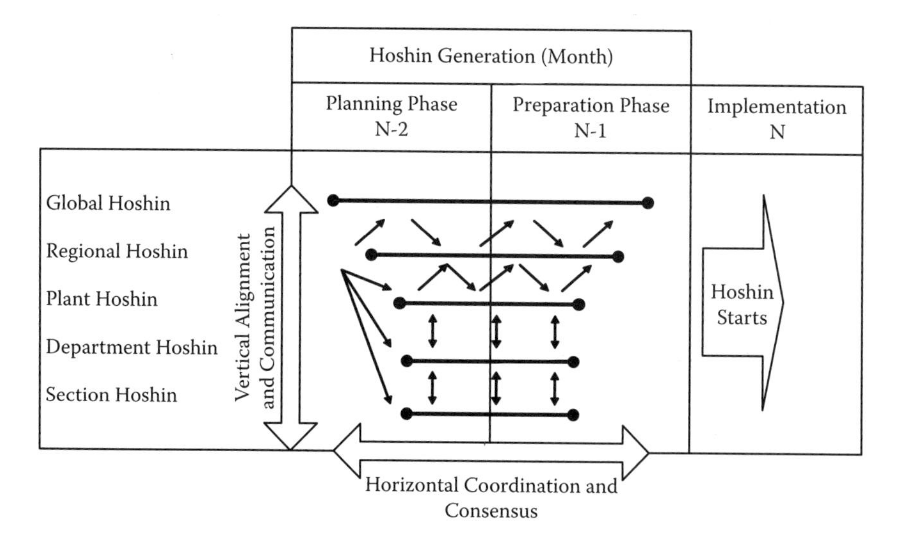

FIGURE 7.15
Hoshin timeline (initial stages only).

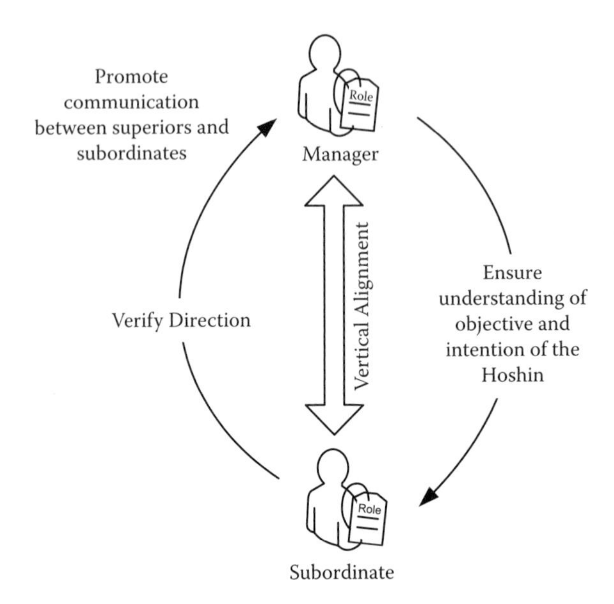

FIGURE 7.16
Representation of two-way communication: vertical.

communicate the company's vision throughout the organization vertically so that each unit is aligned. Vertical communication ensures that both the subordinate and the superior understand the objectives and the intent of the hoshin (Figure 7.16). It is important to note that vertical communication is not the same as downward communication. Vertical communication implies two-way communication. Without two-way communication, subordinates would be unable to participate or contribute their ideas to the hoshin. McCelland (1988) found a number of reasons why upward communication tends to be poor. Employees have fear of reprisal and are generally afraid to speak their minds. Employees also feel that their ideas and concerns are modified as they are transmitted upward. Last, managers generally give the impression that they do not have the time to listen to employees. In the case of hoshin, two-way communication is essential for employee engagement. Employees need to feel comfortable expressing their ideas to management, and management needs to elicit employee's opinions. Without effective two-way communication, hoshin would not enable company participation.

Another essential element in the preparation phase of hoshin is horizontal communication (Figure 7.17). Horizontal communication is information sharing that flows across the organization. In general, organizations

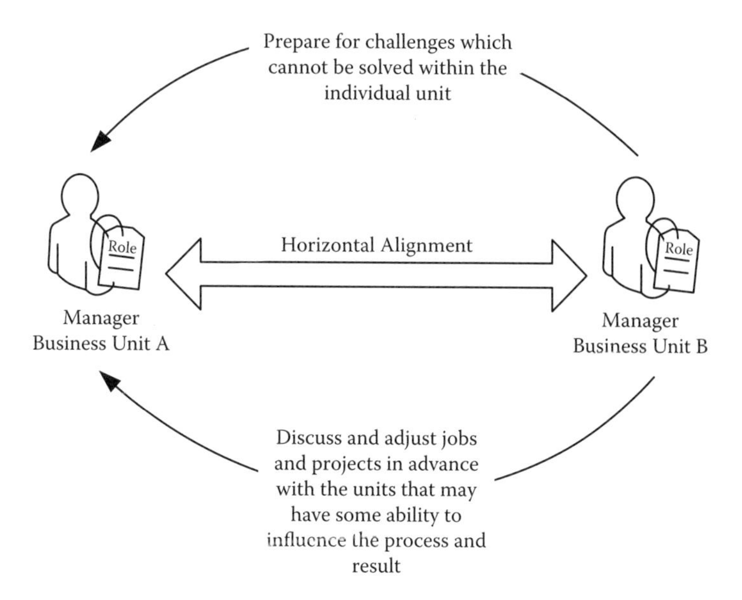

FIGURE 7.17
Representation of two-way communication: horizontal.

have much more horizontal communication than vertical communication. The reason is that there are more employees than managers, and employees generally feel more comfortable conversing with employees at the same level (McCelland, 1988). Units that cannot solve challenges within the unit must seek help from related units. In this phase, units are encouraged to discuss and adjust jobs in advance before implementation. Departments and sections that have an effect on another unit's hoshin should engage in two-way communication to understand each unit's contribution.

A typical problem in annual business planning is that most managers generally do not want to cooperate unless it benefits them. Inherently, managers will do whatever brings them good and avoids problems for them. Unless a manager is receiving resources from a unit, most managers will not cooperate. In this context, most organizations fail because goals are not team goals, and incentives do not promote team work. In the case of hoshin, human resources (HR) play a significant role in encouraging the right behaviors in meeting organizational outcomes.

The HR link in hoshin is a unique aspect in annual business planning. In the first phase of hoshin, each unit lists its departmental role, priorities, and targets. Each employee reviews his or her own role, its work methods, and how the role contributes to the overall goals of the unit. In this phase

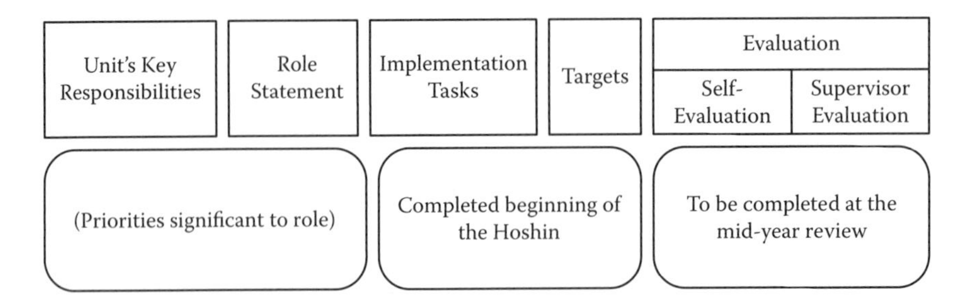

FIGURE 7.18
Performance appraisal factors (adapted example): hoshin form.

of hoshin, HR analyzes and documents each employee's expectations for contributing to the year's hoshin. Each line item of the unit's hoshin is linked to an employee's role; which in turn is linked to each employee's performance appraisal. In this way, each employee understands how he or she will contribute to the goals of the company and how he or she will be evaluated at the midyear and end-of-year review periods. An example of Toyota's initial performance appraisal form for hoshin is shown in Figure 7.18.

An interesting aspect of the performance appraisal process is the concept of target selection. Toyota believes that targets should be measurable, challenging, and achievable. Most companies encourage employees to establish targets without truly understanding the problem domain or current situation. When this occurs, employees have anxiety because the nature of the problem, project, or activity is not truly understood. To make things worse, the manager is also less informed because the nature of the problem was never truly understood. In this context, the speed and quality of management are poor because adequate and sufficient help cannot be provided.

Toyota believes that targets are challenging not because the problem domain is unknown but because the work and effort to achieve the targets have not yet been implemented. An illustration of this concept was shown in a previous chapter relating to Toyota's eight-step problem-solving process. Target selection in Toyota's problem-solving process occurs after step 2. This means that targets should not be selected before defining the problem domain. For targets to be meaningful and achievable, employees must first understand the problem domain.

Once the unit has broken down the company hoshin, the draft of the annual plan is complete. The plan will then go through several approval processes before being finalized. The plan will be approved by management within the unit, discussed and confirmed with related units, and

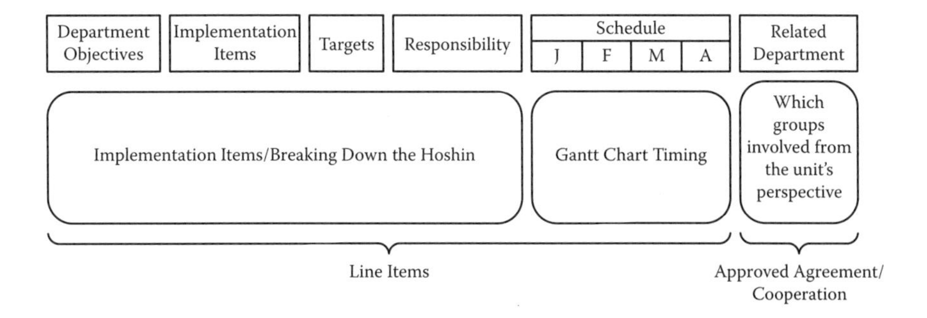

FIGURE 7.19
Annual plan (adapted example): hoshin form.

finally approved by top executives. An example of the annual plan form is shown in Figure 7.19.

7.6.3 Hoshin Implementation Phase

Before hoshin can be implemented, various process management techniques are established to measure and guide the implementation efforts. Tracking ensures that projects and activities of the hoshin stay on course and progress consistently. When processes are abnormal, each unit is responsible for analyzing the causes of the problem and generating countermeasures to get the hoshin back on track. Hoshin is a tool to help managers document and track processes to achieve results. Without sufficient process management, adjustments and corrections cannot be made when problems occur.

A technique that is used by Toyota to monitor the hoshin is plant operational boards (Figure 7.20). Often referred to as visual management boards or key performance indicator (KPI) boards, these displays are used to make information visible by those who need it. Toyota uses these boards to display and track each unit's hoshin at each level. Toyota integrates the hoshin with the operational management indicators, such as safety, quality, productivity, cost, delivery, and human resource development (SQCDHRD). Toyota breaks down each operational management criteria into high-level performance indicators (often referred to as result KPIs), process-level performance indicators (known as process KPIs), and activities (i.e., functions performed by the team). The goal is to visualize the linkage between high-level hoshin goals and lower-level processes. Daily and theme activities are also posted to show how nonroutine projects support the objectives of each unit's hoshin.

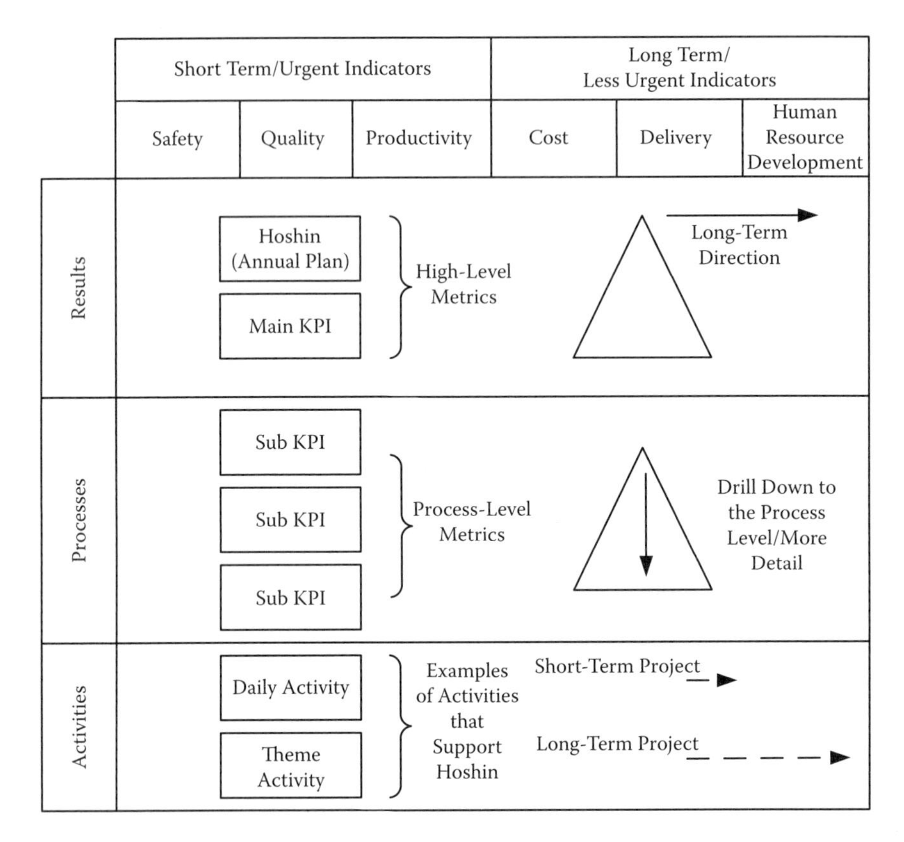

FIGURE 7.20
Operations management board example.

Toyota uses SQCDHRD performance boards for each level of the business unit, namely, the plant, department, section, and group levels. The size of the unit determines whether a board is needed at each level or if it should be combined with other units. Plant-level data is calculated by combining the data from each department. Section data are calculated by combining data from each group. Group, section, and department data are rolled up at the plant level to show the performance of the plant. Figure 7.21 illustrates plant performance tracking for the various levels of the organization.

At the team level, SQCDHRD performance boards are not applied. Toyota uses another form of visual tracking known as the Floor Management Development System (FMDS). Toyota alters the operational management boards because hoshin is a management tool that tracks departmental and interdepartmental activity. At the team level, members are not expected to interact with other teams in other departments. According to system

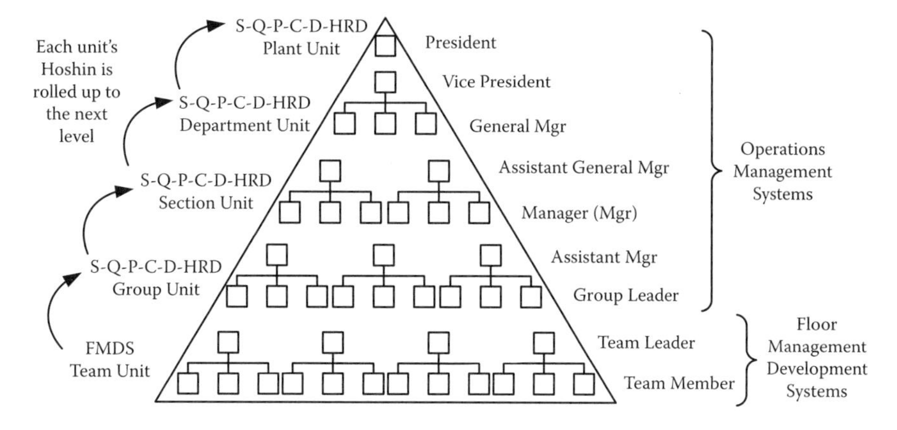

FIGURE 7.21
Division of plant performance tracking for team, group, section, department, and plant level.

theory, this means that hierarchical boundaries at the team level behave as closed systems that require less outside information from the environment (i.e., other teams) to make decisions regarding their work.

The tracking and updating of the FMDS performance boards are completed by the team leader (Figure 7.22). A typical FMDS board contains static and dynamic information relating to the team or work area. Static information is information that does not change over time or as often, such as PPE (personal protective equipment) standards or work instructions (i.e., standardized work). Dynamic information relates to the performance of the team. Examples include productivity charts (i.e., pieces per

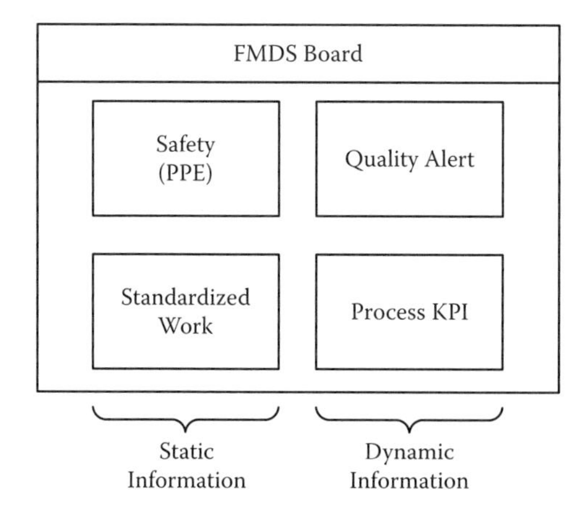

FIGURE 7.22
Floor management development system example: team tracking.

hour) or quality charts (first time through). Overall, Toyota uses process KPIs at the team member level to measure progress for result KPIs at the group level.

7.6.4 Hoshin Evaluation

Toyota's annual business-planning process contains two formal review periods: a midyear review and a year-end review. The purpose of the midyear review is to evaluate from a total perspective the effectiveness of the hoshin. The midyear review forces members to step away from the immediacy of activities and reflect on the overall progress of the hoshin from a larger view. Managers are expected to evaluate the hoshin by prioritizing and adjusting processes due to changing circumstances. The goal is to understand if the initial implementation items are enough to complete the unit's objectives. An example of the hoshin evaluation form is shown in Figure 7.23.

The midyear review is also an accepted way to reinforce successful and repeatable processes. When results are good, managers are looking for signs of standardization. Managers want to know if new processes are sustainable over the long term. If not, the unit has until the year-end review to ensure processes are stable and predictable. In this way, hoshin follows the Shewhart–Deming PDCA (Plan-Do-Check-Act) cycle for each review period and begins a new cycle at the second hoshin period (Figure 7.24).

The midyear review is the first formal performance appraisal process for HR. The appraisal process includes two components: a hoshin component and an employee factor component. The hoshin component looks at how the role was performed in meeting the hoshin goals. This evaluation is concerned with role conformance and the use of established work methods developed by the department to meet its objectives. The idea is to reinforce to each member the role the member plays in each of the unit's

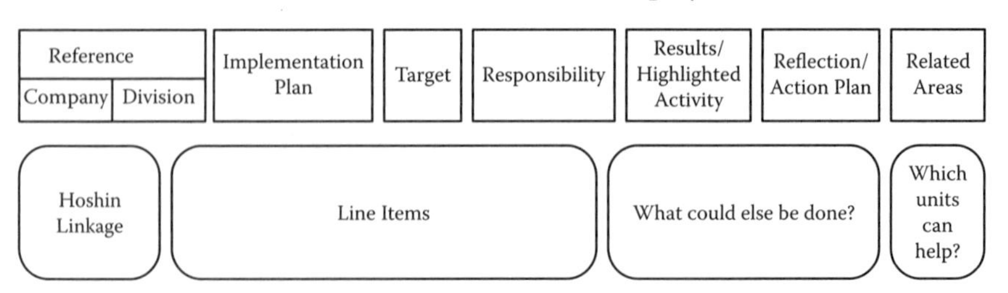

FIGURE 7.23
Annual plan evaluation (adapted example): hoshin form.

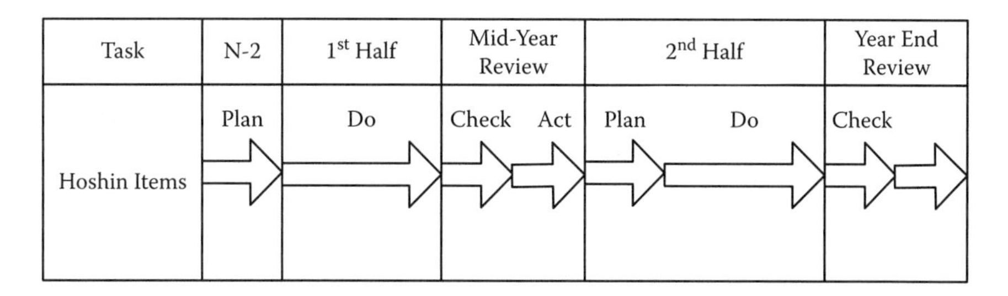

FIGURE 7.24
Hoshin evaluation cycle.

implementation items. This first evaluation ensures that employees are working toward the company goals while fulfilling their role in the unit.

The second appraisal process relates to the employee's capabilities. Toyota uses several employee factors to evaluate role effectiveness. Employee factors include abnormality detection, abnormality management, and relationship and skill development. The concept is that employees should be able to identify issues relating to the hoshin, understand how to manage the issues in the hoshin, and most importantly, conduct themselves in a manner that is conducive to teamwork. Employees are evaluated on how much they have learned and have developed additional skills and training. An example of a performance appraisal factors evaluation form is shown in Figure 7.25.

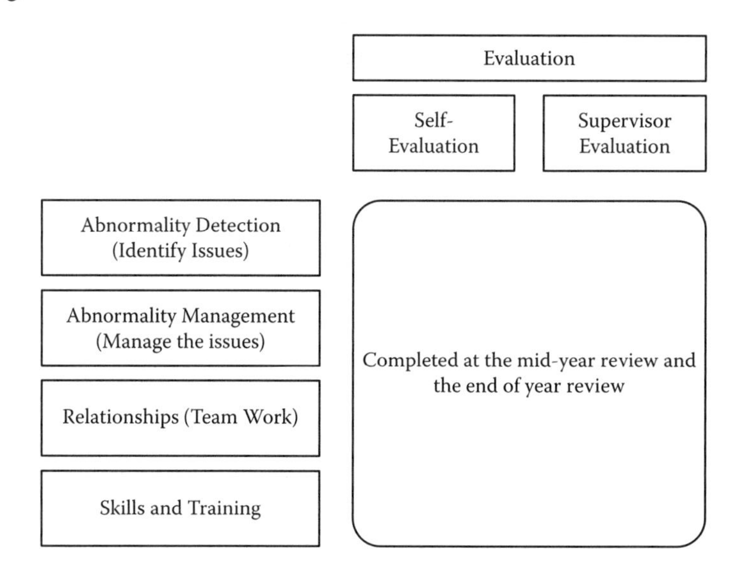

FIGURE 7.25
Performance appraisal factors based on individual capabilities (adapted example).

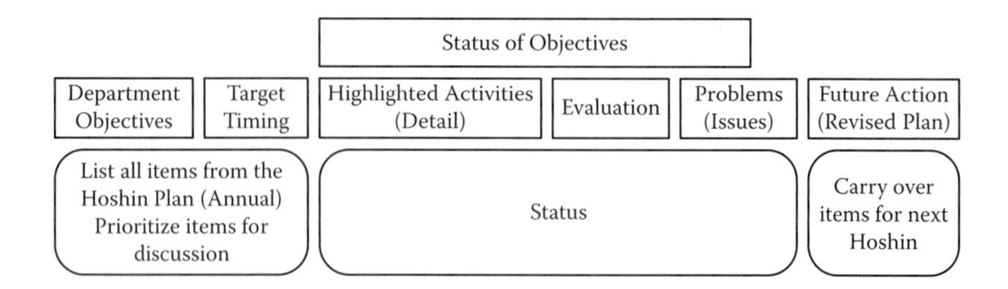

FIGURE 7.26
End-of-year evaluation (adapted example): hoshin form.

7.6.5 End-of-Year Evaluation

The last stage of hoshin is the end-of-year evaluation (Figure 7.26). The purpose of this final step in hoshin is to list all the items from the hoshin plan and formally evaluate how the objectives of the unit were achieved or not fully completed. Implementation items that were completed are evaluated based on their ability to produce long-term and stable results. If targets were achieved yet cannot be sustained, more work is needed. Each manager is expected to highlight and prioritize the hoshin items and to document known problems and issues that have an impact on the hoshin. One of the most essential aspects of the year-end review is to understand which items should be carried over to meet next year's goals. The objective is to use the year-end review to help create next year's hoshin. The more clear and concise efforts organizations spend on the year-end review, the more prepared the organization will be to move forward with a concise plan for the future.

7.7 A THEORETICAL INTERPRETATION OF HOSHIN

In Section 7.5, several strategic planning theories were presented as a way to distinguish the types of thinking used to apply annual business planning. These theories describe elements of the strategic planning process rather than a single strategy that can be applied in all situations and business climates. What is important to understand is how Toyota emulates these different strategies when establishing or carrying out the annual business plan.

Figure 7.27 illustrates the two basic phases of strategic planning and the theories that most closely emulate Toyota's hoshin process. The two types

	Planning Stage	Implementation
Industrial Organizational Theories	*Resource-Based Theory* Hoshin looks at needs within the unit.	*Emergent Theory* Method development is emergent and relies on participation, involvement, and creativity from the unit.
	Deliberate Theory Hoshin is highly structured in the sense of communicating and planning what the organization needs to do.	
	Dynamic Capability Theory Hoshin raises awareness for members to look outside the unit.	
Sociological Theories		*Upper-Echelon Theory* Management controls the approval process for trials.
		Knowledge-Based Theory There is an integrated appraisal review process in hoshin to encourage feedback and learning.
		Agency Theory There is a strong emphasis on vertical and horizontal communication. Hoshin uses roles to manage expectations inside and outside the unit. Tracking is employed to evaluate the line of sight from the top to the bottom to ensure successful implementation.

FIGURE 7.27
Representation of strategic planning theories emphasized during hoshin.

of theories (industrial and sociological) are also highlighted to show when Toyota emphasizes these techniques.

In the planning stage, hoshin emulates more of an industrial approach for establishing what needs to be accomplished for the upcoming year. Hoshin emulates resource base theory and dynamic capability theory to get groups to evaluate what needs to be done outside and inside the unit. Toyota requires each unit to articulate its internal capabilities and to ask for assistance when achieving plantwide goals that cannot be achieved from the unit by itself.

Hoshin also emulates an industrial approach in the planning stages when communicating and planning what the organization needs to do. Mintzberg's deliberate theory (Mintzberg, 2007) defines planning activities as directed, disciplined, and controlling. Hoshin requires target setting to be carried out from the top and cascaded throughout the organization to achieve organizational outcomes.

Mintzberg's emergent strategic planning theory is more closely related to how hoshin is implemented. Emergent strategic planning views that implementation is dynamic and unpredictable, which means that certain aspects of planning should emerge from circumstances. In Toyota's case, kaizen is emergent because it relies on the participation and creativity from employees to generate solutions to meet the hoshin objectives. In hoshin, it is more important that subordinates have their idea first, instead of leadership dictating how to achieve the hoshin. While hoshin is very structured on communicating what needs to happen, Toyota relies on decision making from each employee.

The implementation process of hoshin kanri closely emulates upper-echelon theory and knowledge-based theory. Knowledge-based theory has the view that human capital and learning are primary drivers in strategic planning. Knowledge-based theory closely describes Toyota's hoshin process as it relates to appraisal and feedback. Hoshin requires HR development to be linked to annual business planning so that learning and guidance can be applied throughout the year. Upper-echelon theory describes the ability of leadership to influence and control decisions within the organization. This view suggests that decision making is left up to higher levels of the organization, while lower levels perform as they are told. In the case of hoshin, Toyota follows upper-echelon theory mainly in how management controls the approval process for trials when employees develop work methods. One of the ways that Toyota encourages participation in the workplace is by delegating decision making to the lowest possible level. It is believed that employees closest to the problem have the best chance to solve the problem. Upper-echelon theory describes Toyota's "targeted creativity" approach; employees have the right to test their ideas, while management has the right to approve trials. In this manner, management is not controlling the new method, only the coordination of the trial.

Agency theory is based on the view that groups in isolation have different interests, which means they are less likely to understand system influences and interdependencies. This particular theory is applicable to strategic planning since strategy formation and implementation are often created and achieved in isolation. In the context of hoshin, Toyota employs several mechanisms to make planning and implementation more integrated. In the planning stages, hoshin encourages various forms of vertical and horizontal communication to promote cooperation and teamwork

(Figure 7.27). Roles are also used during planning to help manage expectations from members inside and outside the unit. During implementation, tracking is used to measure progress from a company level to an individual team level. Overall, how Toyota applies these techniques is essential in applying teamwork so that decision making does not occur in isolation.

7.8 SUMMARY

The purpose of this chapter was to explain how TPS exhibits the system property of hierarchies. Hierarchy theory in systems is used to describe how an organism differentiates its layers and subsystems to respond to changing environmental conditions. Permeable boundaries suggest that an organism is more likely to evolve and adapt to its environment than with hierarchical boundaries that are rigid and work in isolation. This chapter described how hoshin exhibits the hierarchical functions within TPS.

The process of carrying out hoshin was also described. To increase the permeability of hierarchies within Toyota, hoshin is used to help units understand what they can accomplish and what they cannot. When units cannot accomplish organizational objectives, hoshin stimulates units to work cooperatively to meet companywide goals. Consequently, role fulfillment within each unit is also increased by integrating the HR appraisal process in the hoshin. The overall effect of hoshin is to move each unit and each member closer toward organizational goals and objectives.

Last, various strategic planning theories were proposed to help explain how hoshin emulates many traditional approaches to annual business planning yet in differing contexts. Hoshin is a management tool that shares characteristics of agency and upper-echelon theory in the selection of targets and goals, yet at the same time draws on many other theories, such as knowledge-based and dynamic capability theories, to help units cooperate and learn. Ultimately, what makes hoshin unique is how these techniques are applied at different phases of the hoshin process.

REFERENCES

Abro, Q., Memon, N., and Shah, R. (2009) Strategic factors for enhancing the innovativeness of the nanotechnology firms. *International Journal of Business Innovation and Research*, 3(6), 596–609.

Anand, N., Gardner, H., and Morris, T. (2007) Knowledge-based innovation: Emergence and embedding of new practice areas in management consulting firms. *Academy of Management Journal*, 50(2), 406–428.

Andrews, R., Boyne, G., Law, J., and Walker, R. (2009) Strategy formulation, strategy content, and performance: An empirical analysis. *Public Management Review*, vol. 11, no. 1, 1–22.

Arthurs, J., Hoskisson, R., and Busenitz, L. (2008) Managerial agents watching other agents: Multiple agency conflicts regarding under pricing in IPO firms. *Academy of Management Journal*, 51(2), 277–294.

Becvar, D., and Becvar, R. (1999) *Systems Theory and Family Therapy: A primer* (2nd ed.). University Press of America, Lanham, MD.

Chaffe, E. (1985) Three models of strategy. *Academy of Management Review*, 10(1), 89–98.

Chandler, A. (1962) *Strategy and Structure: Chapters in the History of Industrial Enterprise*, Doubleday, New York.

Chermack, T. (2004) A theoretical model of scenario planning. *Human Resource Development Review*, 3(4), 301–325.

Cooper, L. (2000) Strategic marketing planning for radically new products. *Journal of Marketing*, 64(1), 1–16.

Crook, T., Ketchen, D., Combs, J., and Todd, S. (2008) Strategic resources and performance: A metal-analysis. *Strategic Management Journal*, 29(11), 1141–1154.

Donohue, J. (2005) *Strategic HR Planning—Theory, and Methods*, Defence R&D Canada, DRDC ORD TN 2005-04. Ottawa, Ontario.

Doving, E. and Gooderham, P. (2008) Dynamic capabilities as antecedents of the scope of related diversification: the case of small firm accountancy practices. *Strategic Management Journal*, 29(8), 841–857.

Faludi, A. (1986) Towards a theory of strategic planning. *Netherlands Journal of Housing and Environmental Research*, 1(3), 253–268.

Freeman, R. (1984) *Strategic Management: A Stakeholder Approach*, Pitman, Boston.

Gigch, J. (1978) *Applied General Systems Theory*, Harper & Row, New York, NY.

Hammer, M. and Champy, J. (1993) *Reengineering the Corporation*, Harper Business, New York.

Holman, P., Devane, T., and Cady, S. (2007) *The Change Handbook*, Berrett-Koehler, San Francisco.

Jackson, T. (2006) *Hoshin Kanri for the Lean Enterprise*, Productivity Press, London.

Kay, J., McKiernan, P., and Faulkner, D. (2003) The history of strategy and some thoughts about the future, in *The Oxford Handbook of Strategy, Volume I: A Strategy Overview and Competitive Strategy*, ed. Faulkner, D. and Campell, A. Oxford University Press, New York.

Kim, W. C. and Mauborgne, R. (2005) *Blue Ocean Strategy*, Harvard Business Press, Boston.

Lavie, D. and Rosenkopf, L. (2006) Balancing exploration and exploitation in alliance formation. *Academy of Management Journal*, 49(4), 797–818.

McClelland, D. (1988) *Human Motivation*. Cambridge University Press, Cambridge, UK.

McLean, G. (2006) *Organizational Development: Principles, Processes and Performance*, Berrett-Koehler, San Francisco.

Mintzberg, H. (2007) *Tracking Strategies*, Oxford University Press, New York.

Morgan, M., Levitt, R. and Malek, W. (2007) *Executing Your Strategy, How to Break it Down and Get it Done*. Harvard Business School Press, Boston, MA.

Overton, W. (1975) General systems, structure and development. *Structure and Transformation: Developmental and Historical Aspects*, 3, 61–81.

Peery, N. (1972) General systems theory: An inquiry into its social philosophy. *Academy of Management Journal*, 15(4), 495–510.

Porter, M. (1980) *Competitive Strategy: Techniques for Analyzing Industries and Competitors*, Free Press, New York.

Rodgers, W., Babcock, R., and Efendioglu, A. (1988) The art of strategic planning. *SAM Advanced Management Journal*, Autumn, 54:4, 26–31.

Scholtes, P. (1988) *The Team Handbook: How to Use Teams to Improve Quality*, Joiner Associates, Madison, WI.

Singh, S. (2009) Knowledge management practices and organizational learning in Indian software company. *International Journal of Business Innovation and Research*, 3(4), 363–381.

Smith, A. and Racic, S. (2009) Strategic recurring traits that contribute to firm loyalty: Case studies of best business practices', *International Journal of Business Innovation and Research*, Vol. 3, No. 6, 610–633.

Smith, R. (2004) *Strategic Planning for Public Relations*, Erlbaum, Mahwah, NJ.

Stavros, J., Cooperrider, D., and Kelley, L. (2003) Strategic inquiry: Appreciative intent: Inspiration to SOAR, *AI Practitioner*, November.

Subramanian, R. and Gopalakrishna, P. (2009) Relationship between market orientation and performance in family owned firms: a context-specific investigation. *International Journal of Business Innovation and Research*, 3(5), 500–514.

Suddaby, R. and Greenwood, R. (2005) Rhetorical strategies of legitimacy. *Administrative Science Quarterly*, 50(1), 35–67.

Tennant, C. and Roberts, P. (2001) Hoshin kanri: Implementing the catchball process. *Long Range Planning*, 34(3), 287–308.

Tischler, L. (2002) Can Kevin Rollins find the soul of Dell? *Fast Company*, 64, November, p. 104.

Wood, V., Pitta, D., and Franzak, F. (2009) Successful strategic alliances with international partners: Key issues for small-to medium-sized enterprises. *International Journal of Business Innovation and Research*, 3(3), 232–251.

Wright, P., Kroll, M., Krug, J., and Pettus, M. (2007) Influences of top management team incentives on firm risk taking. *Strategic Management Journal*, 28(1), 81–89.

8

The System Property of Transformation in TPS

8.1 SYSTEM PROPERTY: TRANSFORMATION

All organized systems contain some form of a conversion process by which elements in a system change inputs into outputs (Gigch, 1978). The conversion process usually adds value and utility to the inputs as they are converted into outputs (von Bertalanffy, 1968). Inputs are quantities acting on a system from the environment, whereas outputs are the results of the inputs acting on the system. A system interacts with its environment by means of inputs and outputs. When an input undergoes some form of a change, a system is often characterized using a transformation model (Figure 8.1). All systems have requirements converting matter and energy. Without some form of a conversion process, a system is likely to run down and die (Miller and Miller, 1992).

The conversion process of a system can be characterized by how it responds to input signals. Signal theory proposes four general system characterization types:

- SISO (single input, single output)
- SIMO (single input, multiple outputs)
- MISO (multiple inputs, single output)
- MIMO (multiple inputs, multiple outputs)

Conversion processes that can handle multiple inputs and outputs are considered to be highly structured systems. The more structured the conversion process is, the greater a system's ability to respond to complex environments. Structured processes can exchange information more

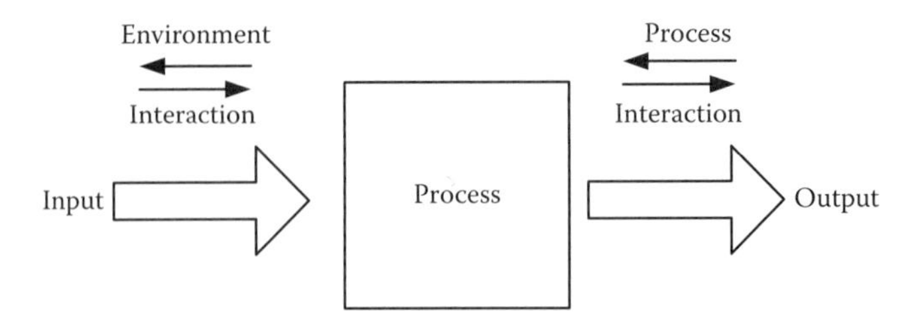

FIGURE 8.1
System characterization of transformation.

readily from their environments and process internal information more easily (Keen and Morton, 1978).

Unfavorable conditions in an environment can force a system out of its normal steady state into an abnormal condition (Miller, 1955). A system can undergo a pathological or unhealthy state when one or more of its critical variables remain beyond the normal steady-state range for a significant period. The factors that can cause a system to enter a pathological state relate to how well a system can maintain its structure when environmental conditions change (Miller and Miller, 1992). Living systems cope with pathological issues at the local subsystem level; however, when these fail other systems become involved. Eventually, higher-order systems become involved when lower subsystems can no longer carry out essential functions (Miller and Miller, 1992). A system can enter a pathological state when (Miller, 1978) there is

1. Lack of matter, energy, or inputs
2. Excess of matter, energy, or inputs
3. Inputs of inappropriate forms of matter or energy
4. Lack of information inputs
5. Excess of information inputs
6. Inputs of maladaptive genetic information in the template
7. Abnormalities in internal matter or energy processes
8. Abnormalities in internal information processes

A theory that describes the types of processes used in the conversion process is living systems theory (LST). For any living system to survive, it must carry out several essential subsystem processes. A sample of those subsystem processes is shown in Table 8.1 and illustrated in Figure 8.2. Without these processes, a system cannot engage in its primary function, which in all systems is its own replication (Maturana and Varela,

TABLE 8.1

Conversion Processes According to Living Systems Theory

#	Living System Theory Essential Process	Description of Process
1	Ingestor	A subsystem that brings matter and energy across a system's boundary from the environment
2	Motor	A subsystem that moves the system or its parts to other areas inside or outside the system
3	Converter	A subsystem that changes certain inputs into forms more useful for the process
4	Producer	The part of the system that forms stable associations for significant periods of time to grow, repair, or replace components of the system
5	Supporter	The part of the system that maintains the proper spatial relationships among components of a system
6	Distributor	A subsystem that carries inputs to other parts of the system
7	Matter energy storage	A subsystem that stores matter or energy in a system location for retrieval at a later time

Source: Adapted from Miller, J. 1978, *Living Systems*, Wiley, New York.

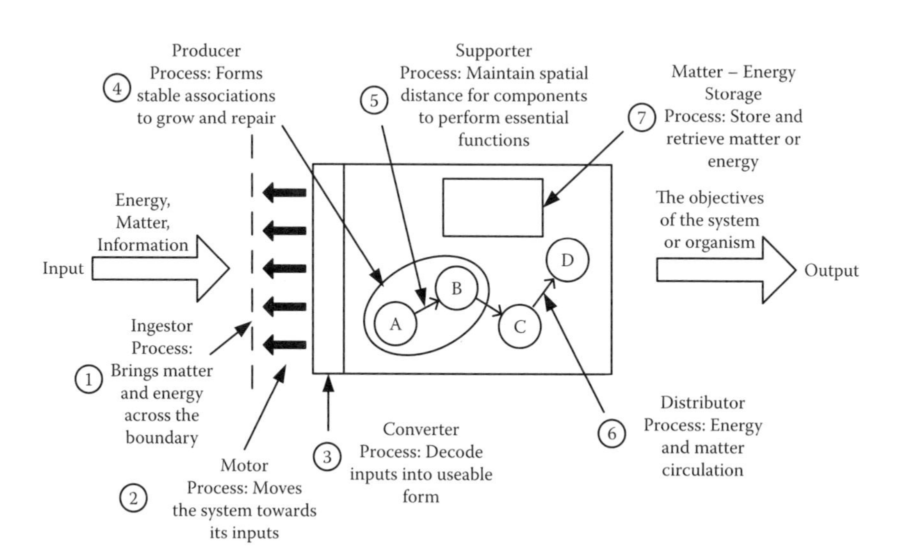

FIGURE 8.2
System characterization of transformation.

1972). A system's first function is to stay alive. Regardless of a system's secondary function, without this first attribute a system would not exist. Conversion processes help systems to maintain their own metabolic processing regardless of environmental challenges. The metabolism of matter and energy is the energetics of living systems and is what gives systems life (Miller, 1978).

8.2 THE CONVERSION PROCESS IN THE TOYOTA PRODUCTION SYSTEM

Conversion processes are similar in all living systems, whether they are simple cells, organisms, countries, or societies. Management systems must also contain conversion processes for them to survive. In the Toyota Production System (TPS), ideas for kaizen must be converted into action; otherwise, TPS will die. The change process that Toyota uses to encourage employees to participate in kaizen can best be described as the metabolism of TPS (Figure 8.3).

Unfortunately, most organizations have a low metabolism for kaizen, with only a few employees expected to be involved in workplace improvement. The primary pathology that organizations experience when implementing Lean is that there are not enough inputs (i.e., ideas) for the organization to improve itself. The second pathological problem is that the inputs or ideas are not in the right form for processing. Using Miller's (1978) biological concept of the motor, there is nothing that drives the

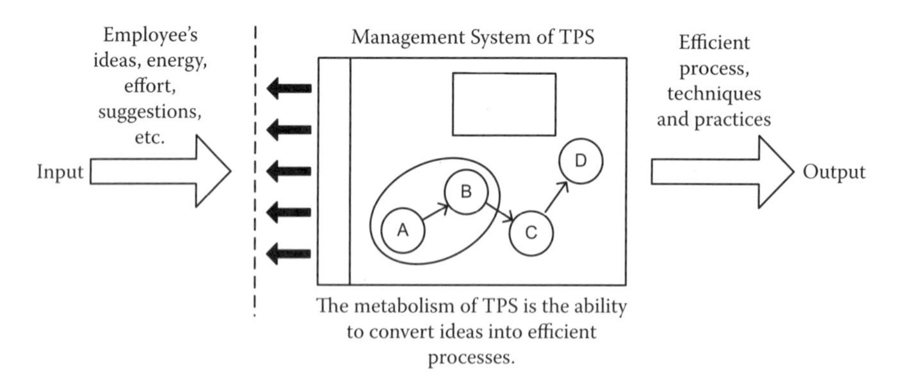

FIGURE 8.3
The metabolism of TPS.

organization toward eliciting ideas from employees. Finally, organizations struggle implementing Lean because there is nothing in the conversion process that allows stable associations to occur for growing and repairing ideas.

At Toyota, the producer process is teamwork. Without teamwork, ideas cannot reach their full potential. When employees engage in problem solving, the quality of decision making improves because ideas are built on another, corrected, refined, and repaired. Organizations that are successful in Lean do well converting ideas into actions, mainly because they understand how to change (i.e., transform).

The purpose of this chapter is to use Miller's (1978) concept of LST to describe Toyota's conversion for TPS. LST is essential for understanding change management in organizations because it can identify the types of conversion processes in systems. While Toyota may employ various forms of change management in TPS, each conversion process has a unique goal and function. LST can organize these types of conversion processes in TPS and provide a holistic explanation of TPS as it relates to change management. Figure 8.4 provides an illustration of LST and the integration of various change management theories.

8.3 LITERATURE REVIEW OF CHANGE MANAGEMENT

The purpose of change management is to alter the basic assumptions currently held in the organization in favor of a more optimal or desired state (Chrusciel, 2008). The main outcomes of change management are learning (Dewey, 1910), improvement (Deming, 1992), and goal achievement (Herscovitch and Meyer, 2002). The techniques used in change management vary, but mostly depend on an organization's particular preference, priorities, and individual circumstances. Some of the main mechanisms that are used to promote change include leadership (Kouzes and Posner, 2007), social management (Lewin, 1947), and planning (Nadler, 1982).

The benefits of change management are that successfully introduced change brings about growth and development of the organization (Stanev et al., 2008). Organizations that are adaptable have increased competitiveness (Neal et al., 1999), improved decision making (Snyder, 2001), and higher performance and profits. An organization that can compensate for dramatic changes can also handle both major threats and major opportunities

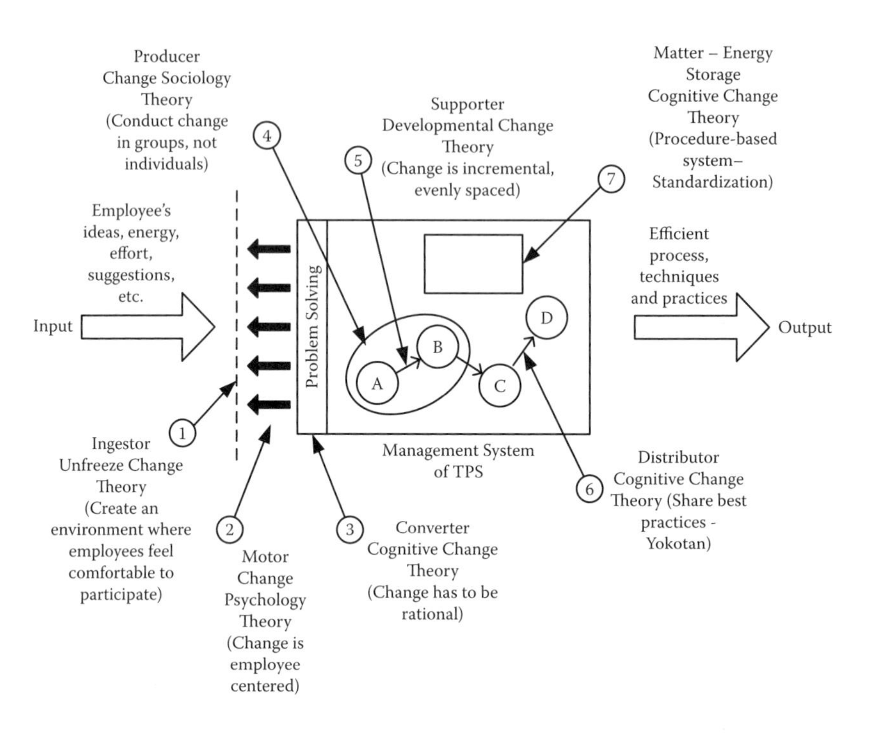

FIGURE 8.4
System characterization of LST and change management theory.

inside and outside the organization (Chrusciel, 2008). Consequently, organizational flexibility also enables companies to better maintain their competitive advantage while stabilizing internal fluctuations (Verdu and Gomez-Gras, 2009).

One of the key elements of change management is dealing with resistance to change (Caldwell et al., 2004). Resistance is defined as employee behavior that seeks to challenge, disrupt, or invert prevailing assumptions, discourse, and power relations (Collinson, 1994). Resistance to change can also be described as a way to restore justice when an individual is at lower authority or power than the source (Folger and Skarlicki, 1999). Inherently, all human behavior exhibits some of these defenses when someone or something alters the status quo (Brenner, 2008). On the other hand, there are schools of thought that disagree with the notion that all employees resist change. When employees are treated poorly or unfairly, employees do not resist change but resist being punished (Harvey, 1975). In most cases, resistance to change is due to psychological issues rather than cognitive or rational aspects. It is often believed that organizations place too much emphasis on the logical aspects of change instead of trust, commitment, and creditability, which is more significant with coping and dealing with acceptance (Morgan and Hunt, 1994).

To date, many techniques and practices exist for implementing change. A summary of these various approaches is shown in Table 8.2. An interesting characteristic in these approaches is that they all view a single dimension as significant when implementing change, which implies that a single methodology fits every company. Although this logic may simplify application, it is unlikely that a particular method can suit all organizations equally or be universal to accommodate a wide range of business conditions. This means that practitioners must choose which method to use by sensing the organization and matching it to the appropriate change technique. To this end, it is not surprising why most organizations improvise change management (Jorgensen et al., 2009).

In addition to standard techniques, many general rules and best practices have been developed to aid in change management. For example, most experts agree that a lack of change should not be ignored. Organizations should be persistent in their techniques and be committed to see change through. Furthermore, organizations should be realistic and implement change according to their timetable rather than some arbitrary number or benchmark. An awareness of the many factors that are out of a company's immediate control can also help an organization to anticipate and manage expectations (Sirkin et al, 2005).

Despite the advancements in change management, most attempts fail, with an average success rate of 25% (Hammer and Champy, 1993). One of the most common problems of change management relate to management's ability to support and embrace the new change. This should not be surprising since the fundamental role of management is to control everything so that things turn out right (Fronda and Moriceau, 2008). When change is introduced, management has the role of stabilizing existing systems, which often counteracts change itself. Change is further complicated by individuals or groups who want to hold on to the satisfactory present (Cutcher, 2009). If the desired state cannot be seen as confronting the concerns of the past and present, there will be little desire to move forward. Last, most organizations fail to manage expectations by not communicating and eliciting opinions by those who are affected by the change itself (Parish et al., 2008). Table 8.3 summarizes some of the more general issues and difficulties relating to change in organizations.

The challenge with change management approaches is that they all view one dimension as significant, which implies that a single methodology

TABLE 8.2

Literature Review of Existing Change Management Approaches

Logical Approach (Brenner, 2008)	Emotional Relationships (Neal et al., 1999)	Social Aspect, Teams and Groups (Axelrod et al., 2004)	Individual (Pullen, 2002)	Transformational Leadership (Bass, 1985)
Provides a common rationale for change, such as pledges that speak to customers, stakeholders, and employees. Often are the "felt needs," that are recognized by managers and leaders currently. Insightful to explore the current philosophy and state before embarking on changing the process.	Technique that is not completely focused on the bottom line but rather a focus on providing emotional support during transformation. Technique is focused on hope, new opportunities, new beginnings, and relationships in promoting employee acceptance and associated with change. Technique also relies on maintaining trust and credibility between employees and management. Examples: Vision, mission, and value-based campaigns	The use of collaboration and teaming for getting groups to set their own change management process. Also emphasis on putting together the right team to deal with the change. Often referred to as self-directed teams or high-performing teams for which a trained facilitator is used.	Change is focused on an individual level within the company. Approach often uses deep structure interviews with individuals to explore areas of misalignment. Technique uses commitment to lead employees to exert effort toward achieving goals for organizational change.	Transformational leadership motivates followers to identify with the leader's vision and sacrifice their self-interest for that of the group or the organization. Leader must possess trust, charisma; recognize the need for change; create the new vision and institutionalize change. This style is better for nonroutine situations and relies on the leader, who understands the politics of the organization.

Involvement in Decision Making: Buy In (Jorgensen et al., 2009)	Success Building (Pullen, 2002)	Rewards and Incentives (Buch and Tolentino, 2006)	Action Learning (Gotsill and Natchez, 2007)	Formal/Technical Approach
Technique that focuses on people and their need to be a part of the process. A common approach is to identify the things that people have lived with for years and get the support of the top team to include them in the change program.	Good way to overcome resistance is to have some early successes, showing that real change is possible. This technique often works with relatively healthy parts of the organization that have the resources and the greatest chance to improve.	Practice that uses rewards to entice participants to change. Rewards can be intrinsic, social, organizational, or extrinsic. Care should be given to rewards or incentives that promote personal gain. In all cases, positive behaviors should be reinforced. A common practice is transactional leadership, which uses rewards in exchange for subordinates' performance. This approach is better suited for event-type pacing, is more mechanistic, emphasizes clarification of goals and follower compliance through incentives and rewards, with focus on task completion.	Action learning is a change management technique that is effective at transferring new skills required in the new environment. Encourages the team to simulate the new process ahead of the change, which is helpful for addressing user anxiety without the risk of making mistakes. Also provides a creative medium to improvise and experiment with a wide variety of low-cost probes. Example: Participative design workshops	Change management approach that emphasizes policy, procedures, and standards. Also, utilizes roles and responsibilities that affect how change is defined, communicated, and enforced.

Continued

TABLE 8.2 (*Continued*)

Literature Review of Existing Change Management Approaches

Change Agent or Champion, Interventionist (Holman et al., 2007)	Appreciative-Inquiry Approach (Cooperrider and Avital, 2004)	Balance Scorecard Approach (Kaplan et al., 1993)	Discussion and Dialogue (Gastil et al., 2005)	Systems Approach (Adams, 1965)
The change agent or champion approach is a technique that assigns a motivator to manage the change process. Role is to manage the in-betweenness from different groups and levels. The reverse of a change champion who gives power, is the change assassin, who takes away power. Approach often resembles external consultancy models. Traits and characteristics of change agents are also studied, including confidence, initiative, ownership, risk taking, sense of urgency, open-mindedness, credibility, flexibility, persistence, and innovation.	A search for the best in people. The positive potential of all employees. Freedom to be heard, to dream, to contribute, to be positive.	Key performance indicators are captured periodically about the change process to adjust, monitor, and evaluate the change process.	A technique used to prompt dialogue and deliberation for increased understanding. Examples include: scenario thinking, town hall meeting, and conversation café.	Systems approach to change management that emphasizes situational factors. Technique is largely based on its circumstances rather than universalistic approaches. Technique evaluates multiple entry points, such as different people, groups, and processes.

Rapid Results Approach (Schaffer, 1988)	Less-Resistance Approach (Alexander, 1979)	Culture Perspective (Jones et al., 2008)	Management Support (Johnson et al., 1990)
Results-oriented change management approach. Often uses a multisystems approach to light many fires and to load experiments for success. Technique often promotes event-based pacing, which is well suited for fostering incremental change.	Work with the forces in the organization that are supportive of the new change instead of those who are defensive and resistant. Most important mechanism for creating a critical mass of people who are solidly behind a change program (Alexander, 1979). Technique relies on a critical mass (20% rule), which means that employees who are solidly behind the idea are likely to promote successful change in the impacted area.	Technique considers cultural perspectives that relate to organizational change. Approach emphasizes member acceptance.	Employees who believe that their managers are supportive tend to be more committed to organizational change.

TABLE 8.3

Common Problems and Challenges in Change Management

Struggle Point	Description of Problem	Reference
Where to start	Not starting with work groups and individuals who already have freedom and discretion in managing their own operations.	Harrison, 1972
Destination point	Lack of shared vision and objectives, no hidden agendas.	Pullen, 2002
Teaming	Group dynamics are not managed (i.e., personalities, attitudes, etc.), group roles are not defined, fear of team destruction compared to team building, friendships should extend beyond the workplace, not willing to spend time with workers outside workplace.	Snyder, 2001
Management blocking	Management does not buy in to the new vision. Blocks change by supporting the existing system rather than teaching the new system. Managers have been dominated by the assumption that the job of the manger is to control everything so that things turn out right.	Carter, 2008
Training syndrome	Belief that short-term training can cause change to occur. Training in the classroom assumes that the ideas agreed to in the classroom will find acceptance on the shop floor.	Alexander, 1979
Mandate change	Leadership has the belief that change can be mandated.	Brenner, 2008
Cannot fire	Some dissenters may need to let go for the greater good of the organization.	Carter, 2008
Poor communication	Understanding and opinions cannot be exchanged because two-way communication is missing.	Pullen, 2002
Territories	Stepping on toes and territories, not-invented-here syndrome; the problems of ownership and transfer, fear of losing status (i.e., the untouchables, the older vested people in the organization).	Burk, 1982
Improvise	Not developing a standard change methodology.	Jorgensen et al., 2009
Paranoia	Under threat, people tend to engage in hypervigilance, in which every social interaction becomes scrutinized for hidden meaning and sinister purpose. Organizational change heightens sensitivity about fairness, can lead to lack of cooperation and industrial sabotage.	Folger and Skarlicki, 1999

TABLE 8.3 (*Continued*)

Common Problems and Challenges in Change Management

Struggle Point	Description of Problem	Reference
Attitudes and behavior	Need to address the way people think and behave.	Pullen, 2002
Firefighting Routine	Change initiatives suffer from fatigue due to firefighting; the solution is to set aside some time from the nonroutine firefighting.	Pullen, 2002
Static culture	More challenging when companies have a static corporate culture that is less willing to take risk.	Brenner, 2008
Change is disruptive	Change is generally disruptive and complex; it involves the unknown and is likely to impact anything from work processes to job security.	Andrews et al., 2008
Cognitive and psychological gap	Bridge the gap between cognitive and psychological. Change management that emphasizes the psychology of change helps individuals to cope with the emotional aspects of change. Change management that is cognitive in nature tries to address the rational and justification of change.	Brenner, 2008
Career arch and risk	Executives may actually be more likely to be risk averse at higher levels of their career arc than earlier in their careers.	De Bono, 2006
Consultant hero syndrome	Organizational development cannot be done by a consultant. Change agents and consultants cannot change anything; the clients can.	Harvey, 1995
Productive conflict	How to surface change conflict and deal with it productively. Overcoming resistance gives the impression that the effort is a win-lose conflict.	Holman et al., 2007
Holding on	Emphasis is placed on the desired state where the problems of the past and present will be solved. Holding on to a satisfactory present, members can resist by drawing on counterdiscourses (counterphilosophies or ideas) of the past. Can be used to validate resistance on a moral authority. Can also disindentify with the new culture. Example: using tradition to be sources for which employees resist change.	Cutcher, 2009
All or nothing	Often, thinking is in either/or terms, which means that there has to be a choice, one-or-nothing approach.	Marshak, 1994

Continued

TABLE 8.3 (*Continued*)

Common Problems and Challenges in Change Management

Struggle Point	Description of Problem	Reference
Inability to measure progress	Inability to measure "progress" therefore becomes a valid reason to question whether change is needed.	Marshak, 1994
Passive resisters	Passive resistance cannot be seen; it looks ordinary. Also known as "discrete resistance," maybe the most widespread today, which can be described as the "go slow" approach. Unlike revolts, which are active and explicit (i.e., strikes).	Alter, 2000
Time	Time is only a statement of an individual's or an organization's priorities. Time is often used to explain failure. Time related to change is irrelevant.	Harvey, 1995

fits every company. It is unlikely that one method is universal or fits all business conditions. To this end, practitioners must sense the organization and match it to the appropriate technique. Because the decision is so daunting, most organizations justify an impromptu approach mainly due to their preferences, priorities, and circumstances.

8.4 CONVERSION PROCESS: INGESTOR

The purpose of the ingestor in living systems is to bring matter and energy from the environment into a system's boundary. This process provides the vehicle for transporting an environment's matter and energy into the system. Without an ingestor, matter and energy, no matter how abundant, would never enter into the system. The ingestor is the system's starting process in conversion.

In TPS, the ingestor can best be described as the open climate that Toyota creates for employees to experiment and test ideas. One of the earliest theories that described how organizations and people go through change is Lewin's (1947) three-step change model. There is a common fascination in change management theory that organizations or individuals go through a series of steps when experiencing change. While it is unknown during which stage companies should exert the most effort in managing change successfully, researchers agree that it is not the emphasis on any stage but that the stage is completed psychologically. Psychologists agree that if a stage is not completed, there is a tendency to revert to a previous stage.

The distinction between various stages or steps of change still remains a valid research point for understanding how change occurs. Change step models suggest that organizations follow and experience distinct periods of change that can be classified and managed. Although modern thinking still supports the concept of linear change management (Tarnas, 1991), other researchers indicated that change is cyclical, not linear (Prochaska and DiClemente, 1984). This view is evident by the fact that many individuals relapse on their change efforts and do not successfully maintain the gains the first time around. Therefore, change cannot be sequenced into an orderly and natural linear progression (Marshak, 1994).

The ingestor phase of conversion is similar to Lewin's initial stage of change management, named the unfreezing stage. At this step, Lewin proposed that organizations and people must be in a free state to experiment and test new behaviors before completely embracing change. Employees must be given opportunities to explore, practice, and learn new behaviors in a psychologically safe environment. One of the ways that Toyota is able to get employees to buy in to the idea of participating in TPS is the kaizen philosophy. The kaizen philosophy is a shared mentality to search constantly for better ways of performing work. Kaizen is about encouraging employees to make many small improvements in the workplace rather than a few and significant ones. The work of Hall (2006) proposed that employees will be more willing to contribute their ideas if the risk for change is small. Small risk and small changes encourage an environment that makes it easier for employees to unfreeze and let loose of their creativity. It is when employees have a lot to lose or risk that they will be more likely to hold back and not become involved.

Toyota's approach to kaizen is to encourage small change and small risk. If an organization is not changing, it is not because it is in a state of equilibrium, but because it is not responding to changes in the environment (Smither et al., 1996). Toyota's view on unfreezing implies that change is not about a destination, but rather a way of facilitating changeability (Michaels, 1994). Toyota views that an open environment is more important than just reaching a destination. The kaizen philosophy is the freedom to keep trying so that ideas can pass through the organization, not necessarily to reach a destination.

Toyota would like for employees to embrace the kaizen concept completely. In the Toyota team member handbook, it is indicated that each

Team Member

Toyota Motto:
There shall not go a day without
one small step
of improvement in my Work life...
Home life...
Community life...

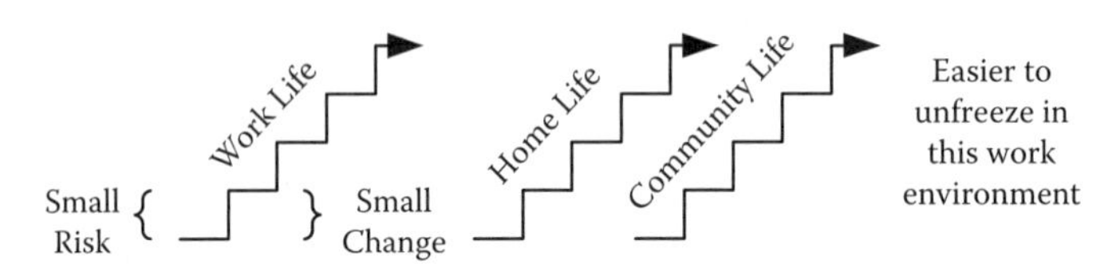

FIGURE 8.5
One small step of improvement. (Adapted from Hall A, 2006, Introduction to Lean—sustainable quality systems design—integrated leadership competencies from the viewpoints of dynamic scientific inquiry learning and Toyota's Lean system principles, published by Arlie Hall, ISBN 0-9768765-0-7.)

Team Member

If team members have a lot to risk to test their ideas, fewer employees will be willing to contribute their ideas

Not one large risk...

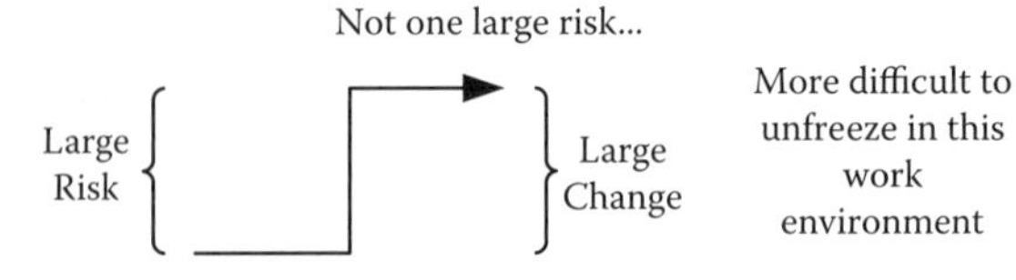

FIGURE 8.6
One large step of improvement, large risk.

employee should take a small of step of improvement in his or her work life, home life, and community life (Figures 8.5 and 8.6) (Hall, 2006). The idea is that employees should share the same type of values, norms, and behaviors of the kaizen philosophy.

8.5 CONVERSION PROCESS: MOTOR

The motor process is responsible for moving the system or its parts in relation to the environment. When environmental inputs are lacking, are in excess, or are abnormal, the motor in the conversion process can help the system to make adjustments. Without a motor in the conversion process, environmental inputs can go unchecked in the delivery to the system. The motor helps to compensate and buffer environmental inputs to keep essential functions of a system regular and consistent.

In TPS, the motor that drives employee participation is the human resource (HR) function. TPS requires a strong HR function for employees to buy in to the idea of workplace improvement. The HR function ensures that employee participation is fair, consistent, and motivating when change occurs in the workplace. The people aspect to change is one of the most critical success factors for promoting kaizen in TPS. The factors associated with people include personal feelings, attitudes, and current thinking. The most challenging aspect of change is that people are less predictable and more difficult to manage.

The HR function helps the emotional and psychological aspects of change by breaking down the resistance to change. Resistance is defined as employee behavior that seeks to challenge, disrupt, or invert prevailing assumptions, discourse, and power relations (Collinson, 1994), also described as a way to restore justice when an individual is at a lower authority or power than the source. Some scientists said that all employees exhibit these defenses (Brenner, 2008), while others suggested that people do not resist change—they resist being punished (Harvey, 1975).

Regardless, change in the workplace can be just as dramatic as real-life change. Workplace change can invoke the same types of feelings that would be encountered in a divorce, bankruptcy, or the loss of a loved one. HR buffers workplace change by creating an emotional and trusting connection between employees, specifically between the supervisor and the subordinate. When employees trust their supervisor, employees are much more likely to have acceptance when change occurs in the workplace. The immediate supervisor plays a critical role in organizations because most people quit their supervisors when they leave a company rather than quitting their job (Figure 8.7). Ultimately, how employees feel about their supervisor is generally how they feel about the company.

HR works with all supervisors so that trust can be built at each level of the organization. The way that Toyota builds trust is through two-way communication. Each supervisor is expected to engage in two-way communication with a subordinate to elicit opinions and understanding regarding workplace conditions. Supervisors are expected to care for each team member and his or her well-being through their actions.

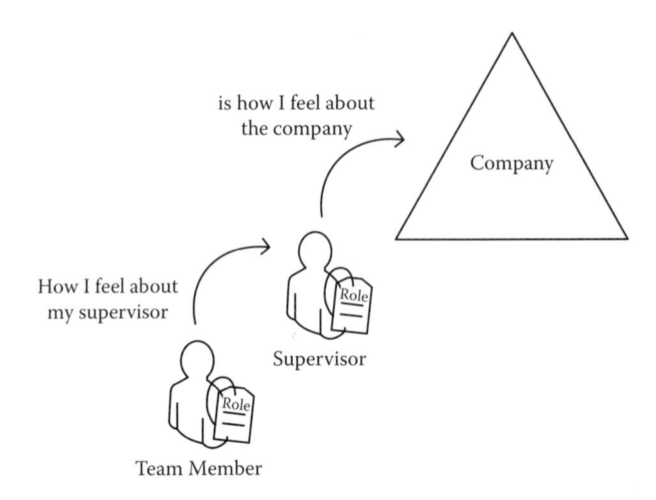

FIGURE 8.7
The importance of the supervisor in change.

Trust between the supervisor and the subordinate is based on three agreements. First, all promises must be kept. If a supervisor miscommunicates the outcome of change in the workplace, employees will be less likely to believe their supervisor the next time change is attempted no matter how appealing it might be. Next, mutually agreed-to items must be put into practice immediately. If the supervisor and team member agree jointly on a set of conditions or terms for workplace change, they must be followed. There is nothing worse than a supervisor agreeing with his or her subordinates and not going forward as planned in an earlier meeting. Last, supervisors must treat everyone fairly and equally at all times under all circumstances. This is especially important when members are struggling through change. While each employee may react differently with change, each supervisor must maintain a fair and consistent approach when making decisions that can have an impact on an employee's work area (Figure 8.8).

8.6 CONVERSION PROCESS: CONVERTER

The converter subsystem has the role of formatting or configuring energy and matter into a useful form before processing. Depending on the level of

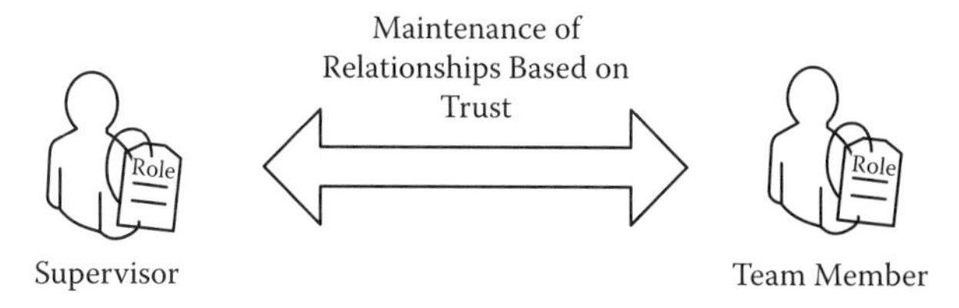

1. Promises must be kept
2. Items mutually agreed upon must be put into practice
3. Everyone must be treated fairly and equally at all times, under all circumstances

FIGURE 8.8
The role of the immediate supervisor in change management.

the subsystem, each converter process variously depends on the requirements needed for the system to function. Without a converter process, energy could enter into the system in an abnormal state, causing inefficiencies and harm to the system. Converter processes in living systems are needed for systems to survive.

The converter process in TPS can best be described as the way ideas are formatted when entering the organization. The decoding technique that Toyota uses to put ideas in usable form is the eight-step problem-solving process. Problem solving provides employees a structured and rational approach for creating change in the workplace. While the motor subsystem appeals to the psychological aspects of change, Toyota's problem-solving process is the basis for rational and cognitive aspects of change. According to management scientists, the cognitive and rational aspects of change are more predictable and easier to manage compared to the psychological aspects of change. Toyota uses problem solving as a logical approach for reasoning and deducing the technical aspects of change.

Problem solving provides a rational and cognitive outlook on change by getting employees to filter out nonessential information, such as opinions, speculations, and assumptions (Figure 8.9). Since problem solving is a way of thinking, it encourages employees to make decisions based on facts and to test their ideas through direct observation. Structured problem solving also prevents employees from wasting time and effort by getting employees to find the true problem and having to re-solve problems already solved. The end result is a conversion of ideas and effort that is effective at finding, fixing, and keeping problems from returning.

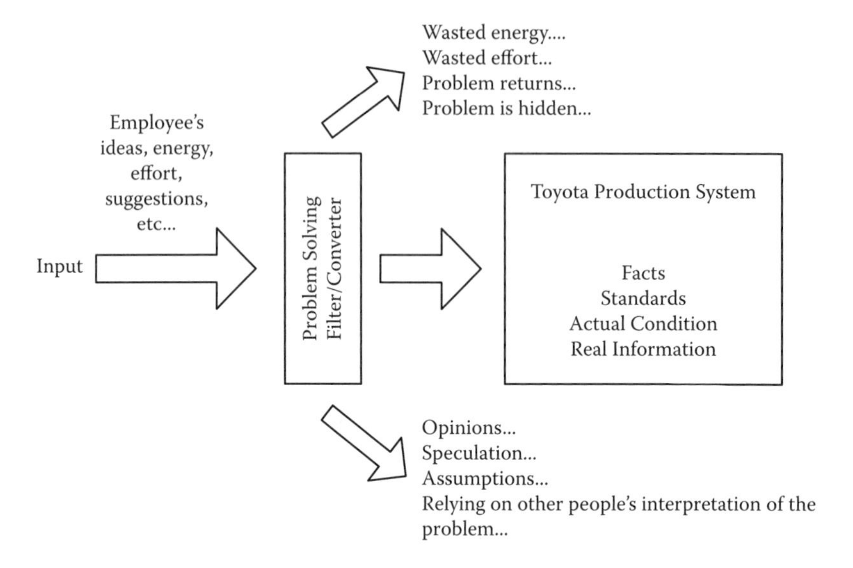

FIGURE 8.9
Problem-solving filter.

The approach that is used to convert employee ideas into usable form is the questioning mindset. Employees are encouraged to think critically during the problem-solving process by asking the following:

What exactly is happening?
What do we actually know?
Can we see the problem when it occurs?
What is normal?
What is the standard?

A technique that is used in problem solving to help employees in understanding and grasping the real problem situation is referred to as the three *R*s (sometimes referred to as the three *G*s). The first *G*, genba, means "real scene" on the shop floor. Employees are encouraged to build detective skills using information at the area where the problem occurs. The second *G*, genbutsu, means "real thing." Managers do not want team members to rely on other people's interpretation of the problem. Team members need to see for themselves the true nature of the problem. Last, there is genjitsu, which means "real fact." In this context, managers stress the importance of gathering accurate information and data when making decisions. These three *R*s are management basics for encouraging employees to think critically when solving workplace problems.

8.7 CONVERSION PROCESS: PRODUCER

The producer process has the role of forming stable associations within a system or components of its subsystem for significant periods. Stable associations exist within systems to grow, repair, or replace certain aspects of the system or subsystem. Producer processes vary at different levels of the system and at areas where the system changes in complexity.

In TPS, the producer process relates to how employees work together to improve and refine ideas. Toyota prefers working in teams because individuals are rarely isolated from group activities. According to change management theory, change at the group level engages everyone in the change process simultaneously (Figure 8.10). The reality of promoting change within groups is that no single member has the same level of readiness. It is rare that members within a group will be at the same level of acceptance for moving forward with change (Figure 8.11).

Other change management theories would argue that change should be focused on an individual level so that personal problems and concerns can be better anticipated and corrected. The main argument between these views is deciding which is more important: starting change at the appropriate level of readiness for each individual or engaging all members to go through the change process simultaneously. Change at the individual level means that individuals can be assisted with change, especially when

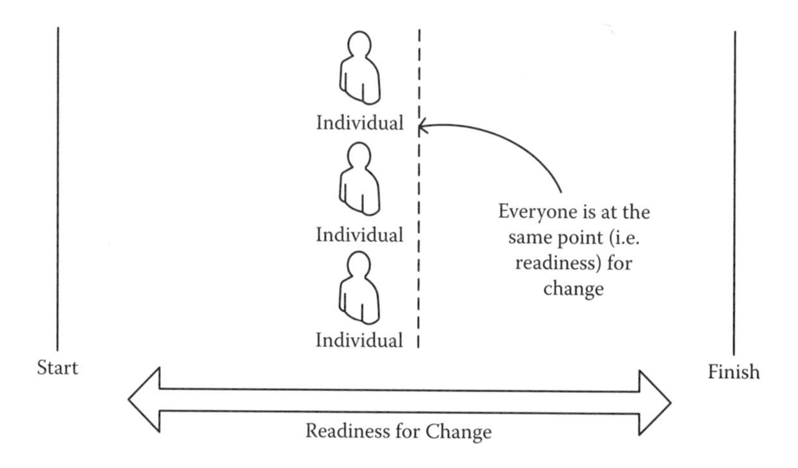

FIGURE 8.10
Member readiness for change (often assumed that change at the group level promotes same-member readiness).

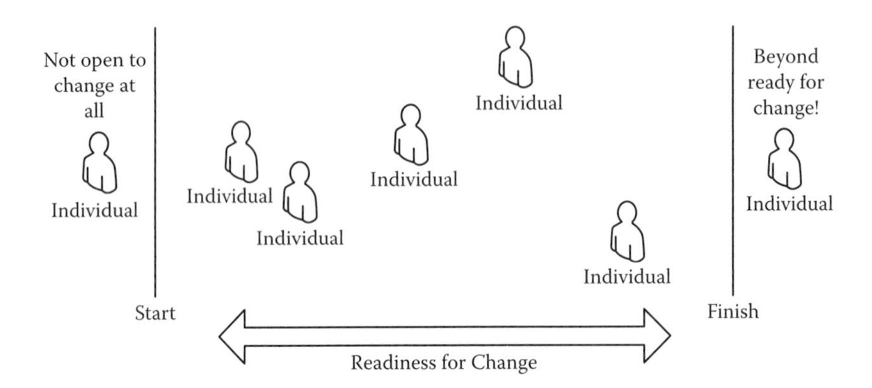

FIGURE 8.11
Member readiness for change (rare that members are at the same readiness level at the same time).

they are having difficulty in coping with change. Change at the group level means that change is more likely to stick because of social dynamics, pressures, and the influence of other group members (Figure 8.12).

In most debates, it can be agreed that it is usually easier to change individuals formed into a group than to change any one of them (Lewin, 1947). According to research, individuals are less likely to transform if group norms, roles, and interactions are not also addressed (Axelrod et al., 2004). At Toyota, the producer process emulates change at the group level. Toyota relies on group activities to encourage employee involvement so that members can learn by watching others. Team activities provide members with opportunities to express concerns and difficulties when applying TPS. Teams also provide a formal network to strengthen communication and relationships among members.

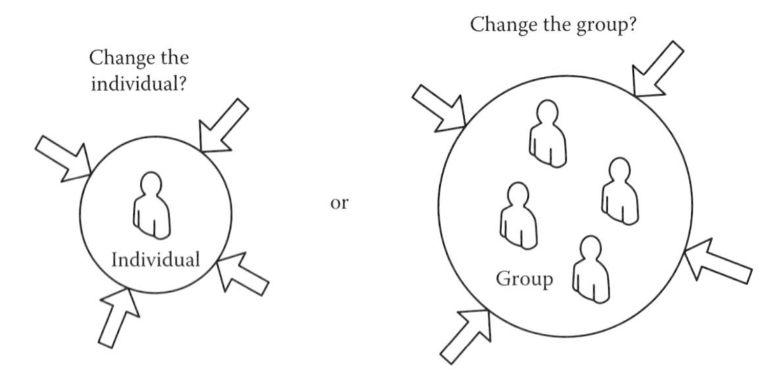

FIGURE 8.12
Two opposing views on change management.

One of the ways that Toyota manages change within groups is the use of roles. Roles provide rights, obligations, responsibilities, and the discretion of how to complete individual and team functions. Toyota uses advisors, coordinators, recorders, planners, leaders, and the most common: facilitators. Roles provide opportunities for individual contribution and a mechanism to evaluate team performance and progress. One of Toyota's main reasons for establishing roles is to stabilize group functions and team dynamics. Because change can cause disequilibrium between members, roles provide a predictable manner for individuals to operate when there is uncertainty and the unknown.

8.8 CONVERSION PROCESS: SUPPORTER

The supporter conversion process in living systems has the function of holding the system together. The supporter subsystem in living systems is the skeletal structure that keeps member components in their proper and physical relationship with one another. Without structure, components of a system would be free to move around without rules or constraints. Higher-order components could cannibalize lower-level components, and boundaries between essential subsystems would become blurred. The order of the system would be jeopardized, mainly because there are would be no DNA template to give the organism the ability to self-organize.

In TPS, the supporter process can best be explained by the type of change that is allowed in the system. The type of change that is used at Toyota is developmental transitional change. Developmental transitional change theory or incremental change theory is a derivative of developmental and transitional change theory. Developmental change theory proposes that change is simply building on what is already known in the organization. Developmental approaches have the least amount of risk because they encourage the organization simply to do more of the things that are already viewed as successful. This approach is sometimes criticized as not a change management approach because there is no transition to manage, just copying and replacing existing systems. This approach tends to be piecemeal across the organization and often encourages groups to work in isolation. Developmental change theory is popular among organizations for two reasons. First, implementation risk is low. Since groups keep doing what they are doing, it is unlikely that any real harm will occur to the rest

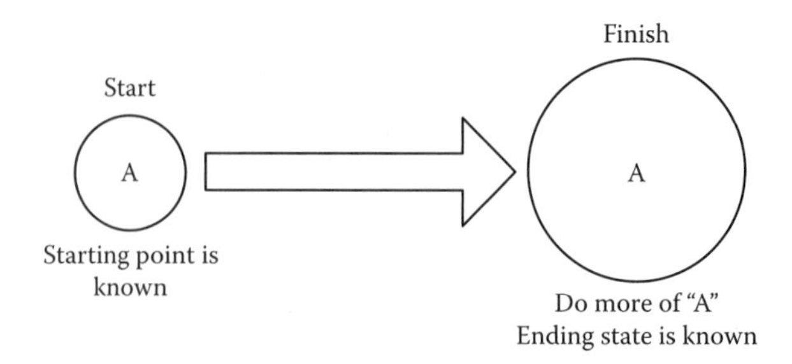

FIGURE 8.13
Developmental change theory.

of the organization. Next, leadership does not have to be charismatic to lead the organization to a new destination. Developmental change theory proposes that the end state is known (Figure 8.13). There is little risk moving the organization to do more of the same thing (Ackerman, 1996).

Transitional techniques cause the organization to move from one state to another. This approach has more risk than the developmental ones mainly because the organization has to manage two different systems, the old and the new, simultaneously. The advantage of transitional change is that the ending and starting point are known. Transitional change theory suggests that the organization should engage in something new, but not unknown (Figure 8.14) (Ackerman, 1996). The challenge is that the organization has to manage through a transition by stopping one type of behavior and transitioning into a new behavior. Implementation risk and leadership skill are higher than developmental change theory, mainly because the organization has to operate using two systems while transitioning to a new system.

Developmental transitional change theory combines both aspects of transitional and developmental change theory (Figure 8.15). The idea is that change should be more interdependent and controlled incrementally. For example, there are aspects of the organization that should take a more developmental approach: do more of something good already in the

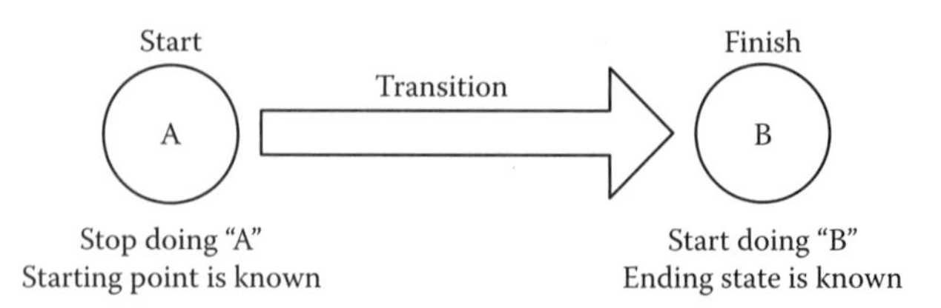

FIGURE 8.14
Transitional change theory.

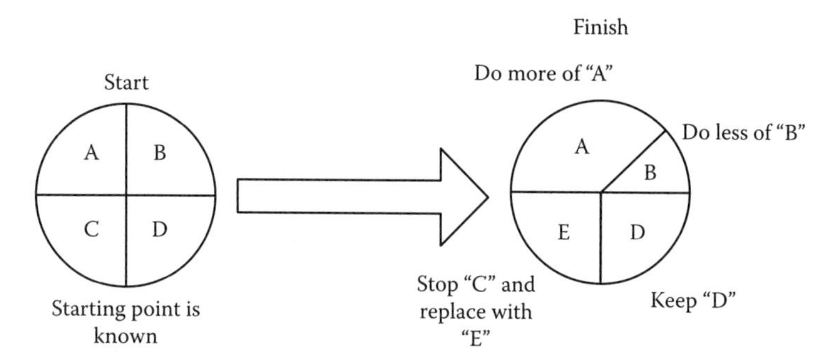

FIGURE 8.15
Developmental-transitional change theory.

organization. The opposite can also be applied: do less of something not good that is in the organization. Consequently, there are some aspects of the organization that should not change at all. In those instances, the new state of the organization is not new at all.

Toyota's use of the developmental-transitional change theory has a lot to do with the kaizen philosophy and the approach to problem solving. Both concepts support the idea that change is small and structured and occurs one by one. The eight-step process breaks large and ambiguous problems into pieces that are more manageable. The kaizen philosophy supports the idea that change should be done incrementally or in stages.

A theoretical perspective that does not emulate TPS as well is Bridges (2003) transformational change theory. This theory proposes that the approach to change is drastic because the ending state of change is completely unknown. Much like a flying trapeze artist who is blindfolded, one must let go to catch the rope on the other side. This type of change assumes that the organization is in a traumatic state: The old ways of doing things must completely stop. This stage represents the coming to terms with endings that must exist for employees to move forward. The quicker that employees are forced to end their current behaviors, the more traumatic the ending will be. After endings, employees enter into a transitional period during which they have the opportunity to practice new ways of doing things. Companies that are successful in the transitional period are ones that take time to let employees sort out their inward and outward feelings. However, employees must not dwell in the transitional state for too long because in this transition employees are pulled in two directions, the old and new ways of doing things. The final

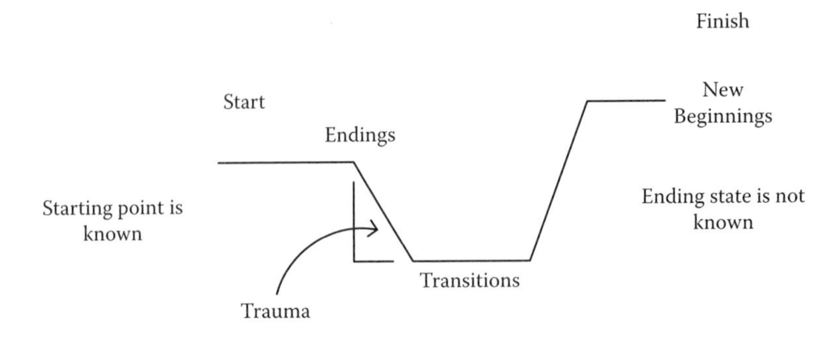

FIGURE 8.16
Transformational change theory. (Adapted from Hall A, 2006, Introduction to Lean—sustainable quality systems design—integrated leadership competencies from the viewpoints of dynamic scientific inquiry learning and Toyota's Lean system principles, published by Arlie Hall, ISBN 0-9768765-0-7.)

state to change is the new beginning. According to Bridges (2003), Burns (1978) and Bass (1985), a highly charismatic leader is required to reshape the new beginning, the future image of the organization, and the new ways of doing things. Transformational change is the most risky because the new image is not known until the old system has died (Figure 8.16). Transformational approaches are for less-routine types of change and require a significant leadership presence that can instill trust in employees to believe in a vision that has not yet been shaped or formed.

8.9 CONVERSION PROCESS: DISTRIBUTOR

The distributor process is responsible for distributing matter and energy throughout a system's structure. Energy and matter are transported from inputs to subsystems and other components as needed for processing. In simple organisms, distributers can represent a channel or a canal. In complex organisms, the distributor may be an intricate vascular system. Regardless of the complexity, without the distributor function, energy and matter would become stagnant and not be able to fuel the system or its components.

In TPS, the distributor function can best be described as the communication techniques employed during problem solving. Toyota's communications are based on three concepts: nemawashi, tataki dai, and

Communication Styles		
	Be personable	Be personable with everyone in the organization; be an ally
	Be a provider	Respond to request in a timely manner
	Be proactive	Seek out request to make communication more effective
	Break for preview	Use informal communication to share and test ideas

Networks		
Existing Networks	Build pyramid	Use and understand the organizational hierarchy
Existing Networks	Begin with peers	Work with peers
New Networks	Bare prospects	Look for potential stakeholders outside the network
New Networks	Build partnerships	Look for potential stakeholders inside the network

FIGURE 8.17
The communication style and network approach in nemawashi.

consensus. Nemawashi is a type of informal communication that is used to discuss and exchange ideas freely (Figure 8.17). In loose terms, nemawashi stands for "preparing the soil." The eight *P*s of nemawashi (be *personable*, build *partnerships*, be *proactive*, build a *pyramid*, be a *provider*, begin with *peers*, break for *preview*, bare *prospects*) describe the approach Toyota employees should use in cross-organization communication. The eight *P*s describe not only the style of communication an employee should demonstrate but also how the employee should communicate inside and outside existing networks.

Tataki dai is another communication technique that means proposals, processes, and problems should be critiqued, rather than people. In other words, Toyota believes that in the exchange of information, employees can sometimes make personal attacks during discussion. Tataki dai is a technique that allows employees to maintain mutual respect during open discussions while directing energy toward the status and the nature of the problem. Consensus in Toyota relates to how ideas are agreed to and supported within teams or groups. Toyota believes that team members should listen from the other's point of view and contribute their ideas, comments, and opinions. Members are allowed to disagree; however, when the team has made a decision, each member should support the team's decision as if it were their own.

8.10 CONVERSION PROCESS: MATTER–ENERGY STORAGE

The matter–energy storage concept in LST proposes that all systems need a way to preserve and maintain energy or matter for later use. Matter–energy storage requires three components: ways to store, maintain, and retrieve energy and matter for later use. Without a storage function, a system is completely reliant on its environmental conditions to maintain a consistent flow of energy. The storage function in conversion processes acts as a way to smooth and buffer internal energy requirements when surrounding condition changes for sustained periods.

The best example of the storage function in TPS is the use of procedure-based systems. Procedures act as a way for TPS to maintain its current functions when environmental conditions vary. Written instructions, protocols, and standards provide the record (i.e., stored energy) of the best way to accomplish work. Without a record, ideas cannot be saved, stored, or retrieved when they are needed. Toyota's approach to saved records can best be explained using the concept of standardization.

Standardization is when groups, sections, or divisions share common work methods. Examples include policies, procedures, or an employee handbook. Toyota uses the principle of standardization to create matter–energy storage devices across the organization to help preserve employee ideas that have been proven to be successful. Without standardization, TPS would not be able to handle varying conditions in the environment. Consider trying to respond to abnormal conditions without a standardized recovery plan. Standardization acts as a way to maintain the knowledge of how the organization should function when the environment varies.

8.11 SUMMARY

The purpose of this chapter was to apply LST to better explain change management inside Toyota. Various change management techniques were presented to describe the conversion processes in organizations. This work illustrated that the metabolism of TPS is strongly dependent on the ability to process employee's ideas, suggestions, and input when improving the workplace. Each conversion process is essential for a living system to survive, which raises the importance of various transformation processes inside TPS.

It was suggested that Toyota manages change by creating an environment in which employees feel comfortable expressing ideas. Toyota uses the immediate supervisor as a way to maintain open communications and a systematic problem-solving process to give rationale and structure to change itself. Toyota sustains change by promoting change at the group level and by standardizing successful practices. Last, Toyota promotes a form of incremental, transitional, and developmental change, rather than the popularized rapid or transformational change approach.

REFERENCES

Ackerman, L. (1996) Development, transition or transformation: The question of change in organizations, *OD Practitioner*, 28:(4), 5–16.

Adams, J. (1965) Inequity in social exchange, in *Advances in Experimental Social Psychology*, Vol. 2, ed. Berkowitz, L. New York: Academic Press, pp. 267–299.

Alexander, P. (1979) The perils and pitfalls of an OD effort. *OD Practitioner*, 11(1).

Alter, N. (2000). L'innovation ordinaire. *PUF Sociologies*, Paris.

Andrews, J., Cameron, H., and Harris, M. (2008) All change? Managers' experience of organizational change in theory and practice. *Journal of Organizational Change Management*, 21(3), 300–314.

Axelrod, R., Axelrod, E., Beedon, J., and Jacobs, R. (2004) *You Don't Have to Do It Alone: How to Involve Others to Get Things Done*, Berrett-Koehler, San Francisco.

Bass, B. (1985) *Leadership and Performance beyond Expectations*, Free Press, New York.

Brenner, M. (2008) It's all about people: Change management's greatest lever. *Business Strategy Series, Northampton*, 9(3).

Bridges, W. (2003) *Managing Transitions. Making the Most of Change*. (2nd Ed). DaCapo Press, Cambridge, MA.

Buch, K. and Tolentino, A. (2006) Employee perceptions of the rewards associated with Six Sigma. *Journal of Organizational Change Management*, 19(3), 356–364.

Burk, W. (1982) Who is the client? A different perspective. *OD Practitioner*, 14(1).

Burns, J. (1978). *Leadership*. Harper & Row, New York.

Caldwell, S., Herold, D., and Fedor, D. (2004) Toward an understanding of the relationships among organizational change, individual differences and changes in person-environment fir: A cross-level study. *Journal of Applied Psychology*, 89, 868–882.

Carter, E. (2008) Successful change requires more than change management. *Journal for Quality and Participation*, Spring.

Chrusciel, D. (2008) What motivates the significant/strategic change champion(s)? *Journal of Organizational Change Management*, 21(2), 148–160.

Collinson, D. (1994) *Strategies of Resistance: Power, Knowledge and Subjectively in the workplace, Resistance and Power in Organizations*, Routledge, New York.

Cooperrider, D. and Avital, M., eds. (2004) *Advances in Appreciative Inquiry*, Elsevier Science, Oxford, UK.

Cutcher, L. (2009) Resisting change from within and without the organization. *Journal of Organizational Change Management*, 22(3), 275–289.

De Bono, E. (2006) Change in management thinking: How to change the way people are thinking in management. Available at http://www.thinkingmanagers.com.

Deming, W. (1992) *Out of the Crisis*, MIT Press, Cambridge, MA.

Dewey, J. (1910) *How We Think*, Heath, Boston.

Folger, R. and Skarlicki, D. (1999) Unfairness and resistance to change: Hardship as mistreatment. *Journal of Organizational Change Management*, 12(1), 35–50.

Fronda, Y. and Moriceau, J.-L. (2008) I am not your hero: Change management and culture shocks in a public sector corporation. *Journal of Organizational Change Management, Bradford*, 21(5), 589–609.

Gigch, J. (1978) *Applied General Systems Theory*, Harper & Row, New York.

Gotsill, G. and Natchez, M. (2007) From resistance to acceptance: How to implement change management. *Fundamentals*, November.

Hall, A. (2006). *Introduction to Lean Sustainable Quality Systems Design: An Integrated Approach From the Viewpoints of Dynamic Scientific Inquiry Learning & Toyota's Lean System Principles and Practices.* Published by Arlie Hall, Lexington, KY. ISBN 0-9768765-0-7.

Hammer, M. and Champy, J. (1993) *Re-engineering the Corporation*, HarperCollins, New York.

Harrison, R. (1972) Strategy guidelines of an internal organization development unit. *OD Practitioner*, 4(3).

Harvey, J. (1975) Organizational development as a religious movement. *OD Practitioner*, 5(3), Winter.

Herscovitch, L. and Meyer, J. (2002) Commitment to organizational change: Extension of a three-component model. *Journal of Applied Psychology*, 87(3), 474–487.

Holman, P., Devane, T., and Cady, S. (2007) *The Change Handbook*, 2nd ed., Berrett-Koehler, San Francisco.

Johnson, M, Parasuraman, A., Futrell, C., and Black, W. (1990) A longitudinal assessment of the impact of selected organizational influences on sales people's organizational commitment during early development, *Journal of Marketing Research*, 27(3).

Jones, J. and Seraphim, D. (2008) TQM implementation and change management in an unfavorable environment, *The Journal of Management Development*, 27(3).

Jorgensen, N., Owen, L., and Neus, A. (2009) Stop improving change management! *Strategy and Leadership, Chicago*, 37(2), 38–44.

Kaplan, R. and Norton, D. (1993) Putting the balanced scorecard to work, *Harvard Business Review* (September-October).

Keen, P. and Morton, S. (1978) *Decision Support Systems: An Organizational Perspective*, Addison-Wesley, Reading, MA.

Kouzes, J. and Posner, B. (2007) *The Leadership Challenge*, 4th ed., Jossey-Bass, San Francisco.

Lewin, K. (1947) Frontiers in group dynamics 1. *Human Relations*, June, 1.

Marshak, R. (1994) *OD Practitioner*, 26(2).

Maturana, H. R. and Varela, F. J. (1972) *Autopoiesis and cognition*, Reidel, Dordrecht, the Netherlands.

Michaels, M. (1994) Chaos theory and the process of change, in *What Is New in Organization Development*, ed. Cole, D., Preston, J., and Finlay, J. Chesterland, OH: Organization Development Institute.

Miller, J. (1955) Toward a general theory for the behavior sciences. *American Psychology*, 10.

Miller, J. (1978) *Living Systems*, Wiley, New York.

Miller, J. L. and Miller, J. G.(1992) Greater than the sum of its parts. *Behavioral Science*, 37(1), 1–38.

Morgan, R. and Hunt, S. (1994) The commitment-trust theory of relationship marketing. *Journal of Marketing*, 58(3), 20–38.

Nadler, D. (1982) Managing transition to uncertain future states. *Organizational Dynamics*, 11(1), 37–45.

Neal, J., Lichtenstein, B., and Banner, D. (1999) Spiritual perspectives on individual, organizational and societal transformation. *Journal of Organizational Change Management*, 12(3), 175–185.

Parish, J., Cadwallader, S. and Busch, P. (2008) Want to, need to, ought to: employee commitment to organizational change, *Journal of Organizational Change Management*, 21(1), pp. 32–52.

Prochaska, J. and DiClemente, C. (1984) *The Transtheoretical Approach: Crossing Traditional Boundaries of Therapy*, Dow Jones-Irwin, Homewood, IL.

Pullen, J. (2002) *Blazing a trail through change management.* Professional Engineering, Bury St. Edmunds, UK. September, 15(17).

Schaffer R (1988) The breakthrough strategy, *Harper Business*, New York, NY.

Sirkin, H., Keenan, P., and Jackson, A. (2005) The hard side of change management. *Harvard Business Review*, 83(10), 108–118.

Smither, R., Houston, J., and McIntire, S. (1996) *Organization Development: Strategies for Changing Environments*, Harper Collins College, New York.

Snyder, M. (2001) *Building Consensus: Conflict and Unity*, Earlham Press, Richmond, IN.

Stanev, S., Krappe, H., Ola, A., and Georgoulias, K. (2008) Efficient change management for the flexible production of the future. *Journal of Manufacturing Technology Management, Bradford*, 19(6).

Tarnas, R. (1991) *The Passion of the Western Mind*, Harmony Books, New York.

Verdu, A. and Gomez-Gras, J. (2009) Measuring the organizational responsiveness through managerial flexibility. *Journal of Organizational Change Management*, 22(6), 668–690.

von Bertalanffy, L. (1968) *General System Theory: Foundations, Development, Applications.* Braziller, New York, NY.

9

The System Property of Entropy in TPS

9.1 SYSTEM PROPERTY: ENTROPY

Entropy is a measure of waste or disorderliness when a system tries to convert energy into some useful or meaningful form of work. Entropy is a property of systems used to understand how processes suffer from inefficiencies. Entropy is a property of energy borrowed from the second law of thermodynamics, which describes conservation. The zeroth law of thermodynamics relates to the equilibrium property (Figure 9.1). When two bodies are in contact and have the same temperature and there is no heat transfer between them, meaning no energy transfer, the bodies have reached a state of thermal equilibrium (Cengel and Boles, 1989).

The first law of thermodynamics relates to the principle of the conservation of energy (Figure 9.2). Energy cannot be created or destroyed; however, it can change forms. During an interaction between a system and its surroundings, the amount of energy gained by a system must be exactly equal to the amount of energy lost by the surroundings. In other words, more energy cannot be retrieved from a system than the amount of energy added to the system.

The second law of thermodynamics relates to the transferability of energy from one form to another (Figure 9.3). When energy goes through a phase change, there is a loss or an amount of energy that cannot be used. Entropy represents that amount of energy that is wasted or cannot be claimed or stored for later use. All processes exhibit entropy. The way that entropy is measured is determined by the reversibility of a process. A reversible process is defined as a process that can be reversed without leaving any by-products to its surroundings. When a system can return to its initial state and all of its surroundings, the process is said to be reversible. In nature, totally reversible processes do not occur.

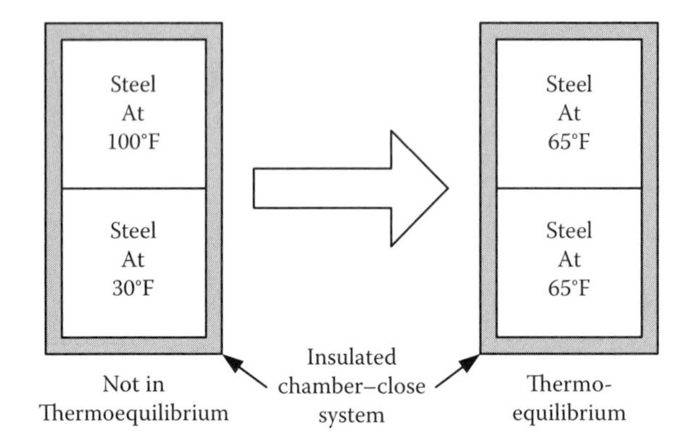

FIGURE 9.1
The zeroth law of thermodynamics. (Adapted from Cengel, Y. and Boles, M., 1989, *Thermodynamics: An Engineering Approach*, McGraw-Hill, New York.)

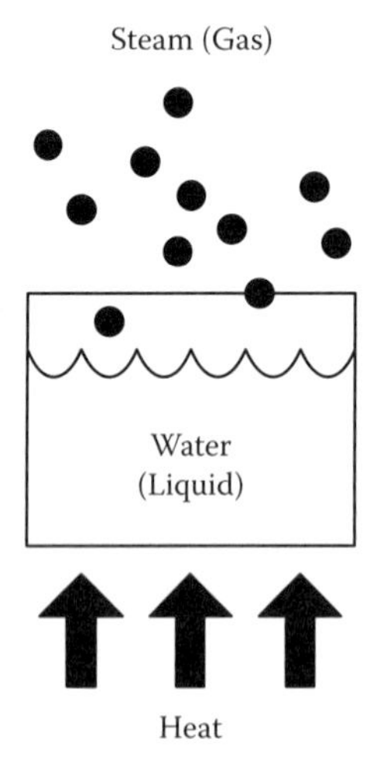

Heat applied to water induces a phase change; however, energy has not been created or destroyed, only transformed into another phase

FIGURE 9.2
The first law of thermodynamics.

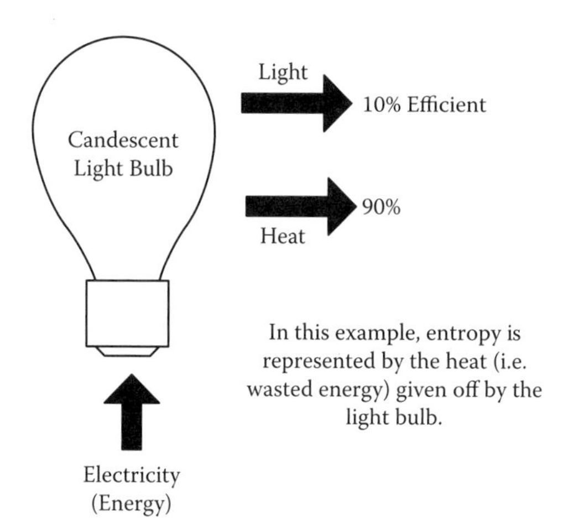

FIGURE 9.3
The second law of thermodynamics.

Processes that are completely reversible are favored by engineers because they are inherently more efficient at converting energy from one form to another. When a process is completely irreversible or one directional, that process is inherently more wasteful (i.e., there is more entropy) at converting energy.

9.2 ENTROPY IN ORGANIZATIONS

The concept of entropy is useful in studying organizations because it can describe the ways energy is lost, managed, and increased in the conversion process (Figure 9.4). While organizations have the ability to resist the trend toward disorderliness, like all living systems, organizations can run-down. From a thermodynamic perspective (Figure 9.5), organizations can experience entropy in three different ways. First, organizations can suffer from entropy when energy is lost and is not replaced. Organizations must be able to import new energy from their surroundings to counteract losses due to conversion and to the environment. Since entropy cannot be conserved, entropy can only be cancelled by new energy.

Entropy can also occur within organizations when energy gradients are opposing and unequal. Differences in energy gradients represent the reversibility of a process. Processes that lack reversibility or balance in

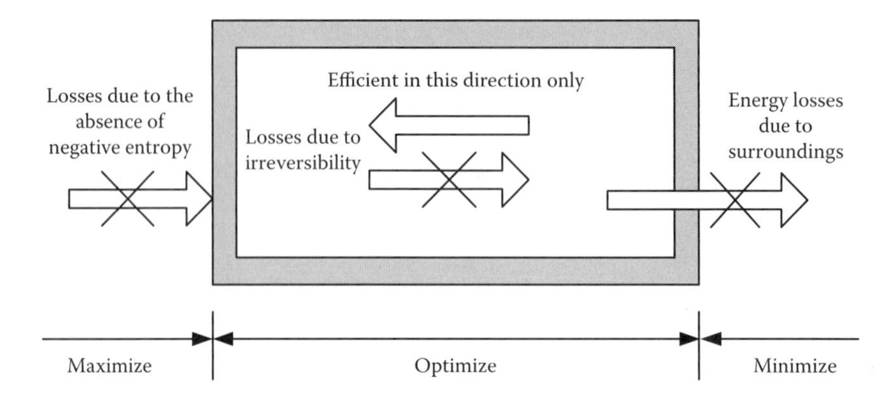

FIGURE 9.4
An example of entropy in organizations: a thermodynamic perspective.

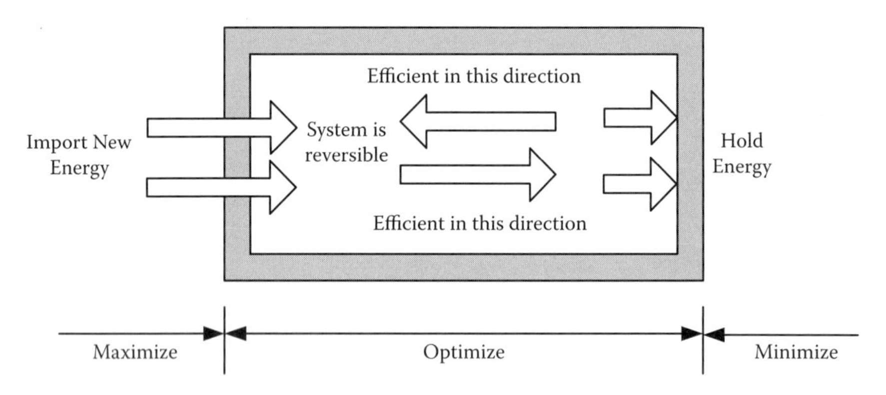

FIGURE 9.5
An ideal thermodynamic state in organizations.

opposing energy gradients create conditions that worsen the effectiveness of the organization. Reversibility drives efficiency in organizations. The less an organization is able to convert energy one to one, the less efficient and reversible it is.

Last, organizations must find a way to limit the amount of energy loss to their surroundings. Organizations are open systems that are highly dependent on environmental conditions. Harsh conditions cause open systems to increase in entropy and their state of disorder. An organization must find ways to minimize losses to their surroundings when converting forms of energy to new ones. Regardless of how well an organization is at converting or importing new energy, if it cannot sustain and hold energy, it will suffer from entropy.

9.3 ENTROPY IN THE TOYOTA PRODUCTION SYSTEM

In the Toyota Production System (TPS), entropy can best be described as the loss of effective management. When TPS can no longer improve the effectiveness of the organization, the system is overrun with entropy. Sources of entropy can include a wide variety of organizational factors; however, as a management system, the most significant source of entropy relates to the use of human resources (HR). In TPS, entropy is the loss of employee engagement (Figure 9.6).

Not until recently have organizations pursued the HR function as a significant factor in implementing Lean. A common pathology in Lean is the loss of employee engagement. Initially, when organizations start Lean, employee morale is high. Employees are engaged, excited, and willing to try new things. After four to six months, morale starts to diminish. There are fewer suggestions and less participation and willingness to experiment in the workplace. Organizations return to their initial state because there are driving factors (i.e., entropy) causing the system to run down. No matter how aggressive or ambitious the Lean initiative, employee engagement is a function of how well the organization manages entropy.

The way that Toyota manages entropy in TPS can best be described using the principles of thermodynamics. The way that energy is added to TPS is through organizational learning (OL). OL helps the organization

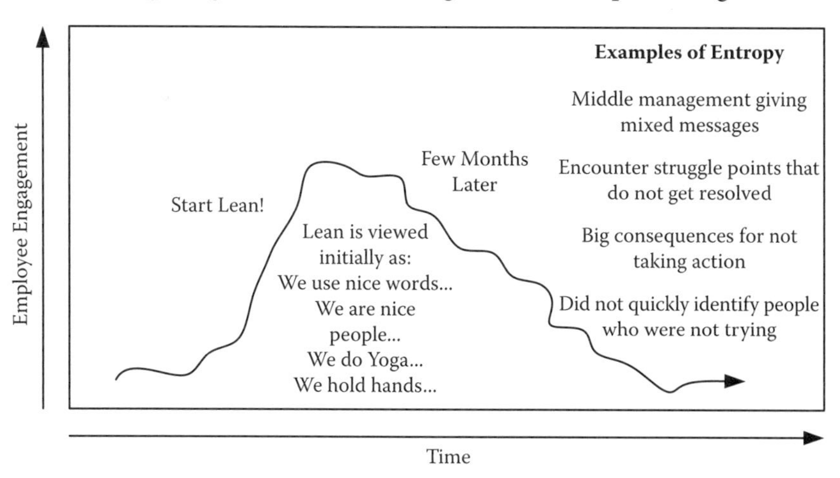

FIGURE 9.6
The entropy of TPS: the rundown of employee engagement.

accomplish new ways of doing things by connecting to each employee in a special way. TPS is not just another participation program looking to raise the number of suggestions in a box; it is a management system that embraces diversity in learning. The best teachers are those who can adapt to each student's particular learning style. Toyota uses OL in the HR function to nurture and cultivate each employee's creative ability.

Toyota utilizes the HR function to prevent energy loss. Energy loss can best be explained by an employee's loss of desire to participate. In more exact terms, this means a loss of faith in and respect for the company. Working in a mass production environment is mentally and physically demanding. Employees are introduced to various industry conditions that challenge their health and well-being. The only way TPS can maintain stable employee relations is through a progressive HR function. Meeting the needs of employees is the first line of defense (i.e., boundary) for preventing entropy.

The last form of entropy in TPS as it relates to organizations is the way management and labor work together. When there is an imbalance in expectations (i.e., energy gradient), the organization becomes less reversible and efficient. Reversibility is not used in this context to mean that management should be able to switch with labor, but to imply that the labor–management equation should balance. Expectations from both parties should be known, not unknown. Both groups should depend on one another and exist with each other (Figure 9.7). When the labor–management relationship is structured, defined, and communicated, opposing energy gradients can be avoided.

One of the ways that the HR function accomplishes this balance is through its historical roots of organized labor. Surprisingly, most

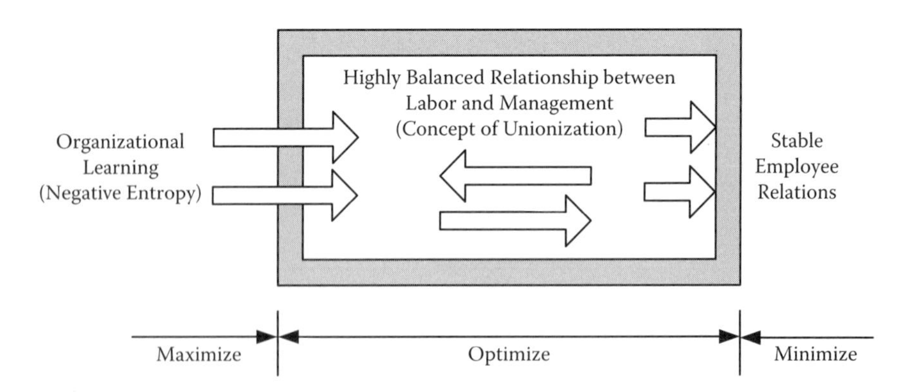

FIGURE 9.7
The ideal thermodynamic state of TPS in organizations.

researchers, consultants, and authors do not acknowledge that TPS grew out of organized labor. TPS is a highly structured management system, not an organic or constantly evolving management system that has fuzzy boundaries. TPS is a disciplined, well-communicated, and highly structured relationship between management and labor. If the system was unstructured, labor and management would constantly be out of balance. TPS resists the entropy of unbalanced processes because the system emulates a highly organized work environment.

The topic of HRs as it relates to entropy and negative entropy is discussed in the following three chapters. Chapter 10 discusses the idea of stable employee relations, Chapter 11 evaluates the topic of organized labor as it relates to the birth of TPS, and Chapter 12 looks at the way Toyota works against entropy using OL.

9.4 LITERATURE REVIEW: THE HUMAN RESOURCE FUNCTION

The concept of human resources management (HRM), often referred to as personnel management or labor relations, has been increasingly used to refer to philosophies, policies, procedures, and practices related to the management of people within an organization (French, 1986). The purpose of HRM is to develop processes that unleash expertise to improve organizational performance. The benefits of applying effective HRM have been well researched and have improved many areas, including employee relations (Palmer, 1987); knowledge management (Parise, 2007); strategic alignment (Wright et al., 2000); and most importantly organizational performance (Guest et al., 2003). One of the most appealing aspects of HRM is that effective practices can provide many competitive advantages that are hard for competitors to imitate (Figure 9.8). In fact, sophisticated technologies and manufacturing innovation can do little to improve operational performance unless HRM systems are effective.

A significant body of research suggested that a specific set of HRM practices is key to enhancing organizational performance and competitive advantages (Theriou and Chatzoglou, 2008; Ahmad and Schroeder, 2003). Unfortunately, the selection of effective HRM practices and their combinations is still under debate. There is work that suggested that HR practices should be employee centered and emphasize employee's attitudes

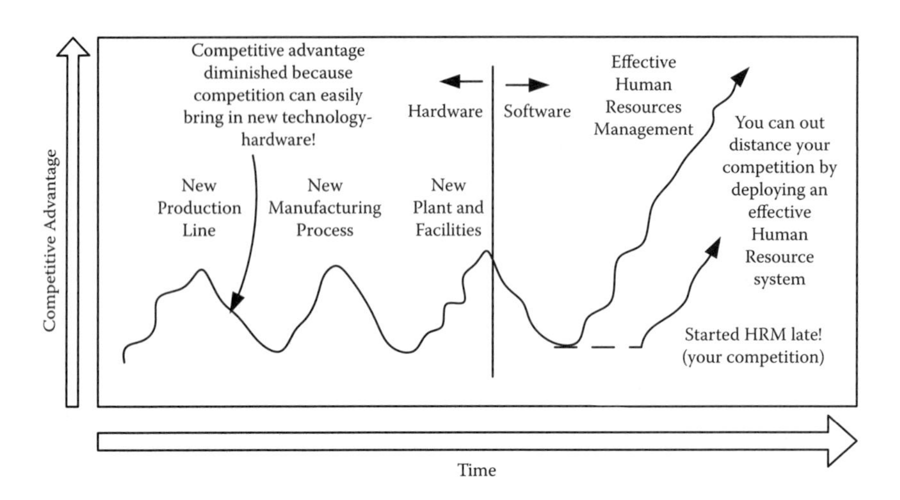

FIGURE 9.8
Using human resource management as a competitive advantage compared to technology.

and well-being (Applebaum et al., 2000), while other work showed that HRM should be more extrinsic and value employees through rewards and compensation (Johnson, 2000). There is also work that suggested that HRM should play a more significant role in the organization by aligning organizational practices with the business strategy (Fombrun et al., 1984).

The most common approach in applying HRM is the use of "best practices." Researchers have questioned the basis of these universal claims and have shown that HRM practices vary substantially in different organizations and countries (Guest, 1999). Consequently, best practices are static and only describe what to do under one circumstance. Another problem in applying HRM practices is the belief that individual components of a system can be applied in "pieces" or "bundles" that provide a concerted impact on organizational performance (Hendry, 2003). Therefore, understanding the various relationships in HR systems is crucial if the total system is to offer any net benefit (Figure 9.9).

9.4.1 Japanese Management Techniques in Human Resources

A trend in HRM is the study of Japanese management techniques, which in the 1980s were acknowledged as key in Japan's economic rise. Since then, numerous books and articles have been written on Japanese management systems examining the differences between Western and Asian techniques. Some research showed that Japanese HR practices tend to

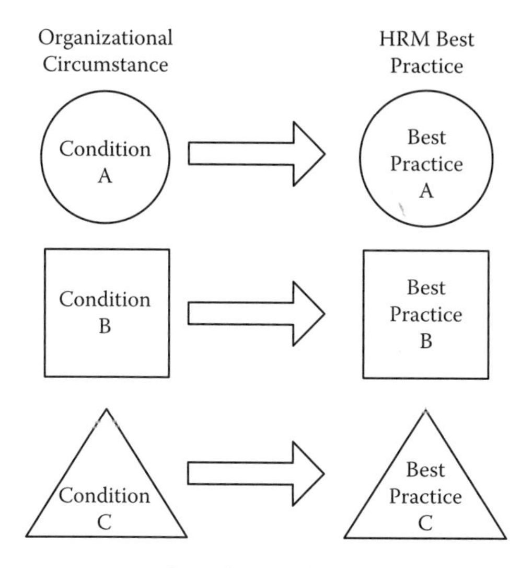

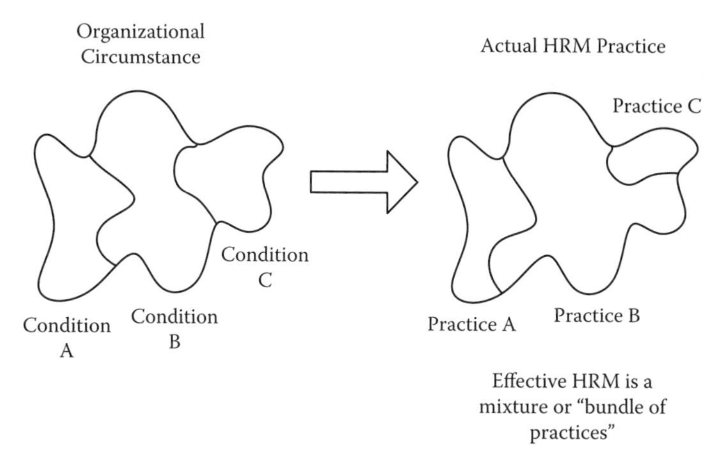

FIGURE 9.9
Comparison of traditional and progressive HRM approaches.

reinforce lifetime employment, generalist career paths, and automatic pay raises and promotions (Beechler and Bird, 1994). Other research suggested that Asian cultures have been reported to have higher commitment to preserve harmony through intricate social rituals, to cooperate with workers, and to have strong employee cohesion (Selmer, 2001). In addition, Japanese practices in HR showed that they rely on informal controls such as values and inward employee beliefs rather than formal controls such as rules and policies in influencing employee behavior (Lincoln and Kalleberg, 1990).

One of the most appealing characteristics of Japanese management techniques is the ability to make better use of human capital through training and worker involvement. American HRM practices seem to struggle with adopting the right blend of techniques to engage shop floor workers to contribute their tacit knowledge without fear of losing their status or employment in the company. Currently, most companies are coupling Japanese HRM techniques to offset these fears in an attempt to maximize worker involvement and engagement.

Unfortunately, more damage than good can result in the implementation of Japanese management practices if the nature of these systems is not understood. For example, some attempts to imitate Japanese HR systems have been to engage in activities that positively affect the bottom line of the organization (Edgar, 2003). This approach tends to lead to worker intensification and to generally greater exploitation of employees within the workplace (Fucini and Fucini, 1990). Other research suggested that Japanese HRM techniques are aiming at lower management density, which has a dampening effect on employee expectations of upward career mobility (Selmer, 2001). There is also the issue of compensation; employees are paid a lower starting salary but can look forward to automatic pay raises with increasing age and seniority until retirement (Lincoln and Kalleberg, 1990). These types of systems have worked because the company has the incentive to invest continuously in their employees without risk of them leaving or taking with them proprietary knowledge due to lifelong employment. Last, Japanese management systems introduce many conflicting individual dynamics; pay is based on team performance, yet stronger team members are expected to assist weaker ones for the good of the team as a whole (Debroux, 1997). Thus, there is a need to study Japanese management techniques more completely to understand how such practices can be used effectively.

9.4.2 Japanese Management Practices and Toyota

Not surprisingly, Toyota's view of HRM draws on many traditional practices of Japanese management and Asian cultural traits. Toyota's HRM system has been examined and is said to be superior due to its strong emphasis on people development, two-way communication, and teamwork (Liker and Hoseus, 2008). Toyota's HR system is believed to be unique compared to Western views due to its organizational priorities, which stress stable employment, fair and consistent policies, and rewards for cooperation. Toyota's culture emphasizes group centeredness, lifelong employment, long-term planning horizons, generalist career paths, and authoritarianism (Beechler and Bird, 1994). Other ways Toyota's HRM practices are similar include the emphasis on in-house training (Dennis, 2007), open communications (Besser, 1996), and long-term planning (Kochan and Dyer, 1993). From the outside, it would appear that Toyota's practices in HR are similar to traditional Japanese management techniques.

Other work suggests that Toyota's practices are unique compared to traditional Japanese management techniques. Research shows that Toyota's view on HRM may be unique as it relates to problem solving, method development (Liker and Hoseus, 2010), worker motivation (Hall, 2006), and organizational culture (Badurdeen et al., 2009). These examples offer many interesting insights that Toyota's HR system is both unique and similar compared to traditional Japanese management systems.

9.5 MINIMIZING ENTROPY USING THE HUMAN RESOURCE FUNCTION IN TPS

A common theme is the separation between the hard and soft characteristics of HR (Pascale and Athos, 1981). The hard version of HR is primarily concerned with operational performance, tasks (Black et al., 1992), capability (Stalk et al., 1992), profitability, compensation (Kochan and Dyer, 1993), and the well-being of the organization. These models have little concern for the worker and mostly emphasize strategy, structure, and systems (Sparrow and Hilltrop, 1994). A classic example of a hard HRM model is the Michigan school model.

The main challenge in HR is that of better aligning people and organizations. It has been shown that if people feel that the organization is responsive to their needs and supportive of their goals, managers can count on their followers' commitment and loyalty (Bolman and Deal, 1997). Consequently, organizations can benefit by applying effective HRM practices by simply being attentive to the basic needs of workers. However, if the organization does not make an attempt to meet the extrinsic and intrinsic needs of their workers, the organization will most likely suffer from an unreliable labor pool and a loss of energy and talent (i.e., entropy). Therefore, the HR function has the role to seek the interest of both the organization and its employees in fulfilling needs on both sides.

In contrast to the hard approach is the soft-sided view of HRM. These practices (such as the Harvard school model) show equal concern for staff (i.e., employees), skills, and the style of management. This view emphasizes a relationship between the organization and employees while considering the interest from all affected stakeholders (Beer et al. 1984). The soft approach emphasizes the need for achievement (McClelland, 1986), affiliation, opportunities for learning (Luthans, 1995; Wright et al., 2000), teamwork (Wageman, 1995), and behavior (Gilbert, 1978). Overall, this type of thinking suggests a softer approach to HRM, which means that both the organization and the employees can benefit from HRM practices.

In TPS, employees cannot be expected to completely participate and give all their energy and talent if their needs are not met. According to Herzberg's hygiene and motivation theory, employee engagement and willingness are dependent on two factors (Herzberg, 1987). The first factor relates to hygiene or extrinsic aspects of the work environment. This factor has been shown not to increase worker motivation, but without it, worker dissatisfaction can occur. Often described as a form of Maslow's lower-level needs theory, hygiene theory relates to the basic needs that employers must fulfill for their employees, such as stable employment, stable compensation, safe working conditions, and consistent and fair management practices. A worker will not be willing to do more than the bare minimum according to hygiene theory. The second factor relates to the intrinsic or internal aspects of the job, such as achievement and opportunities to advance and learn. This factor relates to the inward aspects of work and has been well documented as the most important element of worker motivation. Both of these factors prevent employees from losing interest in TPS. These factors act as barriers or insulation to prevent the loss of employee engagement (Figure 9.10).

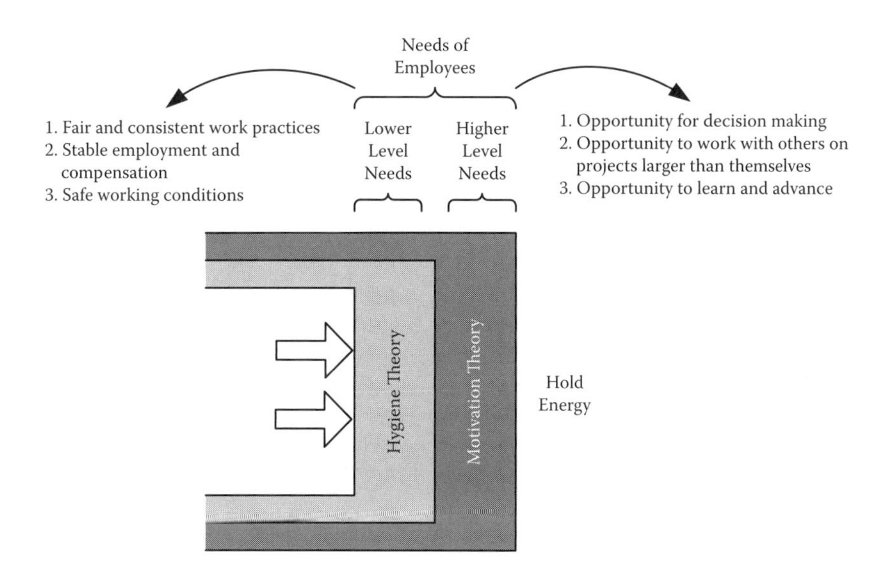

FIGURE 9.10
Minimizing entropy by meeting the needs of employees.

9.6 HYGIENE THEORY AND TPS: THE ESSENTIAL NEEDS OF THE INDIVIDUAL

For employees to support the company's management systems and practices, basic needs must be fulfilled (Figure 9.11); otherwise, employees will do the bare minimum to keep their supervisor off their backs. When employees do the bare minimum, they are not working with you, but not necessarily working against you either. These employees are known as the isolated wage earners. They are in it for themselves and will avoid problems by being a passive resister, meaning they are not outwardly expressing their discontent but doing it behind the company's back. This is a dangerous situation for an organization because these employees are toxic. Because horizontal communication is more prevalent than vertical communication, these employees can spread discontent quickly around the organization.

To prevent employees from doing just the bare minimum, Toyota utilizes a set of hygiene factors to motivate employees. Hygiene factors are often labeled as a form of extrinsic motivation because they require a drive for resources that exist outside the employee (i.e., food, water, shelter, etc.). Because hygiene factors often are related to external factors, they are also compared to formal controls. Formal controls include both disciplinary

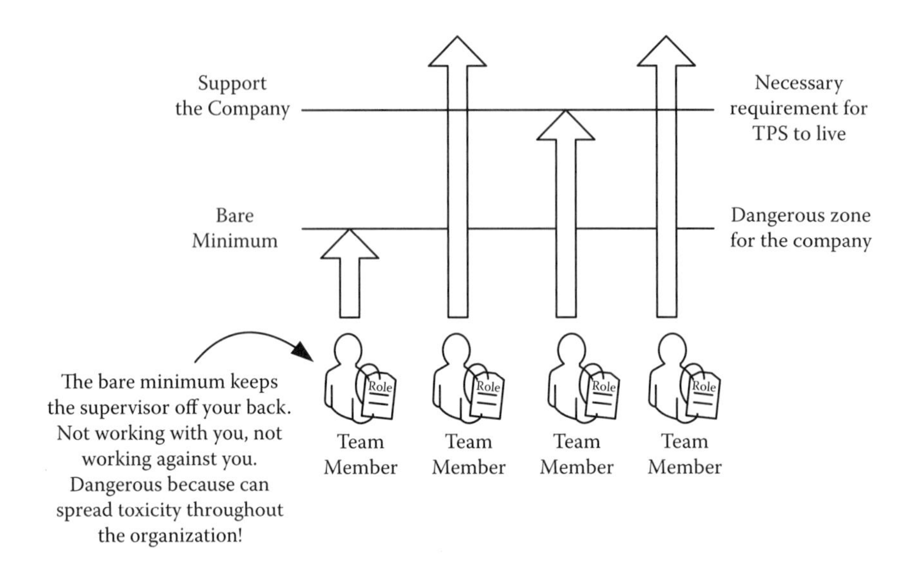

FIGURE 9.11
A comparison of employee engagement.

systems and incentive systems. Formal controls relate to the mechanisms used by an organization to influence and maintain employee behavior (Deming, 1986; Katz and Kahn, 1978; Gardner, 2009). Without hygiene factors, Toyota would rely completely on employees' internal motivation to encourage them to participate in TPS (Figure 9.12). While hygiene factors do not lead to motivation, without them they can cause TPS to deteriorate (Herzberg, 1987). Toyota divides hygiene factors into four HR functions: stable employee relations, fair appraisal, safe working conditions, and communication.

9.6.1 Stable Employee Relations

Stable employee relations relate to the way the company and the employees make an effort to maintain a stable and long-term relationship. Toyota's view of stable employee relations is based on the idea of long-term mutual trust and respect. The concept of trust was developed by Toyota through a cooperative effort by the employees and the company in 1949 during their most widely known labor disputes. The dispute started after the collapse of the Japanese economy; Toyota was forced to lay off 1,500 workers, at that time 25% of the total workforce. To calm labor disputes, the president at the time, Kiichiro Toyoda, son of Sakichi Toyoda, stepped down. Kiichiro assumed responsibility for the company's difficulties despite the postwar

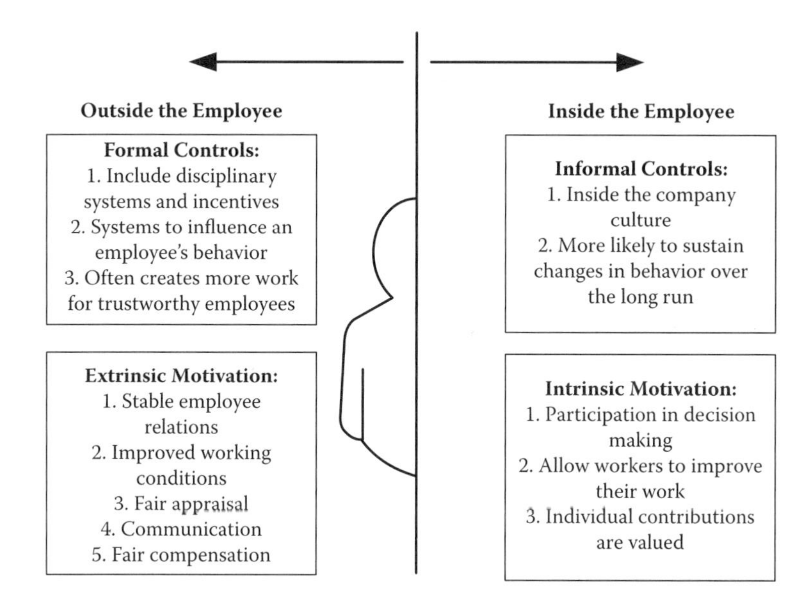

FIGURE 9.12
A comparison between internal and external factors for human resources.

confusion. After two months of an intense labor dispute, Toyota's first union was born: the Toyota Motor Works Union. Ten years later, Toyota and the union signed the Joint Labor and Management Declaration.

After the layoff, the company vowed to do three things: first, never lay off people again; second, never borrow money from the bank again; and third, always build in quality because quality sells in good times and bad times. It is important to distinguish that these concepts are philosophies, not necessary policies. Toyota does not have a policy that states that it cannot lay off people. Toyota's philosophy implies that a layoff is the company's absolute last resort.

9.6.2 Fair Appraisal and Treatment

The idea of stable employee relations requires more than just a no-layoff philosophy; the organization must also have fair and consistent management practices. One example is the role of HR as it relates to fairness and favoritism in the workplace. The HR function has the ultimate decision to fire and promote employees. In most organizations, that power resides with the immediate supervisor. When supervisors have the right to fire, it assumes that the supervisor can do no wrong, and that all of the responsibility lies in the performance of the subordinate. One of Toyota's sayings,

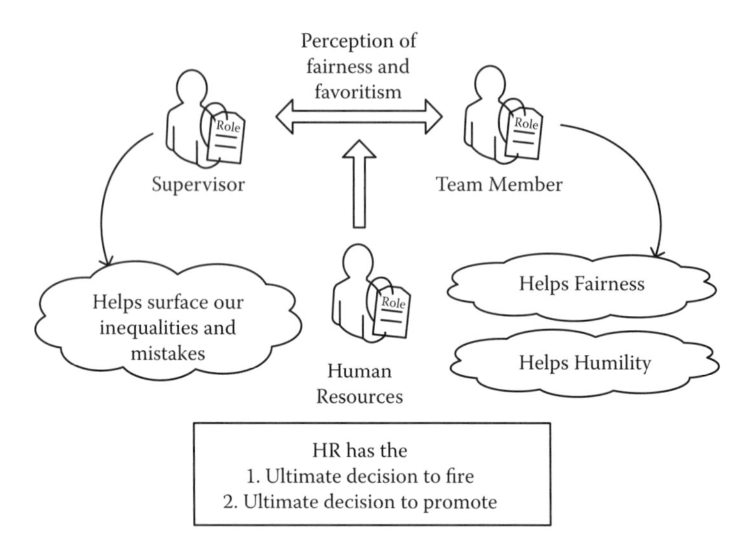

FIGURE 9.13
Illustration of HR's role for fairness and favoritism.

"the student shall never fail," implies that Toyota looks within rather than outward if an employee is not doing well. The HR function provides a nonbiased role that enables inequalities and mistakes to surface from both sides. When HR controls the status of employees, fairness and humility can improve decision making (Figure 9.13).

Organizations may follow a similar practice as Toyota; however, a prerequisite for HR's ability to be neutral is the reliance on roles. To achieve day-to-day trust, each employee needs to understand his or her role in the organization and how to perform it. Role clarity, as discussed previously in this book, provides a natural way to manage the expectations of employees and managers. Again, the reliance on a known condition, or known operating parameter, allows the HR function to be fair and consistent. Without roles, supervisors and employees will not understand how to participate and, most importantly, how to contribute.

Another example that demonstrates fairness and consistency at Toyota is the processes for corrective action (Figure 9.14). Traditionally, disciplinary actions are negative and do not contribute to a productive work environment. Most programs emphasize the negative aspect of an employee's performance rather than the improvement aspects that draw attention to the positive. Consequently, a corrective action process can be humiliating. Toyota would like for employees to retain their respect and dignity rather than decrease their self-confidence. Toyota utilizes a series of steps

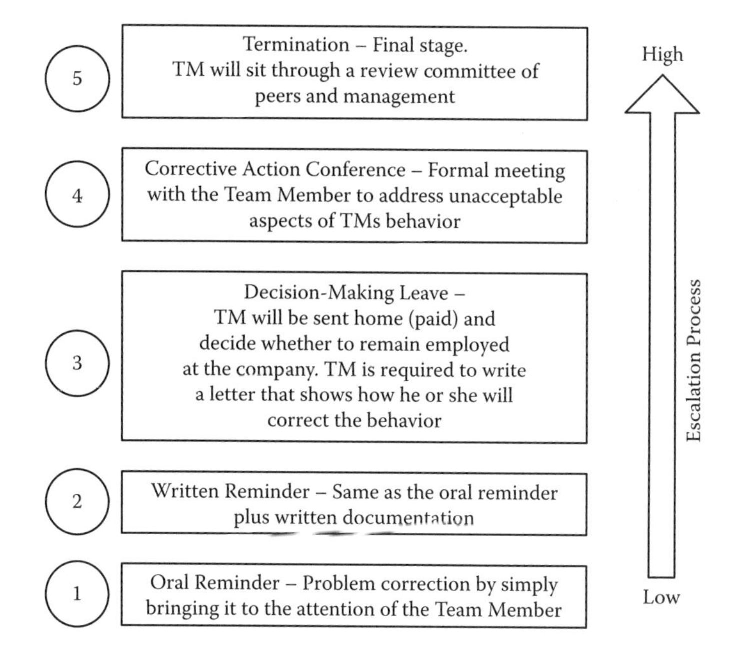

FIGURE 9.14
Example of corrective action process in HR. TM, team member.

or stages in the corrective action that is believed to bring consistency to the corrective action process. The ultimate goal is to change behavior, not to punish. While Toyota reserves the right to skip steps depending on the circumstances, Toyota never wants to be put into a position that leads to terminating an employee.

The appraisal process is another way that the HR function ensures fair and consistent work practices. The appraisal process starts when each role defines how he or she will contribute to the hoshin. The contribution of each employee is determined by the role and the established work methods connected to each role. Toyota evaluates employees based on the hoshin and their ability to apply the work methods in the role. Work methods represent the process used by the role to achieve a result. While most companies are looking to connect results to people, Toyota is looking to connect roles to process. Toyota evaluates employees mostly on how well they apply the work methods in their role rather than the results they can achieve on their own. The evaluation process has two components, process and results. The ratio that Toyota uses to evaluate employee's performance is 60% process and 40% results. Toyota's reliance on process ensures that employees will be evaluated fairly based on the process they follow.

9.6.3 Improve the Terms and Conditions of Employment

Another source of entropy due to hygiene factors relates to the terms and conditions of employment. Employees expect that work conditions should improve over time. Work conditions and environment are the basic elements for a stable and satisfying life. One of the prerequisites for a satisfying work environment is safety.

A process that is used to raise awareness of potential workplace problems when a team member performs their job is the daily team member check. Every day, the team leader inquires about the health and well-being of those team members in the rotation zone by asking such questions such as

> How are you feeling?
> Are there any processes that are having an impact on you?
> Are you experiencing any discomfort when you perform your job?
> Do any other team members experience these pains?
> Do you have any stiffness in your back?
> Are there any ongoing aches or pains in your hand or arm?

A check sheet is used to track when team members have difficulty or discomfort on a process. Daily checks are essential for early detection of problems and for improving employee relations and morale. To ensure effective two-way communication, an ergonomic questionnaire is also used for team members to rate how they feel. Instead of relying only on the team leader's subjective assessment of a team member's physical condition, team members are expected to participate in weekly or monthly ergonomic assessments.

One of the most essential aspects of workplace safety is taking action when early symptoms appear (Figure 9.15). Toyota employs an organizational approach for conducting investigations when team members start to develop or experience ergonomic symptoms. Various levels of the organization are expected to carry out aspects of the investigation within a time frame. Toyota's early symptom investigation process spans over a month, with intervals starting when a symptom first appears. The idea is that everyone has a plan and a job for managing safety. Toyota's process for ergonomic safety is just one example of how hygiene factors such as work conditions are managed to prevent entropy from building up in TPS.

Known Safety Issues for a Job	We ask you: How are you feeling?	You tell us: How you are feeling	Early Response of Issues
Standardized Work	Daily Team Member Check	Team Member Ergonomic Questionnaire	Early Symptom Investigation
1. Includes key points for safety 2. Visual control and training aid	1. How are you feeling? 2. Any difficult processes? 3. What process/steps cause you to have pain?	1. Assess process difficulty 2. Assess risk factors 3. Assess burden 4. Rate discomfort factor	1. Early intervention 2. Standard for escalating safety issues and for checking follow up items

FIGURE 9.15
An example of process management for workplace safety.

9.6.4 Workplace Communication

One of the last hygiene factors that Toyota employs to prevent workers from losing interest in TPS is the concept of workplace communication. Workplace communication is essential for keeping team members informed of current happenings. Workplace communication is necessary for creating an environment in which team members can feel comfortable voicing their opinions, suggestions, and ideas. Toyota believes that the best conduit for workplace communication is the supervisor. The challenge for HR is to set leaders at all levels to be successful at workplace communication. As stated, the way employees feel about their supervisor is how they feel about the company (Figure 9.16). Most employees quit because of how they think and feel about their immediate supervisor. When supervisors engage in two-way communication with employees, employees are more likely to have a positive image of their supervisor.

One of the most delicate areas for supervision is the ability to resolve team member concerns and issues (Figure 9.17). Supervisors have to create an open climate of trust by caring for team members through their actions. Toyota believes that supervisors can create an open climate of trust by sharing management's thinking for making decisions. Ultimately, supervisors have a role to instill all of their best thinking into their

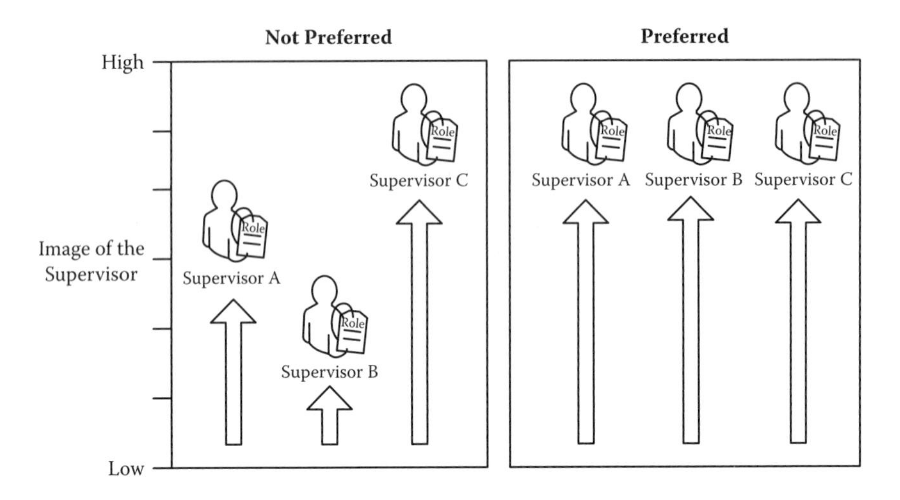

FIGURE 9.16
A comparison of how supervisors can be viewed differently in organizations.

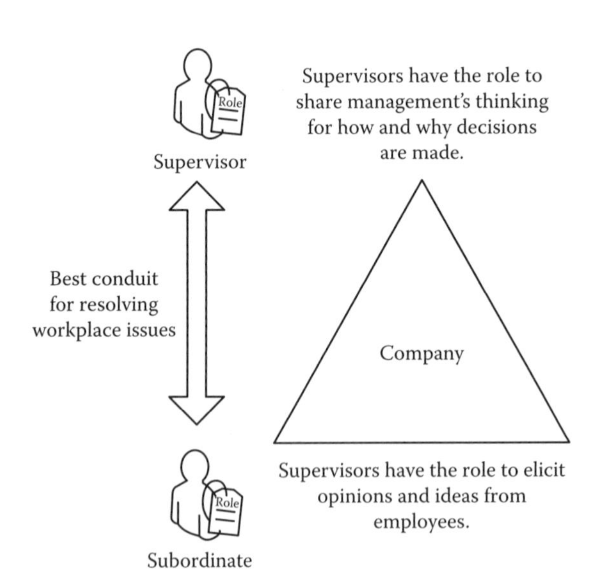

FIGURE 9.17
The best conduit for resolving workplace issues.

subordinates. This can only be accomplished if supervisors can send clear messages to their subordinates, especially relating to the conditions of the company and how decisions are based on customer needs. Team members will feel comfortable to express their concerns if their immediate supervisor walks the talk.

While Toyota believes that the supervisor role is the best conduit for resolving team member concerns, there are times when an employee may not feel comfortable expressing concerns to their immediate supervisor. In this setting, Toyota utilizes a twenty-four-hour hotline so team members can leave messages, anonymously if they do not feel comfortable identifying themselves, to voice concerns or issues. Regardless of how the issue is raised, all concerns are treated seriously and important for the company to address.

Finally, Toyota has initiated various other programs to foster open communication. The frequency, length, and purpose of these communication mechanisms are illustrated in Table 9.1. What is interesting about Toyota's approach to communication is that it does not rely on a single channel or HRM practice. Toyota's approach is to bundle a variety of communication channels to provide a more holistic approach to effective HRM.

9.7 MOTIVATION THEORY AND TPS: THE INTRINSIC NEEDS OF THE INDIVIDUAL

Intrinsic motivation is defined as interest in or enjoyment of an activity for its own sake (Lepper, 1981; Ryan, 1993). When employees are intrinsically motivated, their quality of work increases, and they are more likely to contribute toward organizational outcomes. Employees have more job satisfaction, higher performance, greater persistence, greater acceptance of organizational change, and better psychological adjustment (Baard et al., 2004). Often, it is not enough for employees to be externally motivated (i.e., by a paycheck) for employees to want to stay at their jobs. The idea is that people perform better if they like what they do, and the challenge creates a climate in which employees want to do more than the just the bare minimum. Not intentionally, organizations create conditions so employees have minimum opportunities for involvement (Figure 9.18). When this occurs, a contract of minimum effort exists between employees and the organization.

TABLE 9.1

Examples of Organizational Communication at Toyota

Meeting Type	Frequency/ Trigger	Length	Who Chairs the Meeting	Purpose
Daily communication meetings	Daily (mandatory)	5 min	Group leader/ or team leader	Operations meeting to communicate safety or quality issues for the day
Lunch box meetings	4 to 10 a year. (voluntary participation from team members)	1 h	Section management	Off-shift or shift communication meetings between section management and team members; intended to unify team spirit of the section or department
Executive roundtable meetings	1 to 3 times a year (random invitation to team members; voluntary participation)	1 h	Assistant general manager and above	Relaxed environment to communicate with senior management and executives of the company; information meeting where questions and answers are addressed
Opinion surveys	1 every 2 years (strong voluntary participation)	15 min	Questionnaire provided by HR	Opportunity to provide feedback and raise concerns

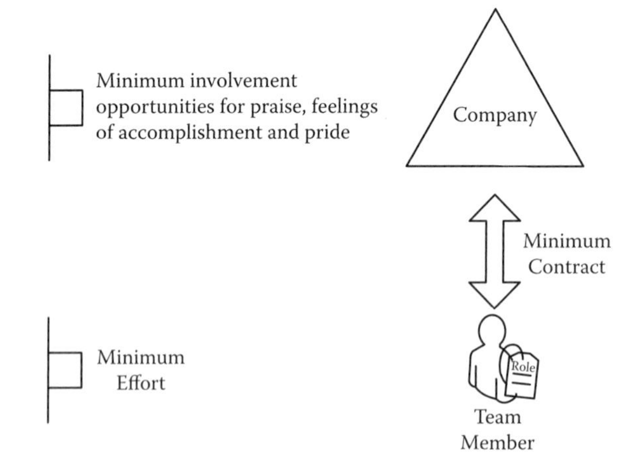

FIGURE 9.18
The minimum contract between employees and the organization.

A theory that describes the types of conditions that an organization must create to establish a mutual understanding between employees and the organization is self-determination theory (SDT). SDT suggests that employees have three central psychological needs: relatedness, effectance, and autonomy (Deci, 1975; Deci and Ryan, 1985). Relatedness can be described as the idea that people are inherently motivated to feel connected to others. Effectance is defined as the self-efficacy an employee feels when confidently working in his or her environment. Autonomy is the sense of personal initiative to work freely in one's environment. Several studies have shown that competence, autonomy, and relatedness are in fact true needs (Deci and Ryan, 1985). When organizations consider intrinsic forms of motivation in the work environment, employees are more likely to exert maximum effort in meeting organizational outcomes. The end result is more of a maximum agreement than a minimum one between employees and the organization (Figure 9.19).

SDT practices are generally associated with informal means of control since internally driven behaviors are regulated by internal norms, values, and behaviors. Informal controls are typically easier to manage because they exist within the employee (Katz and Kahn, 1978; Gardner, 2009). Informal controls are also more likely to endure over the long run because they are internalized inside the individual, meaning they do not require an external stimulus to sanction behaviors.

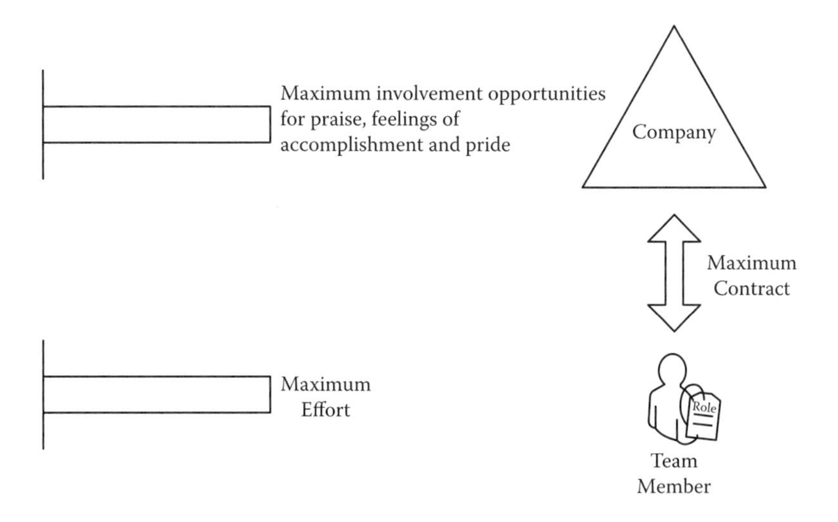

FIGURE 9.19
The maximum contract between employees and the organization.

Toyota's approach to intrinsic motivation mirrors that of SDT. Toyota has divided motivational characteristics into three HRM functions: decision making, teamwork, and workplace competence. Toyota has developed these concepts to help the organization operate efficiently under changing business conditions. Even if an organization has excellent technical systems, the company will not necessarily achieve good results unless the people who work in the system are competent and involved and work together. In this way, the HR function plays a critical role in the company's success and in encouraging employees to be highly motivated.

9.7.1 Decision Making

Toyota believes that employees should be given opportunities to be involved in decision making, especially when it means improving their work. Toyota believes that the kaizen mind should be promoted in all activities to improve productivity and quality. Toyota recognizes that the employee's mind is the best resource for kaizen (Figure 9.20). When employees find problems in their work, they are the ones who know best what to do. It is at this point that it is critical for organizations to let employees think or leave some leeway for people to think on their own. Toyota believes that the workplace should allow employees to use their own initiative. If employees are only allowed to do maintenance or keep conditions stable and not engage in kaizen (i.e., improvement), workers will lose interest in their work. In this way, kaizen is important in decision making because it encourages employees to think of ways to improve their job and not just make repairs.

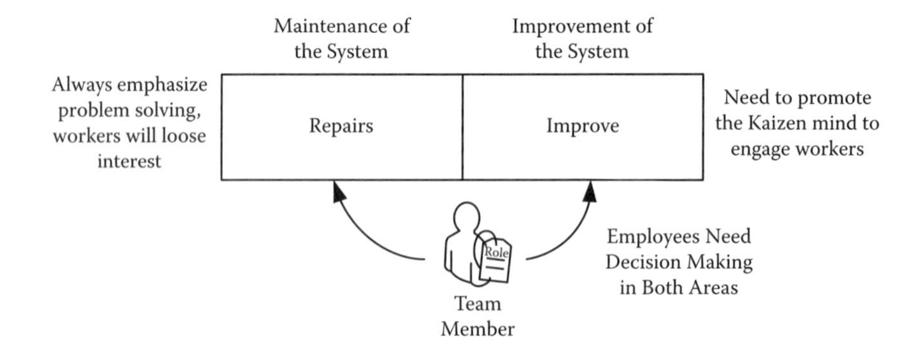

FIGURE 9.20
The importance of the kaizen mind for engaging workers.

9.7.2 Teamwork

Teamwork is important in organizations because no single person or group can accomplish all of the organization's functions. If an organization does not have good relations and cooperative activities, the organization will not be able to achieve its business goals efficiently. TPS ensures teamwork at all levels by encouraging team members to share ideas and work together in achievement of common goals. Teams encourage the organization to be aware of other groups and stakeholders. Teams promote organizational members to capture and document so that they can discuss larger issues with impacted groups. When a unit cannot cope with departmental challenges, help from other units can create a sense of unity.

Toyota believes that working in teams is satisfying and supportive of social relationships. When employees connect with others, they have the opportunity to feel socially valued. Employees who feel part of a team often express their work-related and personal troubles to other members. The less organizations emphasize teams, the more likely employees will feel lonely and not confide their troubles with employees at work. Working in teams encourages employees to share a sense of mutual respect with one another, to care and to rely on each other.

Teams are encouraged at Toyota in a variety of different ways. One of the most fundamental concepts of teams at Toyota is the organizational structure. Toyota utilizes a highly dense management structure that encourages a high number of hierarchies and a small span of control (Figure 9.21). According to system theory, the more hierarchies a system contains, the better it can react to environmental conditions. Fewer hierarchical levels create a larger power distance index within the organization, which is the extent that power is distributed within the organization (Hofstede, 2001). This would suggest that this type of organizational structure would encourage decision making to be delegated at the most appropriate level. Consequently, the dynamics of small teams create opportunities for affective-based trust, which is the kind of trust found in personal relationships and family. Affective-based trust is social–emotional trust that helps members psychologically when there is conflict or when they need support. Coupled with the use of roles, small teams can also benefit from cognition-based trust, which is trust based on a person's faith to carry out their duties. Small teams rely more on the social pressures of the group to sanction behaviors rather than deterrence-based trust or calculative trust, which uses politics or punishment to drive behaviors. While both types

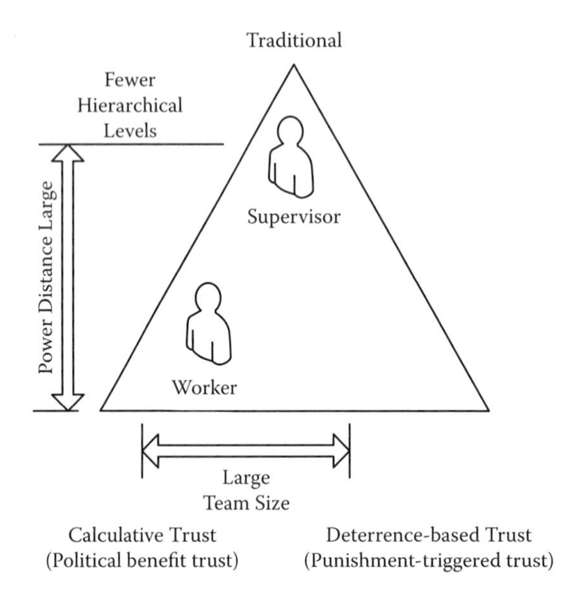

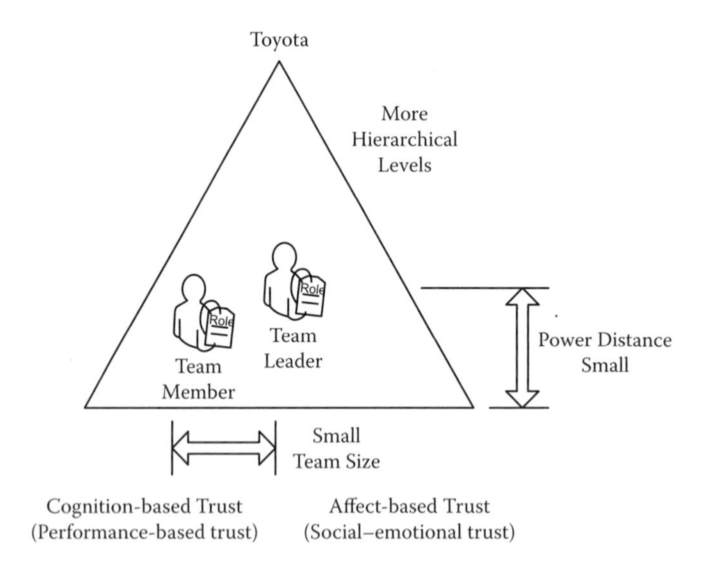

FIGURE 9.21
A comparison of hierarchy and team size in organizations.

of positive or negative forms of trust are present in all teams, larger-size teams tend to create a climate in which a strong or dominant personality has to emerge to structure the decision-making process. Since these individuals do not necessarily hold the rank or position, other forms of power are employed. Simply, small teams are more effective in the workplace because each member has a more equal and participative opportunity for decision making.

Toyota's approach to teaming shares many similarities with social learning theory (SLT). SLT indicates that people learn by watching others (Bandura, 1977). The main advantage of SLT is reinforcement. When members experience support from their team, they gain positive attitudes about implementing new skills. They increase their own capability, confidence, and most importantly like-mindedness of other members. Social learning reduces the risk of learning new behaviors because actions are guided by following good examples. Because Toyota generally involves a learning component in team activities, members are more likely to adopt the new behavior because it results in outcomes the team values. Toyota's approach to team learning encourages a climate conducive to learning and reinforcement because new skills have functional value and worth for the team.

9.7.3 Workplace Competence

SDT states that a worker is satisfied when he or she feels competent when working in an environment. Toyota's job instruction training process, modeled from Training Within Industry (TWI), describes one way that workers can feel competent when learning new jobs. SDT implies that the more competent a worker is, the more satisfied the worker becomes. Toyota's view on lifelong learning, like many other Japanese organizations, closely resembles SDT theory.

It is argued that what makes Toyota's approach to SDT theory unique is how it achieves competence. This can best be illustrated by Toyota's concept of worker flexibility. The automobile industry is subject to various market conditions that make it difficult for the organization to predict customer preferences. An organization needs to have the ability to respond to changing conditions so that it can be profitable. One of the main functions of HR is to maintain a flexible organization in response to changing conditions. For this reason, Toyota simplifies job classifications so that workers can move around in the organization to meet changing demands of the customer. For example, team members are classified into four divisions: production, general maintenance, tool and die, and skilled trades. Fewer job classifications allow workers to move to areas of the plant that are overloaded with work.

One of the ways that Toyota is able to move workers around is their emphasis on in-house training. Job rotation is one approach that allows employees to be trained in a wide range of skills in a job category. After employees master their operation, they are developed for another operation. This approach requires a mid- to long-term staffing plan that hires employees based on their potential to contribute to the company. Consequently, this system allows employees to continue to expand their skills throughout their working life.

An economic theory that describes Toyota's implementation approach to acquiring knowledge is human capital theory (Becker, 1993). Human capital refers to the knowledge and expertise an employee accumulates through education and training (Becker, 1993). This theory highlights two critical dimensions for HRM: general training and specific training.

Specific training refers to organizational training (typically in-house training or on-the-job training [OJT]) that increases productivity of employees working in an organization. This type of training is only beneficial for the organization. An example of specific training is Toyota's concept of job rotation, which is often related to a multiprocess or multifunctional worker, a worker performing more than one type of related job in his or her work area.

General training refers to raising the skills and productivity of people valued by the same amount in other organizations. Examples of this type of training include leadership training, problem solving, and communication skills. Because competitors can benefit from hiring trained employees from another organization, some organizations may be less likely to offer general training (Becker, 1993). Therefore, the incentives for an organization to invest in training are different for these two types of training.

The types of general training that Toyota provides is illustrated in Figure 9.22. General training is separated between team leader and supervisory training. Team leader training is composed of meeting facilitation, problem solving, standardized work, kaizen, and a variety of quality circle training. Group leader training emphasizes work site communications, leadership, job relations, conflict management, and suggestion system training. Talent management and quality circle doctrine training are reserved for assistant manager and manager training. A variety of electives

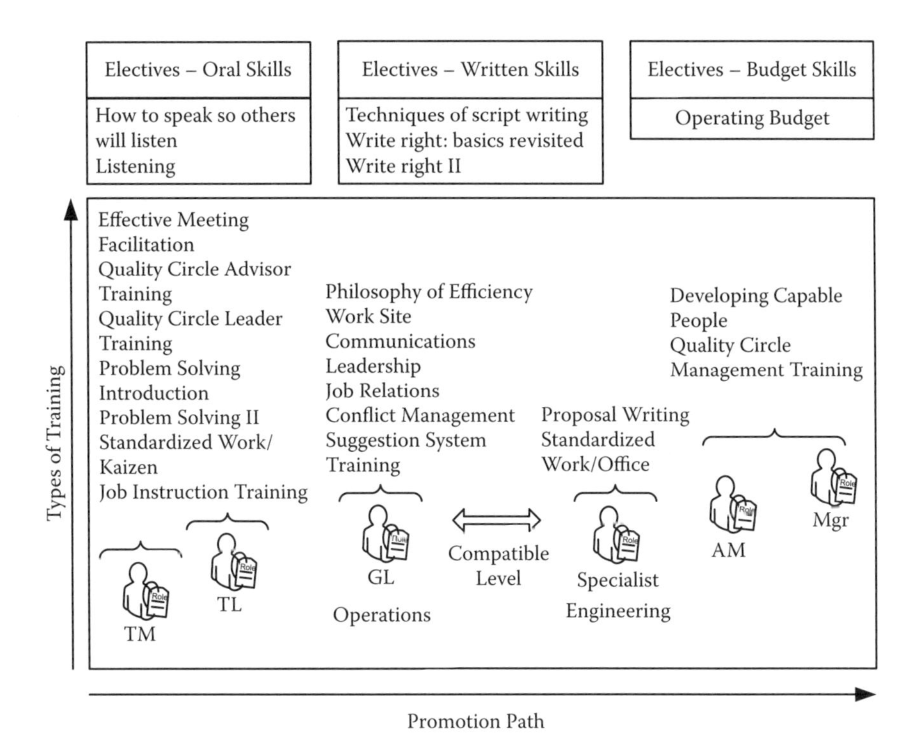

FIGURE 9.22
Example of general training at Toyota. TM, team member; TL, team leader; GL, group leader; AM, assistant manager; Mgr, manager.

is also provided in Toyota's curriculum of general training. They include communication and written skills along with budget development skills.

9.8 SUMMARY

This chapter presented the idea that entropy can be generated in TPS when employees no longer feel open or engaged in workplace improvement. Hygiene and motivation theory was used to describe the needs of employees in the workplace. Low-level needs or hygiene needs do not encourage employee engagement, but without job security employees will never feel like participating in improving jobs that may lead to their dismissal. When this occurs, TPS suffers from entropy or deterioration or simply runs down. The HR function must stabilize the basic employment terms and conditions to place employees in a free state, absent of fear and full of trust.

Higher-level needs include a set of HR functions that relate to teamwork and workplace learning. These represent motivating factors that encourage employees to maintain a certain level of job satisfaction that is instrumental in keeping TPS from returning to a state of disorder. SLT showed that Toyota's team concept creates an environment conducive to learning among members. SDT was presented to help explain the view that Toyota shares in lifelong learning, and human capital theory was used to describe two specific approaches to lifelong learning. Toyota's use of specific and general training provides a barrier from entropy by creating conditions that keep workers engaged in TPS throughout their life cycle.

REFERENCES

Ahmad, S. and Schroeder, R. (2003) The impact of human resource management practices on operational performance recognizing country and industry differences. *Journal of Operations Management*, 21(1), 19–43.

Applebaum, E., Baily, T., Berg, P., and Kalleberg, A. (2000) *Manufacturing Advantage*, Cornell University Press, Ithaca, NY.

Baard, P. P., Deci, E. L., and Ryan, R. M. (2004). Intrinsic need satisfaction: A motivational basis of performance and well-being in two work settings. *Journal of Applied Social Psychology*, 34, 2045–2068.

Badurdeen, F., Marksberry, P., Hall, A., and Gregory, B. (2009) No instant prairie: Planting lean to grow innovation. *International Journal of Collaborative Enterprise, Special Issue: Technological Innovation*, 1(1), 22–38.

Bandura, A. (1977). *Social Learning Theory*, General Learning Press, New York.

Becker, G. (1993) *Human Capital: A Theoretical and Empirical Analysis with Special Reference to Education*, 3rd ed., University of Chicago Press, Chicago.

Beechler, S. and Bird, A. (1994) The best of both worlds? Human resource management practices in U.S.-based Japanese affiliates, in *Japanese Multinationals*, ed. Campbell, N. and Burton, F. London: Routledge.

Beer, M., Spector, B., Lawrence, P., Mills, Q., and Walton, R. (1984) *Managing Human Resource Assets*, Free Press, New York.

Besser, T. (1996) *Team Toyota: Transplanting the Toyota Culture to the Camry Plant in Kentucky*, State University of New York Press, Albany.

Black, J., Gregersen, H., and Mendenhall, M. (1992) Toward a theoretical framework of repatriation adjustment. *Journal of International Business Studies*, 23(4), 737–760.

Bolman, L. and Deal, T. (1997) *Reframing Organizations: Artistry, Choice and Leadership*, 2nd ed., Jossey-Bass, San Francisco.

Cengel, Y. and Boles, M. (1989) *Thermodynamics: An Engineering Approach*, McGraw-Hill, New York.

Collins, W. A., ed. (1981) *Aspects of the Development of Competence: Minnesota Symposium on Child Psychology*, Vol. 14, Erlbaum, Hillsdale, NJ, pp. 155–213.

Deci, E. L. (1975). *Intrinsic Motivation*, Plenum, New York. Japanese Edition: Seishin Shobo, Tokyo, 1980.

Deci, E. L. and Ryan, R. M. (1985). *Intrinsic Motivation and Self-Determination in Human Behavior*, Plenum, New York.

Deming, W. (1986) *Out of the Crisis*, MIT Press, New York.

Dennis, P. (2007) *Lean Production Simplified*, 2nd ed. CRC Press, Boca Raton, FL.

Edgar, F. (2003) Employee-centered human resource management practices. *New Zealand Journal of Industrial Relations*, 28(3), 230.

Fombrun, C., Tichy, M. and Devanna M. (1984) *Strategic Human Resource Management*, John Wiley, New York, NY.

French, W. (1986) *Human Resource Management*, Houghton Mifflin, New York.

Fucini, J. and Fucini, S. (1990) *Working for the Japanese: Inside Mazda's American Auto Plant*, Free Press, New York.

Gardner, M. (2009) Managing people in a Lean environment: The power of informal controls and effective management of company culture. *Journal of Business Case Studies, Littleton*, 5(6), 105–111.

Gilbert, T .(1978) *Human Competence: Engineering Worthy Performance*, McGraw Hill, New York.

Guest, D.E. (1999). Human resource management: The workers' verdict. *Human Resource Management Journal*, Vol. 9, No. 3 5–25.

Guest, D., Michie, J., Conway, N., and Sheeman, M. (2003) Human resource management and corporate performance in U.K. *British Journal of Industrial Relations*, 41(2), 291–314.

Hall, A. (2006). *Introduction to Lean Sustainable Quality Systems Design: An Integrated Approach From the Viewpoints of Dynamic Scientific Inquiry Learning & Toyota's Lean System Principles and Practices*. Published by Arlie Hall, Lexington, KY. ISBN 0-9768765-0-7.

Hendry, C. (2003) Applying employment systems theory to the analysis of national models of HRM. *International Journal of Human Resource Management*. 14(8), 1430–1442.

Herzberg, F. (1987) One more time: How do you motivate employees? *Harvard Business Review*, 65(5), 109–120.

Hofstede G (2001) *Culture's Consequences: Comparing Values, Behaviors, Institutions, and Organizations Across Nations*, 2nd ed., Sage, Thousand Oaks, CA.

Johnson, E. (2000) The practice of human resource management in New Zealand: Strategic and best practice. *Asia Pacific Journal of Human Resources*, 38(2), 69–83.

Katz, D. and Kahn, R. (1978) *The Social Psychology of Organizations*, 2nd ed., Wiley, New York.

Kochan, T. and Dyer, L. (1993) Managing transformational change: The role of human resource professionals. *International Journal of Human Resource Management*, 4(3).

Lepper, M. R. (1981) Intrinsic and extrinsic motivation in children: Detrimental effects of superfluous social controls. In *Aspects of the Development of Competence: the Minnesota Symposium on Child Psychology*, Volume 14. Collins, W. A. (ed.), Erlbaum, Hillsdale, NJ.

Liker, J. and Hoseus, M. (2008) *Toyota Culture—The Heart and Soul of the Toyota Way*, McGraw-Hill, New York.

Liker, J. and Hoseus, M. (2010) Human resource development in Toyota culture. *International Journal of Human Resources Development and Management*, 10(1), 34–50.

Lincoln, J. and Kalleberg, A. (1990) *Culture, Control and Commitment: A Study of Work Organization and Work Attitudes in the US and Japan*, Cambridge University Press, Cambridge, UK.

Luthans, F. (1995) *Organizational Behavior (7. Auflage)*, McGraw-Hill, New York.

McClelland, S. (1986) The human performance side of productivity improvement. *Industrial Management*, September–October, 14–17.

Palmer, G. (1987) Human resource management and organizational analysis, *Asia Pacific Journal of Human Resources* July Vol. 25, No. 2, 5–17.

Parise, S. (2007) Knowledge management and human resource development: An application in social network analysis methods. *Advances in Developing Human Resources*, 9(3), 359–383.

Pascale, R. and Athos, A. (1981) *The Art of Japanese Management*, Simon & Schuster, New York.

Ryan, R. M. (1993) Agency and organization: Intrinsic motivation, autonomy and the self in psychological development, in *Nebraska Symposium on Motivation: Developmental Perspectives on Motivation*, Vol. 40, ed. Jacobs J. Lincoln, NE: University of Nebraska Press, 1–56.

Selmer, J. (2001) Human resource management in Japan: Adjustment or transformation? *International Journal of Manpower*, 22(3), 235–243.

Sparrow, P. and Hilltrop, J. (1994) *European Human Resource Management in Transition*, Prentice Hall, Englewood Cliffs, NJ.

Stalk, G., Evans, P., and Shulman, L. (1992) Competing on capabilities: The new rules of corporate strategy. *Harvard Business Review*, 70(2), 57–69.

Theriou, G. and Chatzoglou, P. (2008) Enhancing performance through best HRM practices, organizational learning and knowledge management: A conceptual framework. *European Business Review*, 20(3), 185–207.

Wageman, R. (1995) Interdependence and group effectiveness. *Administrative Science Quarterly*, 40, 145–180.

Wright, P., Geroy, G., and MacPhee, M. (2000) A human resources model for excellence in global performance. *Management Decision*, 38(1), 36–42.

10

The System Property of Reversibility in TPS

10.1 SYSTEM PROPERTY: REVERSIBILITY

According to the first law of thermodynamics, energy can neither be created nor be destroyed. The second law of thermodynamics states that entropy cannot be destroyed, but it can be created. An interesting aspect of the second law is that it says something about the efficiency of converting energy from one form to another. The second law states that processes are not perfect (Lay, 1963). Simply, processes have a natural tendency to go in one direction than in another. For example, it is easier to mix gases than to unmix them.

It turns out that the most efficient processes for converting energy from one form to another are processes that are reversible. In a reversible process, the net entropy change of the system and the surroundings is zero. For example, this could be a process that does not contain any friction, like an electric current flowing through zero resistance, or the mixing of two ingredients that are exactly the same and at the same state, such as two cups of water at the same temperature.

The measure of entropy or the reversibility of a process determines the best that can be done in converting energy. The examples illustrate low or net-zero entropy generation; which means they are more likely to be reversed efficiently. In this case, the natural tendency to convert energy can be in either direction.

In reality, no thermodynamic process is entirely reversible (Figure 10.1). No mechanical process is entirely frictionless, and a conductor does not exist that has zero resistance. No two cups of water are exactly the same in composition or state. The study of reversible processes is important because reversible processes are more efficient than irreversible ones.

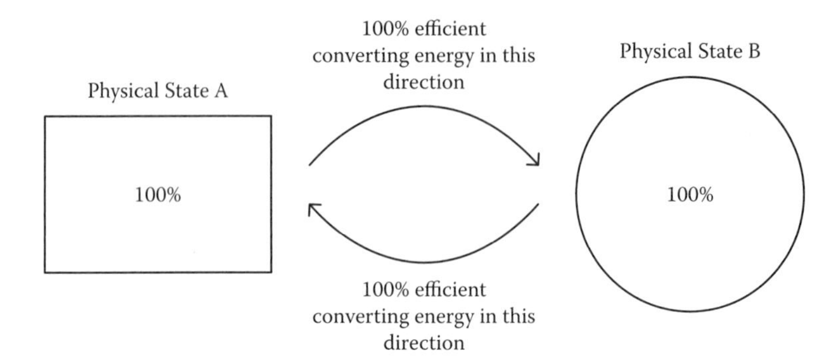

FIGURE 10.1
Reversibility in a thermodynamic process.

Reversible processes are considered limits of ideal or theoretically efficient processes (Cengel, 1994). To this end, the better the design, the higher the reversibility and the higher the second law efficiency. It is when processes are irreversible that entropy is generated (Figure 10.2) (Lay, 1963).

While no process is completely reversible, engineers still use the idea of reversibility to bring insight into studying how systems behave. One approach is the classifying parts of a system as either internally and

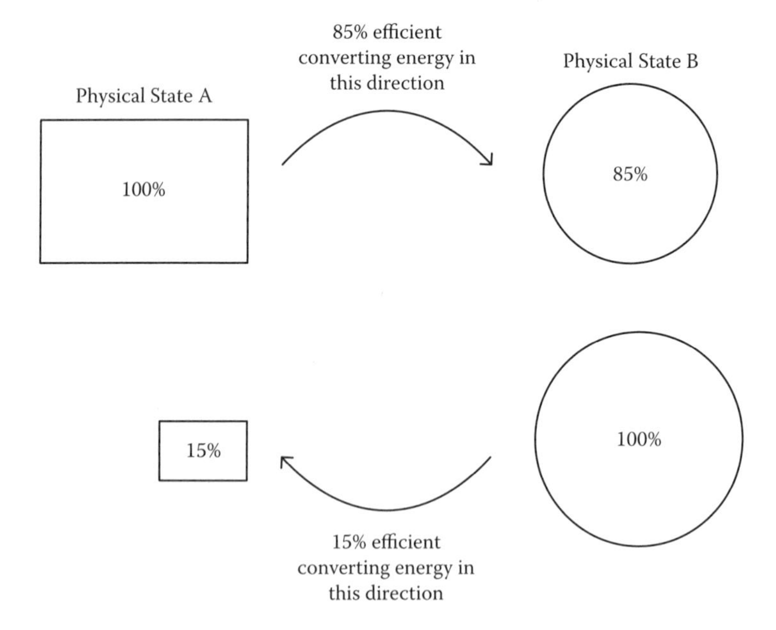

FIGURE 10.2
Irreversibility in a thermodynamic process: most efficient conversion process only occurs in one direction.

externally reversible. An internally reversible process is one that contains no irreversibilities within the boundaries of the system as energy is converted from one form to another. An externally reversible process occurs when no irreversibilities occur outside the system. In other words, restoration is made without leaving any net changes to the surroundings.

Consider a cup of hot coffee. Once the coffee cools, it will not heat up by retrieving from its surroundings the heat it lost. If the coffee was able to recover its lost heat from its surroundings and the surroundings were able to be restored to the original condition, this would be both an internally and an externally reversible process. It could be argued that the coffee can return to its original state by heating its surroundings; however, this does not guarantee reversibility. The surroundings and other systems must also be restored to their exact original state. Reversibility implies that a process that can bring about change in a system and bring the surroundings back to their original state. Irreversible processes involve a change in a system and its surroundings that is permanent (Lay, 1963).

10.2 MINIMIZING ENTROPY IN THE MANAGEMENT AND LABOR RELATIONSHIP (REVERSIBILITY)

The idea of using the theory of entropy in organizations is appealing because it represents the best possible exchange of energy a system can experience. In the Toyota Production System (TPS), the second law represents the best possible conversion of employee ideas, energy, and effort by the company. When the effort of employees cannot be converted completely, entropy is created, which is a measure of irreversibility. Ideally, TPS should behave as a reversible system, one that has few losses due to its internal and external conversion processes. In other words, the way that employees give their ideas inside the system should be efficient, as well as that of management, who is responsible for creating favorable external conditions of the system itself. In thermodynamic terms, labor is represented by the internal conversion process, and management is represented by the external conversion process. For TPS to be completely reversible, both the internal and external conversion processes must be efficient (Figure 10.3).

An example of irreversibility (i.e., entropy) in TPS is when there is friction between labor and management (Figure 10.4). In this manner, employees do not feel that working to the system is the best way

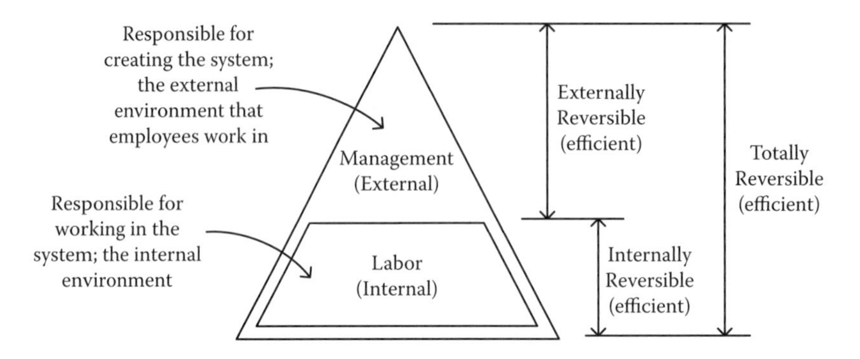

FIGURE 10.3
The external and internal systems of an organization according to the laws of reversibility.

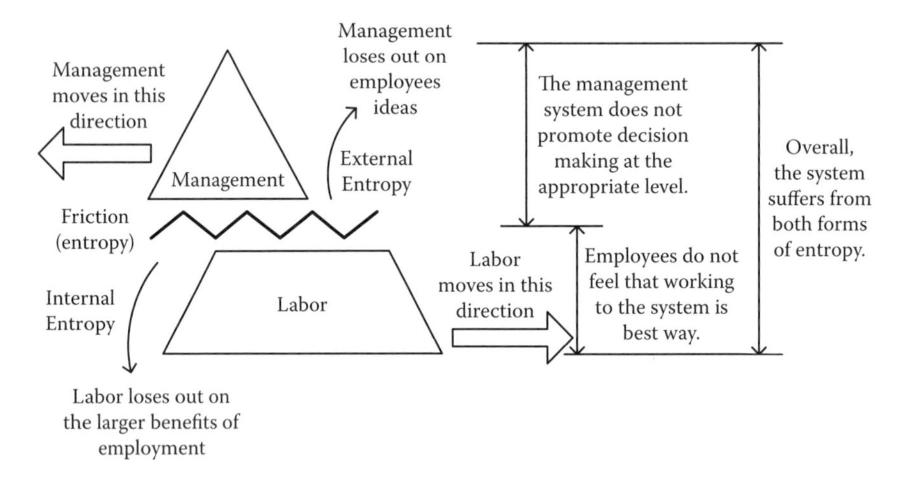

FIGURE 10.4
Illustration of entropy (i.e., friction) between labor and management.

for employment, and management does not feel that employee ideas can improve the organization's performance. When this occurs in organizations, the system suffers from both internal and external entropy. When internal irreversibilities (efficiencies and entropy) become increasingly high in organizations, unions are often established to smooth the effects of the management–labor relationship. The idea of unions from a thermodynamic perspective is to return the internal environment back to its original and healthy state. An organization is said to be reversible if the surroundings (i.e., management) can also return to a previous state of restoration.

The concept of using unions to describe entropy in organizations can bring insight in how worker's rights are increased to the point they are in equilibrium with the rights of management. It is argued that TPS balances irreversibilities associated with management and labor so that the total system suffers less from entropy, meaning that neither management nor labor is responsible for all the ideas. Neither management nor labor has all the rights. The sections that follow give an introduction to how unions have evolved in the automotive industry. The literature review also covers how Lean has been applied in organized environments. The review shows that even with the most progressive human resource (HR) functions, internal and external irreversibilities in organizations can still occur, which makes the study of unions an interesting equalizer in the workplace as it relates to TPS.

10.3 LITERATURE REVIEW: THE BIRTH OF THE TOYOTA PRODUCTION SYSTEM AS A UNION

In 1949, Toyota announced that it would dismiss 25% of its workforce (~1,500 employees) due to an ailing Japanese automobile market. While the president of the Toyota Motor Corporation, Kirrichro Toyoda, claimed responsibility for the company's failure, his resignation could not stop the company's most historical labor dispute, which lasted nearly twelve years. In 1962, Toyota and its employees signed the Joint Labor and Management Declaration, which started the first organized labor movement, known as the Toyota Motor Works Union (TMWU). Today, Toyota has over 64,000 union members in Japan and is part of the Federation of Toyota Workers Union, the Confederation of Japan Automobile Workers Union, and RENGO (Japanese Trade Union Confederation).

Most of what is known about Toyota's HR systems grew out of the TMWU (Figure 10.5). Toyota's first labor–management agreement represents the beginning of TPS, which is a formal and structured cooperative declaration based on mutual trust. The agreement represents the rights and obligations between two parties with the goal to establish peaceful and stable relations. The starting point for TMWU and the company was to create a clear understanding of each other's perspective. Various venues for discussion, including councils and conferences, were established to promote the exchange of opinions on problems at each level of the organization (Figure 10.6). Ideally, each party could address issues in a timely manner and incorporate views into the company policy. Both parties

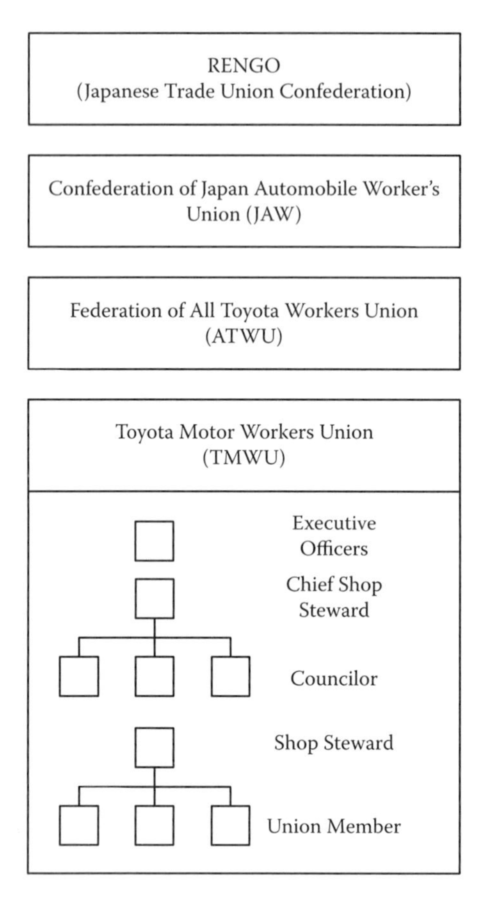

FIGURE 10.5
General organization of the Toyota Motor Workers' Union.

agreed to work hard to increase productivity and improve the working conditions of the organization. Ultimately, the company will do its utmost to improve the terms and conditions of employment, and each employee will cooperate in realizing the company's policies.

Toyota's human resource management (HRM) practices have been widely benchmarked and studied and are contributed as the primary factors for successfully achieving the TPS (Hall, 2006). Most research attempts described Toyota's HRM practices as a collection of progressive techniques that aim to raise respect and humility at the workplace, simultaneously raising profits and productivity (Liker and Hoseus, 2010; Mann, 2005). Toyota's safe work environment encourages employees to participate in kaizen by voicing their ideas without fear of losing their job in the process.

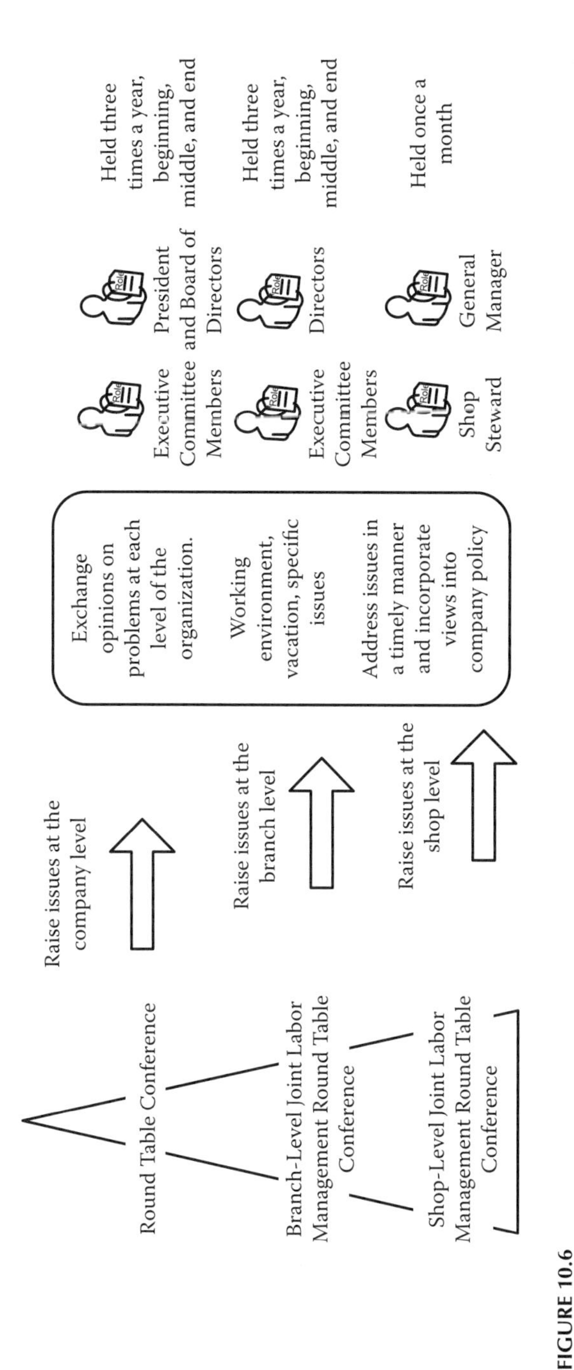

FIGURE 10.6
Illustration of management labor declaration: an example of a structured exchange of issues and problems.

Interestingly, researchers viewed Toyota's HRM practices from a nonunion perspective, yet the development of their core HR philosophies and practices grew out of organized labor.

Toyota's wholly owned transplant companies, such as Toyota Motor Manufacturing Kentucky (TMMK), have been a point of study in HR mainly because they have remained nonunion despite several attempts by the United Automotive Workers (UAW). Somehow, Toyota has been able to create a "healthy tension" between the workforce and management that has workers participating and contributing, which was never thought possible. This "management-in-conflict" stance in HR is possibly one of the most unusual aspects of TPS and one of the most difficult aspects to emulate.

What is unknown is how TPS can create an environment in which its members have solidarity within the workplace. If TPS does share more of a unionized approach for increasing solidarity among its members, which union labor theories best describe TPS? Can TPS be considered a specialized form of unionization because it shares the same types of participation theories found in an organized environment? If TPS is a new form of unionization, to what degree does the power of workers increase compared to management? Possibly, the reason American cultures have done so poorly implementing TPS-like systems is because they tend to oversimplify Toyota's flexible workplace environment compared to a narrowing and aggrandizing system inherent in most unionized environments.

It is speculated that TPS is a unique form of unionism since it shares many of the same narrowing and restricting characteristics of a traditional unionized labor–management model. In other words, there is not unlimited freedom in how members are expected to participate in workplace improvement. While most studies of Toyota's HR practices tend only to highlight the autonomous aspects of the work environment, the theoretical perspective of unions paints a more holistic and representative picture of how both sides, management and labor, "struggle cooperatively," never quite satisfied, in raising profits or making the company a better place to work (Figure 10.7). For example, unions can negatively restrict and aggrandize or positively improve and protect the organization. The negative effect of unions relates to a lop-sided monopoly situation in favor of labor. The positive effect of unions describes a balanced and democratic situation in which a mutually benefiting relationship exists between the company and labor (Kaufman, 2010).

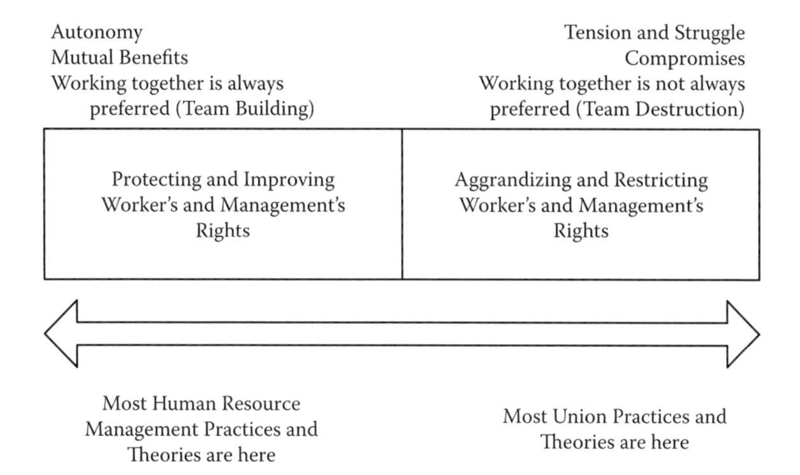

FIGURE 10.7
Human resource management and union viewpoints compared.

10.3.1 The Role of Trade Unions and the Human Resource Function in the United States

In 1935, the National Labor Relations Act (NLRA) legalized the formation of labor unions and collective bargaining in the protection of worker's rights in the United States. Unions have historically been able to increase their members' wages in the range of 10% to 30% higher than the wages of nonunion workers and provide a mechanism for members to voice opinions regarding workplace conditions through collective bargaining (Kaufman, 2007). Unions offer an opportunity for individuals to participate above the levels their jobs might afford them (Hoell, 2004), the security of stable employment (Weather and North, 2010), and the protection of worker rights (Commons, 1905).

One of the primary deterrents of organized labor are the implementation of progressive HRM practices (Ichniowski et al., 1997). When employees feel fairly treated by their company, most employees will reciprocate the gift of hard work and loyalty without seeking representation (Hoell, 2004). On the other hand, if workers feel unfairly treated, they react by punishing the company through greater absenteeism, loafing on the job, and a bad attitude toward customers (Akerlof, 1990).

The leading challenges in unionized environments have been to effectively apply HRM practices to enhance organizational performance while maintaining balance between productivity and worker input (Bhatti and Qureshi, 2007). Research showed that satisfied employees tend to be more productive, yet few have made employee involvement (EI) a top priority in organizations (Webster, 2008). The most common HRM approach to raise satisfaction among the workplace has been the use of participation programs. These "new social agreements" have been a way to motivate workers without necessarily increasing wages in union environments (Ariovich, 2007; Kochan et al., 1986).

10.3.2 The Use of Lean in Human Resource Development

In the 1970s, a new form of worker participation, named Lean production, became one of the most dominant and preferred HRM practices in both unionized and nonunionized environments. In Lean environments, workers not only have the opportunity to participate in workplace decisions but also have the opportunity to think actively when solving workplace problems (Womack et al., 1990). Lean offers opportunities for workers to be involved in how the workplace operates and greater autonomy in learning workplace jobs. Workers are generally more satisfied in a Lean environment because they are more empowered to have an impact on their work (George and Hancer, 2003).

Unfortunately, implementing the soft side of Lean is not easy. Most companies choose to apply Lean by changing a single HRM practice rather than a group or cluster of practices, which has little to no effect on productivity or participation (Ichniowski et al., 1997). One of the most significant struggles in HR is choosing how much power and flexibility should exist in a Lean environment. Regrettably, the implementation of Lean production methods in a union environment have fared poorly, mainly due to misinterpreting the amount of freedom in the workplace.

10.3.3 A Difficult Implementation of Lean in Union Environments in the United States

Historically, participation programs in union environments do not perform well mainly because every social interaction becomes scrutinized for hidden meaning and sinister purpose. Participation programs, like Lean production, often heighten sensitivity about fairness, which leads to no cooperation and industrial sabotage. Total quality management (TQM) and EI programs are often viewed by unions as additional efforts to make employees work harder (Eaton, 1995). Union activists believe that workers

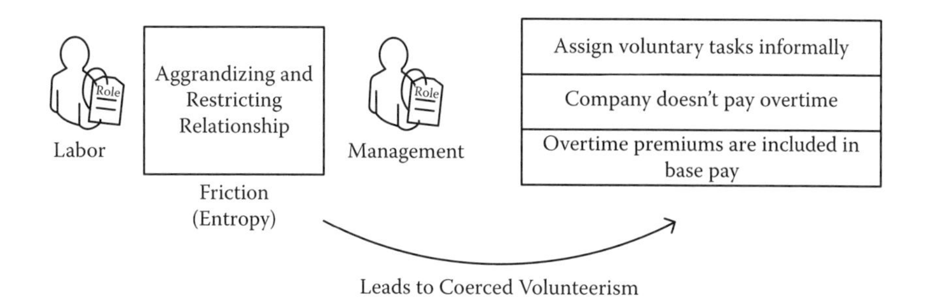

FIGURE 10.8
Example of Lean when it is aggrandizing and restricting: coerced participation.

who do not volunteer or participate in quality circles (QCs) or who fail to meet the expected quota of suggestions pay a heavy price for their lack of commitment (Babson, 1993). Labor experts have described some participation initiatives as coerced volunteerism (Figure 10.8) (Makoto, 1996). In this situation, the organization assigns voluntary tasks informally or denies or does not pay overtime. In extreme cases, organizations manipulate wage allowances or claim that overtime premiums are included in base pay (Weather and North, 2010).

Participative programs and teamwork initiatives are also viewed as antiunion activities intended to penetrate and influence small work groups. Unions are concerned that small groups will convince employees to believe that the company is a great place to work and management really cares, despite lousy pay and rotten conditions (Grenier, 1988). In addition, workers may feel obligated to go around union contracts, not to file grievances, and to start viewing the union as an outside third party (Banks, 1984). Consequently, unions are less likely to support participative programs because no union official wants to go out on a limb with the membership and endorse something that is later discontinued (Figure 10.9) (Cole, 1984).

One of the biggest complaints about Lean in union environments is the overstated freedom to control one's work (Figure 10.10). In practice, team members cannot alter their standardized work without supervisory approval, and casual deviation is strongly discouraged. "If you follow your standardized work all the time, every day, regardless of how you feel—quality will follow" (Babson, 1993). Some authors suggested that when the job cycle is described in minute detail, including specific tasks, sequence, and number of seconds allotted for each task, a highly standardized short work cycle is a complex extension of Taylorism (Berggren, 1992).

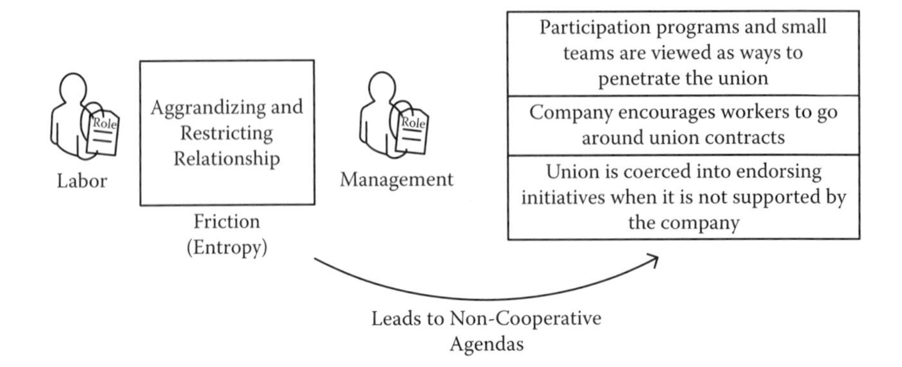

FIGURE 10.9
Example of Lean when it is aggrandizing and restricting: noncooperative agendas.

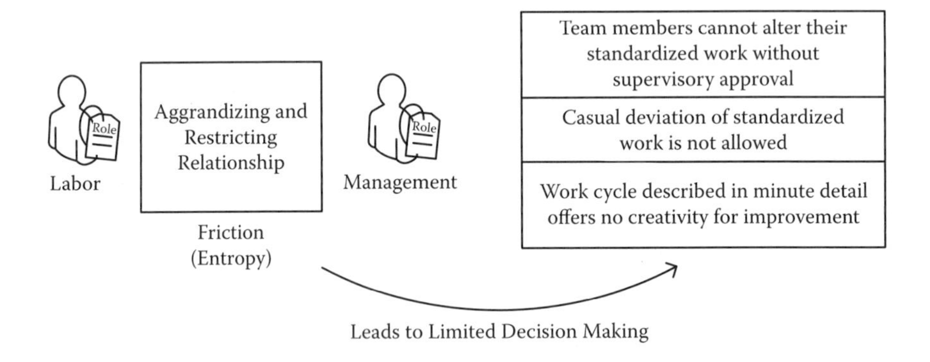

FIGURE 10.10
Example of Lean when it is aggrandizing and restricting: limited decision making.

10.3.4 Union Problems and Lean Production in the Automotive Industry

The application of Lean in the automotive industry has been less than successful in organized environments. Lean production methods have been attempted by American original equipment manufacturers (OEMs), by Japanese transplant companies in America, and by American–Japanese partnerships in America. Most approaches to emulate Lean in union environments tend to take on a wide interpretation of how workers are expected to participate. This is surprising since most union labor agreements tend to be narrowing rather than open in regard to how workers are expected to be involved in areas such as problem solving, decision making, and workplace improvement.

One of the first examples of Lean applied in an automotive union environment was that of General Motors (GM). The Saturn project in 1985 was intended to beat the Japanese at their own game by producing a small car using the same Lean production methods. Other features of Japanese production management were also incorporated, such as extensive training, protection against layoffs, team concept, and problem solving. The twenty-two-page labor agreement eliminated the status differences between labor and management by enabling the union to be a full partner in scheduling, budget, housekeeping, safety, ergonomics, maintenance, inventory control, training, job assignment, repairs, and absenteeism (Saturn, 2000). The contract also eliminated the management rights clauses of GM's paragraph 8, which had been established since the 1930s (Yanarella and Green, 1994). The new contract reduced the number of job classifications from over 100 to 6 and added 92 hours of formal training per year.

It was not until 1998 that the labor–management agreement at Saturn was fully tested. GM's second-largest work stoppage since the 1970s caused twenty-seven of twenty-nine assembly plants to shut down. Saturn's refusal not go on strike triggered a diverse set of responses among union members. A faction inside local 1853, named the Concerned Brothers and Sisters, finally forced the union to strike and reclaim seats in a subsequent election by running on a more traditional GM–UAW platform. In 1999, the union and GM made an attempt to recommit to the original memorandum laid out in the original Saturn philosophy. In 2004, the union reestablished its linkages to the national UAW agreement, which ended most of the Saturn philosophy that emulated Lean production methods. In 2010, GM discontinued the Saturn brand due to falling sales.

In 1986, GM made another attempt at adopting Japanese management techniques by partnering with Suzuki. Canadian Automotive Manufacturing, Incorporated (CAMI) was established in Ontario, Canada, to emulate Lean production methods using the CAMI Production System (CPS). CPS shared many of the traditional Japanese management features, such as teamwork, standardized work, quality circles, multiskilled training, and a truncated job classification structure. In 1992, the CAW (Canadian Auto Workers) rejected the Japanese production methods and held concessions to neutralize the worker suggestion system that led to unfair and inconsistent payouts. Management was accused of inflating estimated cost savings and cooking the books in exaggerating reward-worthy suggestions. In 1996, the company laid off 250 workers and in

2001 another wave of 500 workers, thus contradicting the idea that under Japanese management systems loyal workers could expect permanent, lifelong employment. In 1998, the union won the rights to select team leaders based on seniority and purging existing team leaders with lesser seniority for those with more years on the job (Yanarella and Green, 1994). In 2009, Suzuki ended the partnership with GM, yet GM continued to use the CPS in manufacturing the Chevrolet Equinox and Terrain.

In 1987, Mazda and Ford opened a new plant in Flint Rock, Michigan, to produce coupes and sedans. Mazda's Lean production methods were criticized early by the UAW mainly due to the emphasis on teamwork and exaggerated ideas of worker flexibility. The Mazda philosophy endorsed cross training; however, union workers did not have the freedom to seek alternative jobs within the plant. Most workers would like to have access to less physically demanding work and jobs that are less monotonous. Workers also favor jobs that provide opportunities for overtime. The union viewed Lean as negative because management did not provide access to all jobs within the unit and would not train workers in learning these types of jobs (Babson, 1993). In 1991, the UAW won concessions to amend Mazda's Lean production system. The union added a new agreement that established a temporary assignment pool for workers whose jobs were eliminated by kaizen, new technology, model changes, and production cutbacks. There was also mandated training to upgrade workers to make them able and capable for favored jobs in their unit. When a worker was offered overtime and was not able or capable, the company was obligated to give both the low-hour worker and the truly able worker overtime assignments to train (Babson, 1993).

10.3.5 A Successful Case of Lean Production in the Automotive Sector in a Union Environment

In 1984, GM and Toyota Motor Corporation (TMC) created a joint venture named NUMMI (New United Motors Manufacturing) in Fremont, California. Its mission was to produce small cars while introducing the Japanese-style "Lean production" to the U.S. auto worker (Jenkins, 2009). The Fremont plant had been closed in 1982 by GM due to the worse quality and productivity levels ever recorded in GM history. The workforce had absenteeism over 20% and was believed to have the worst labor relations of all the GM plants (Adler, 1996). When NUMMI opened, Toyota accepted representation by the UAW union mainly because of its history as a GM plant with a union workforce (Holusha, 1985). Consequently, a nonunion GM plant would also disrupt labor at other GM plants. By 1986, with the same workforce and comparable equipment, productivity was almost twice that of GM–Fremont in its best years and close to TMC's sister plant in Takaoka, Japan. NUMMI was producing the highest level of quality in the industry (Adler, 1996). Absenteeism decreased to around 3%, and participation in the suggestion program climbed to 90% in 1991. Job classifications went from over eighty to around five.

The UAW cautiously embraced the new labor–management model offered by Toyota. One of the main areas that Toyota focused on at NUMMI to engage the union was the use of teams (Figure 10.11). Both parties agreed that the company would utilize teams of five to ten members. All members of the team share the responsibility for the work performed by the team and participate in the quality/productivity improvement programs, such as QCs and kaizen. Also, team members are expected to rotate jobs within the team (NUMMI, 2001).

Teams were successful at NUMMI mainly due to the role of the team leader, which was also part of the bargaining unit. The process for team leader selection was extremely structured, including criteria such as hands-on job evaluation, job performance, safety, quality, and productivity. The selection was based on each employee's record, attendance, behavior, and overall attitude. The selection process evaluated how employees

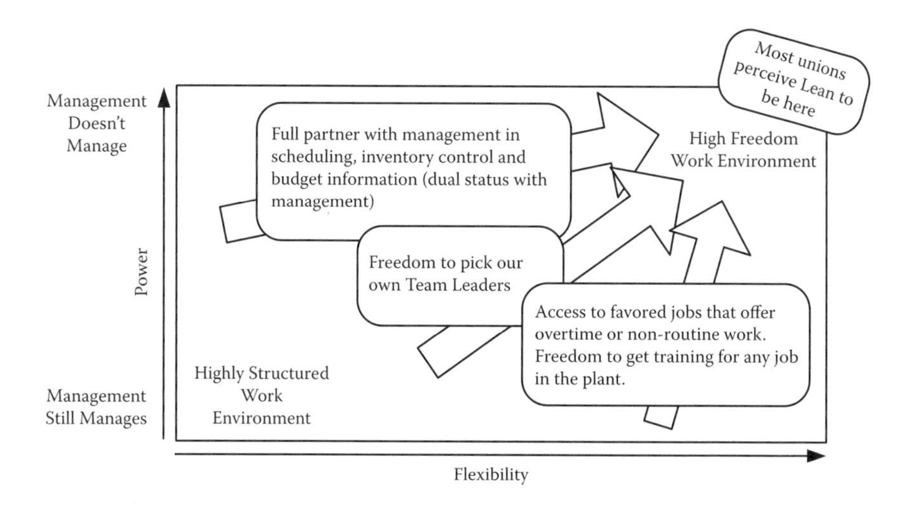

FIGURE 10.11
Most automotive unions perceive Lean to offer more freedom and autonomy in the workplace.

performed in people handling, how they motivated team members, and how they settled problems that related to trust and communication. Finally, NUMMI's team leader role would be entirely selected from a pool of team members who would enroll in a voluntary pretraining program, unpaid, and on their own time (NUMMI, 2001).

Other criteria in the NUMMI labor agreement emphasized how union workers could review problems or difficulties in achieving or developing standardized work. The labor agreement stated how problems with standardized work can be resolved and the escalation process when resolution exceeds a given time frame. Issues that escalate are handled by a committee consisting of both management and the union. The committee will review the problem and conduct a fact-finding investigation that will evaluate all aspects of the job, ranging from quality, tooling, and physical capabilities to training. If the problem is not resolved in five days, issues are escalated to a standardized work board (SWB), consisting of the vice president of manufacturing and HR, the international representative of the UAW, and the chairperson of the bargaining committee of the local union for resolution (NUMMI, 2001).

Last, the labor agreement provided conditions for how employees can challenge ergonomics, how to feel comfortable pulling the andon (i.e., a cord that team members can pull to call for help), and how to participate in kaizen. NUMMI's contract provides a detailed plan that illustrates how team members can transfer from areas within the unit and outside the unit when there are reductions or changes in manpower. In this situation, union workers know how they would change jobs and the types of jobs they could perform (NUMMI, 2001).

Possibly the biggest trust point for the union was in 1987. Just three years after the plant opened, capacity dropped to under 60%. It was at this time that most union workers believed that a layoff was inevitable. Instead, workers were put into extra training programs and put to work on kaizen projects and facilities maintenance jobs previously contracted out (Adler, 1996). It was at this critical juncture that the no-layoff philosophy and mutual long-term trust and respect became real for Local 2244.

10.3.6 The Influence of Labor Unions in Japan on Lean Production

Historically, the Japanese labor movement has been described as a harmonious work environment populated with loyal and satisfied workers

(Kuruvilla et al., 1990). In actuality, labor relations in the pre-postwar era resembled Western ones more at the outset of the twentieth century in the United States (Gordon, 1987). Japanese unions were strong in the prewar era and fought lengthy, and sometimes violent, battles over the independence of management's need to control their labor (Turner, 1995; Moore, 1983).

In the 1950s, the practices of scientific labor and technology allowed management to regain control over labor. Japanese workers and work groups became deskilled and were separated through automation. While management had seized control over productivity, there was the threat of opposition due to the isolated wage earner. To counteract commitment issues from workers, Japanese management used the nenko system to increase wages and guarantee employment. The nenko system is a class system in labor used to describe a worker's seniority and performance or merit in the company. Merit in Japan has always been related to performance and latent ability. It has been suggested that Japan's constantly evolving use of the nenko system has in effect been a substitute for the functions of a labor union (Makoto, 1996). The nenko class system was also based on the promise that any worker could have an open-ended career with the company. But, for a system like this to be plausible, it meant that the numbers to whom the promise was made had to be limited. Thus, the beginning of the team leader role started to take shape, which brought with it a highly selective process.

To develop the latent ability of workers and to remobilize work groups, a new form of work groups, named quality circles, was initiated. These new work groups were used to reinforce equality and solidarity among workers by challenging the self-centered behavior of the isolated wage earner. Interestingly, the QC movement and the changes to the nenko system emerged just when union activism started to center on the struggle for workplace control (Makoto, 1996). Historians suggested that workplace participative programs acted as a deterrent to offset changes in the Japanese labor movement.

10.4 INTERNAL REVERSIBILITY IN TPS: A UNION PERSPECTIVE

A theory that describes the two basic functions of unions is institutional economic (IE) theory. IE theory states that unions have two primary functions: economic and political (Freeman and Medoff, 1984). The economic function relates to the monopolistic view of unions; unions raise wages above competition levels by threat of strike. The political function of unions represents the larger employment aspects of work, such as a worker's status within a company and the worker's involvement in decision making (Kaufman, 2007).

Toyota's compensation program reflects a monopolistic view similar to that of unions mainly because of how the organization values its workforce. Toyota considers that wages should motivate and reward team members for full contribution and high performance. Annual wages for a Toyota team member are on average about 10% higher than the national UAW union worker. A wage theory that best describes Toyota's approach to compensation is adverse selection theory (AST). Unlike traditional fair wage models that try to retain employees by minimizing compensation, Toyota's monopolistic approach to high wages is based on recruiting employees based on competence. AST proposes that with higher wages, a larger pool of applicants can be attracted so that the company can more carefully select the types of employees is wants. Toyota's approach as a positive wage monopoly acts as a strong deterrent against unions because it relies on a compensation system not driven by turnover or training cost (Steel, 2002), shirking (Gintis, 1976), or effort (Akerlof, 1982).

Figure 10.12 illustrates several circumstances in which a company may want to pay a worker's wage that is above the market level. Wage efficiency models suggest that higher wages lead to more productivity, but not all components of human capital are observable by the company. All models make assumptions about employees and make cases for global optimality on local rationality. The mentality is if it works there, it must also work here.

If Toyota acts as a union in offering high wages beyond the average, then the question is whether workers differ in relevant ways or production processes give leeway for existing differences to be translated into greater productivity. The same argument could be applied toward Ford's $5-a-day wage in 1914. Why did Ford offer such high wages?

One argument is that Ford had turnover rates as high as 400% the year before the $5-a-day wage minimum. According to turnover theory, Ford would pay $5 to avoid the cost of retraining employees. Turnover theory proposes that the main factor in setting wages has to do with the time and cost of retraining employees. At Ford, an operator could be trained in about five to ten minutes from one or two demonstrations by the foreman

Turnover Theory	Shirking Wage Theory	Rent/Profit Sharing Theory	Gift-Exchange Theory	Adverse Selection Theory
Fear of retraining	Fear of slacking on the job	Fear of Collective Action	Effort Based	Competence Based
Workers will be less likely to quit their jobs if they are paid wages higher than they would get at another company. If the organization can pay wages above this minimum, workers will be less likely to quit their jobs.	This theory proposes that workers will be motivated to work harder (and not shirk – slow down) if they are given a wage higher than the average. The idea is that workers will be in fear of losing their job and will not slack because the wage is a decent wage.	Rent sharing theory proposes that an organization will have to pay higher than average wages because workers have exerted collective pressure. Higher wages act as quasi-rents.	Gift-Exchange Theory is a positive view of wage compensation. The theory states that higher wages are seen by workers as a gift and will return the favor in the form of hard work and effort.	When wages are well above the labor market equilibrium, more workers can be drawn into the organization. More workers allow the organization to be more selective. The goal, is to recruit workers that have the types of preferred skills favored by the organization.

Negative Wage Monopoly ⟺ Positive Wage Monopoly

Fair Wage Effort or Efficiency Models
(Compensation based on priority and preference)

FIGURE 10.12
Continuum of wage efficiency theories.

or a worker who was already trained in doing that job. Ford's radical division of labor does not support the ideas presented in turnover theory; however, turnover did drop to about 25% after wages increased, which made the threat of dismissal painful (Meyer, 1981).

Another argument is that Ford's high wage was meant to prevent shirking or slacking on the job. Some believed that because Ford's new production methods gave too many opportunities for shirking and discretion, high wages might serve as an incentive to individuals to refrain if there were some chance of getting caught. This argument is not probable mainly because of Ford's moving assembly line concept, which encouraged pace. Supervisors could look down the line and tell at a glance if there was trouble, mainly due to parts backing up (Raff, 1988). Solutions to problems were swift. Ford workers could not shirk even if they wanted to do so.

The most plausible case presented by the work of Raff (1988) is that Ford offered high wages because there was a threat of collective action. Collective action would interfere with the company's means of dealing with shirkers or anyone who wanted to interfere with the pace and coordination of work (Raff, 1988). Ford would have more difficulty dealing with groups rather than individuals, who could be instantaneously replaced and would do the work without hesitation. Researchers proposed that buying the peace (rent or profit-sharing theory) is a much more plausible economic wage theory model than turnover cost theory or shirking wage theory. Unionism was in the air, and Ford Motor Company was the most vulnerable in the industry, producing profits three times that of its competitor, GM, in 1913.

Toyota's approach to high wages, while holding some similarities to Ford, is much more focused on recruiting a certain type of worker. Toyota's approach to AST proposes that certain workers have more leeway in impacting organizational performance than others. An example of AST as it applies to the hiring process for a production worker at Toyota is shown in Figure 10.13. The selection process contains components for both hygiene and work motivation factors. Motivation factors that make TPS achievable relate to learning, roles, teaching, teaming, and decision making. Hygiene factors relate to the basic conditions of the work environment, how much a worker stands, the presence of mechanical noise, and a willingness to work overtime when needed. The more selective Toyota is at meeting these criteria, the more TPS is likely to be successful.

Toyota's approach to screen for certain worker traits bears questioning on the types of worker characteristics needed to keep TPS alive. For example, the more an employee favors hygiene factors compared to motivating factors, the less an employee is committed to the organization (Parkes and Razavi, 2004). Fukami and Larson (1984) have shown that role thinking or job autonomy is directly and positively related to organizational commitment. If workers are given opportunities to learn different types of jobs in different places every day, workers are generally more satisfied and more committed to the organization. Mathieu and Zajac (1990) have shown that workers who are more competent also report higher levels of organizational commitment, as do workers who are allowed to participate in decision making (Meyer and Allen, 1997). While these personality factors tend to be reported as deterrents to union ideas, they are not new to progressive HRM practices. The goal in HR, like TPS, is to get more employees involved and engaged in their job.

All team members and team leaders are compensated the same mainly because employees are never intended to compete horizontally for wages (Figure 10.14). Wages should be based on each employee's ability and encourage a spirit of challenge. In Japan, the wage system is similar to

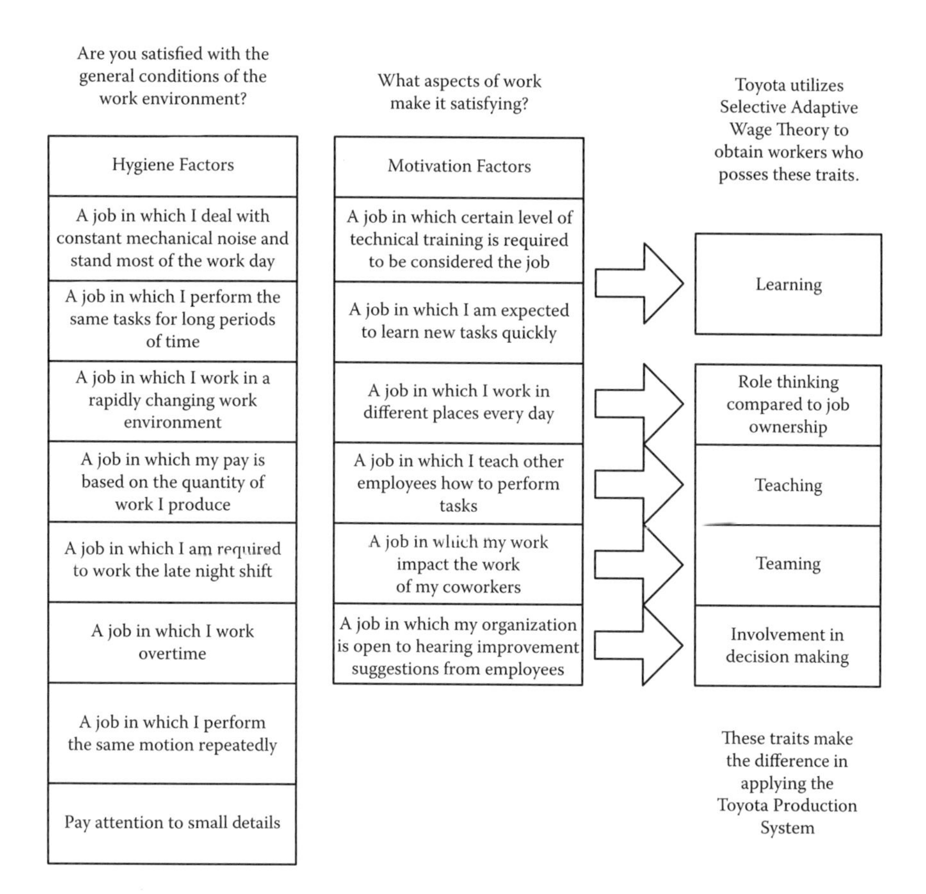

FIGURE 10.13
Selective adaptive wage theory: example of worker traits that make TPS possible.

that of the United States except there is a factor for age. The nenko system promotes seniority because in Japan it is a common belief that older workers should earn more than younger workers. Bonus wage allowances are the same for all team members regardless of the area where they work at the plant. If bonuses were different for each section or department, team members would switch to more favorable payouts. In that context, the compensation system would promote competition among members.

10.4.1 Larger Employment Issues

The second political function that unions represent is the idea that the governance of the system should be replaced with one that is more democratic and representative. When workers have a high degree of autonomy and decision making in the workplace, their rights are increased.

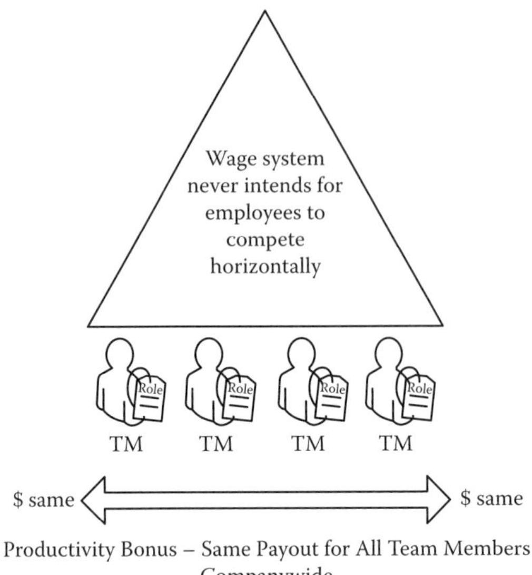

FIGURE 10.14
Wage system concept: no horizontal competition among members. TM, team member.

Similarly, when workers have a pathway to increase their rights (i.e., increased decision making in the workplace), employees can benefit from larger employment.

One of the ways that Toyota increases employees' rights is how they are promoted. The team leader role is unique because it represents a pathway for any worker to increase his or her rights as an employee. The team leader role provides growth opportunities for employees and the opportunity to develop their full potential. The team leader process is voluntary and is open to anyone who is interested in advancing within the company. Because the guarantee is open to anyone, the selection process is highly selective (Figure 10.15).

The team leader selection process has the goal to identify the most capable individual from the company. To be a team leader, a team member has to attend team leader training after work, without pay, to be put into an eligible pool. Team leader opportunities are communicated to interested and qualified employees. Employees with attendance rule violations within one year are not eligible to apply. Consideration is given to the group with the vacancy. This is important because the team leader has to know all the jobs in the area. Other areas of evaluation used to select team leaders are as follows:

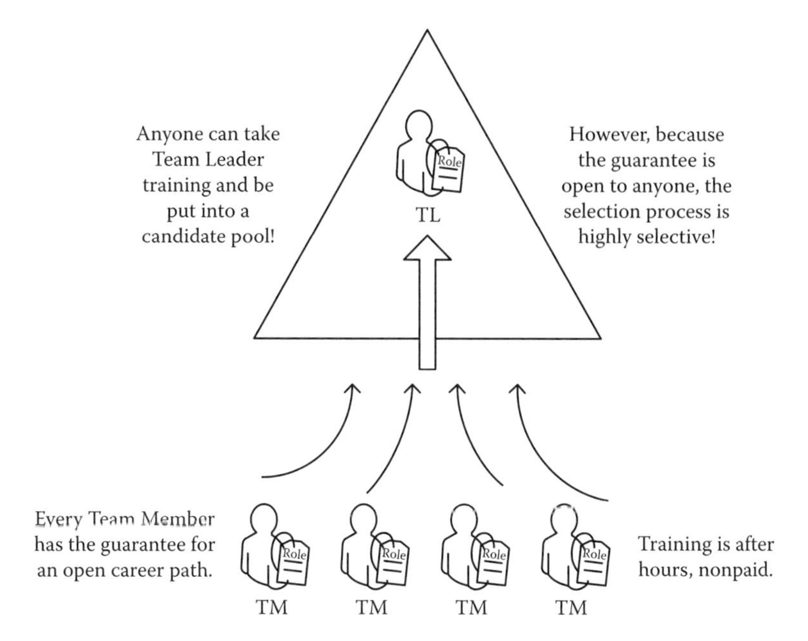

FIGURE 10.15
The open pathway to increase worker's rights.

1. Completion of a leadership skills evaluation
2. Job evaluation on all work jobs in the work area
3. Evaluation of standardized work
4. Assessment of attitude and behavior
5. Evaluation of attendance
6. Participation in the suggestion program
7. Role playing in terms of job assignment
8. People handling
9. Job knowledge
10. Training record, completed courses in standardized work, safety rules, kaizen concept, and problem solving

Team leaders who are promoted have a four-month evaluation period. If the team leader is unable to perform the role, the employee will be returned to his or her group as a team member (Figure 10.16).

The team leader role (Figure 10.17) is a successful medium to increase the rights of workers mainly because of the way the work environment is organized. The team leader has a span of control of four team members, which allows participation to be increased among the group. The ratio

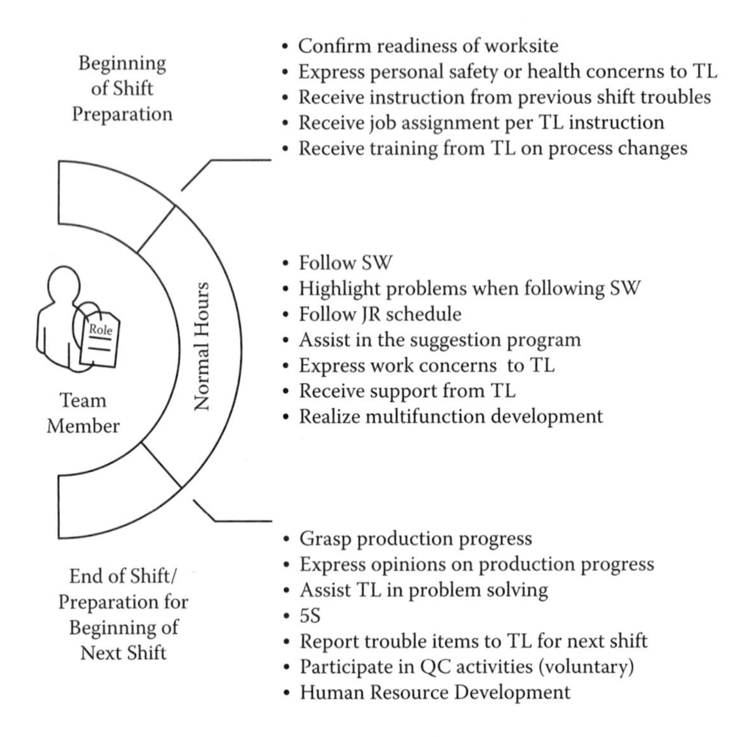

FIGURE 10.16
Role of the team member.

encourages a wide variety of desirable small team dynamics, one of which is improved communication. Team leaders are responsible for a variety of job functions, including team member training, emergency assistance, and initiating problem solving when issues arise. Because the position is quasi-tied to the production line, the team leader has the autonomy to analyze the larger aspects of the work environment. While all team members have the right to stop the line, the team leader is the only labor position that can establish the best way to perform the job (Figure 10.18).

Even if wages are high (which allows the organization to select an achievement-oriented worker) and there is a pathway to increase the workers' rights, the organization will still not function effectively if those ideas cannot be collected and aligned to the targets and goals of the company. Wages and rights only keep employees engaged at best to follow the organization's systems and practices, meaning that employees will buy in to the idea that working to the company's system is the best way to go. Wages and rights keep employees in a restoration state at which they

- Check log book
- Confirm trouble items
- Prepare worksite/report readiness to GL
- Report attendance to GL (make recommendations)
- Conduct warm-up exercises
- Start-up communications
- Perform Job Instruction Training (new people or new processes)

- Write Standardized Work (SW)
- Respond to andon calls
- Perform problem solving
- Meeting facilitation
- Check/give guidance on SW
- Assist in job rotation
- Perform checks: part quality, equipment, etc.
- Encourage Team Member (TM) contributions
- Maintain high morale between TMs

- Support Group Leader (GL) in communications
- Prepare for next shift
- Coordinate/support 5S
- Make shift log
- Lean QC meeting
- Promote QC themes

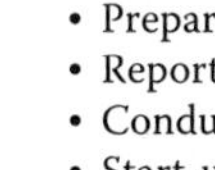

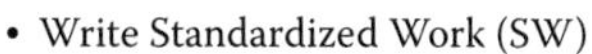

FIGURE 10.17
Role of the team leader.

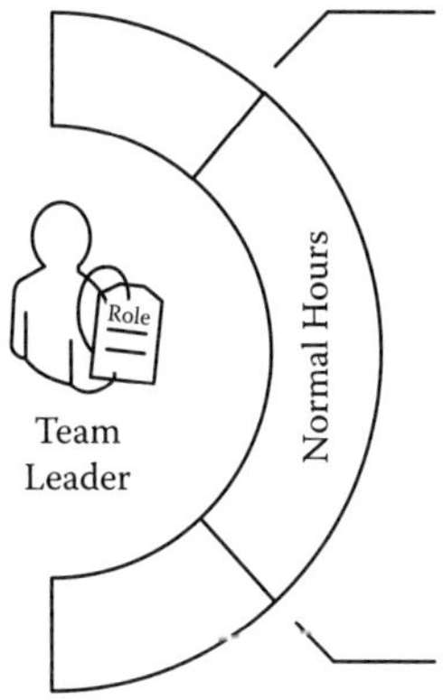

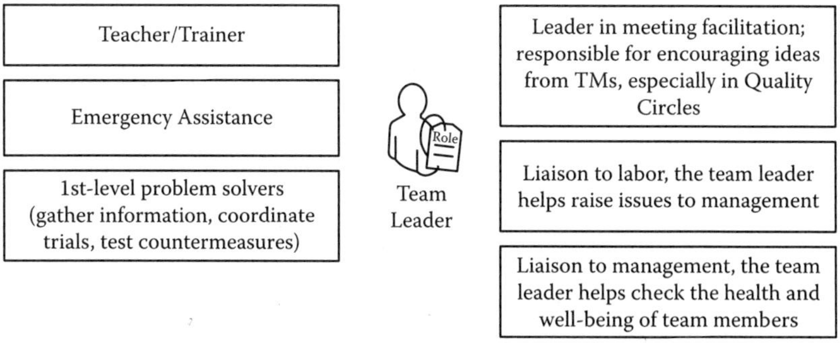

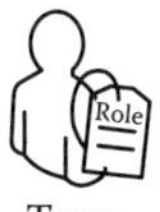

FIGURE 10.18
Summary: role of the team leader.

feel comfortable participating in workplace improvement, but it does not mean that the organization can benefit from the effort. While the system may be internally reversible, people willing, the entire system may still suffer from entropy.

10.5 EXTERNAL REVERSIBILITY IN TPS: A UNION PERSPECTIVE

The external aspects of reversibility in organizations relate to how well ideas from labor can be collected and converted into organizational outcomes. Ideas could represent how to solve a problem or how to improve a process. A mechanism that is used by Toyota to elicit worker's ideas for organizational outcomes relates to the role of the team leader. The team leader is responsible for eliciting opinions and suggestions from team members. Team leaders are a great conduit for bringing issues to management's attention and for checking the health and well-being of team members' attitudes and morale. One of the reasons why the team leader role is effective at establishing a conduit between labor and management can be explained by reviewing union participation theories.

Various participation theories have been proposed by social scientists to explain the behavior of union members (Fullagar and Barling, 1987). Participation is the extent to which members involve themselves and devote energy to the operation of their work group or to the union (Tannebaum and Kahn, 1958). Two common theories of union participation relate to the rationale and convenient aspects of participation. The theory of reason action (TRA) states that human beings are rational and make systematic use of information, meaning they will consider the implications of their actions before they decide to engage or not engage (Kuruvilla and Sverke, 1993). On the contrary, while someone may want to participate, there is still a question of how that can be achieved when there are so many obstacles barring a member from participating. Some researchers suggested that a more appropriate theory is the theory of planned behavior (TPB) (Conner and Armitage, 1998). TPB is the individual's perception of the extent to which participation is easy or difficult. The theory implies that employees will participate if their ideas can be easily tested or implemented.

Even if employees are willing to give their ideas, not all ideas are ready for implementation. In some circumstances, ideas may need to be refined before they can be applied. Trials may be need to be performed to test the feasibility of an idea, or information may need to be gathered to prove that the idea is effective. In all cases, not every idea can be implemented or should be implemented. Simply, management reserves the right not to approve trials if needed.

In terms of external reversibility, management needs a way to convert ideas without appearing as aggrandizing or limiting. If management interferes with kaizen and problem solving at the team member level, employees will feel less willing to participate. The question is, how can the rights of management be increased without limiting the employees' participation to kaizen and problem solving?

One way that Toyota has increased the rights of management is through a highly structured protocol defined in the role of the team leader. The protocol uses two techniques for filtering information flow from team members. First, before an idea can be generated there has to be a trigger or mechanism that signals employees that the state of the system needs improvement. Toyota accomplishes this by separating normal work and abnormal work. Normal work is the way the job should be performed when there are no problems. Toyota uses the term *standardized work* to describe how the job should be performed. At Toyota, team members perform normal work. Abnormal work, such as andons or repairs, is work that is unplanned. At Toyota, the team leader performs abnormal work. When an employee performs both types of work (normal and abnormal), all work looks to be productive. Toyota separates normal and abnormal work so that jobs can be easier managed and evaluated.

Normal versus abnormal thinking is also a technique that helps to limit and focus problem solving and kaizen. Normal and abnormal thinking draw attention to standardized work and help to target creativity for improving consistency of operations at the team member level. The protocol encourages team members to follow a system (i.e., normal work) and to highlight problems when there is abnormal work (Figure 10.19).

Another technique that is used in combination with abnormal work is the concept of early detection, early resolution (Figure 10.20). The andon is the simplest and easiest way to call for help when normal work cannot be completed. Team leaders have the role to respond to andon pulls, to grasp what is going on, and to find the causes that made the system go from normal to abnormal (Figure 10.21). At this point, the system is officially alerted, which engages a series of steps to diagnose and troubleshoot the abnormality. Andon tools are effective when they give warning to the

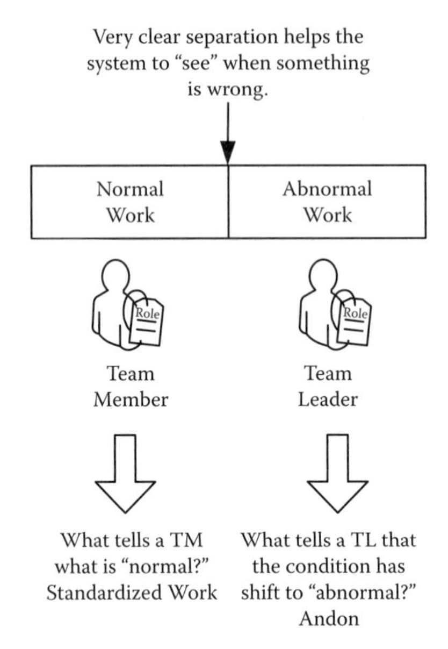

FIGURE 10.19
Separating the roles of normal and abnormal work.

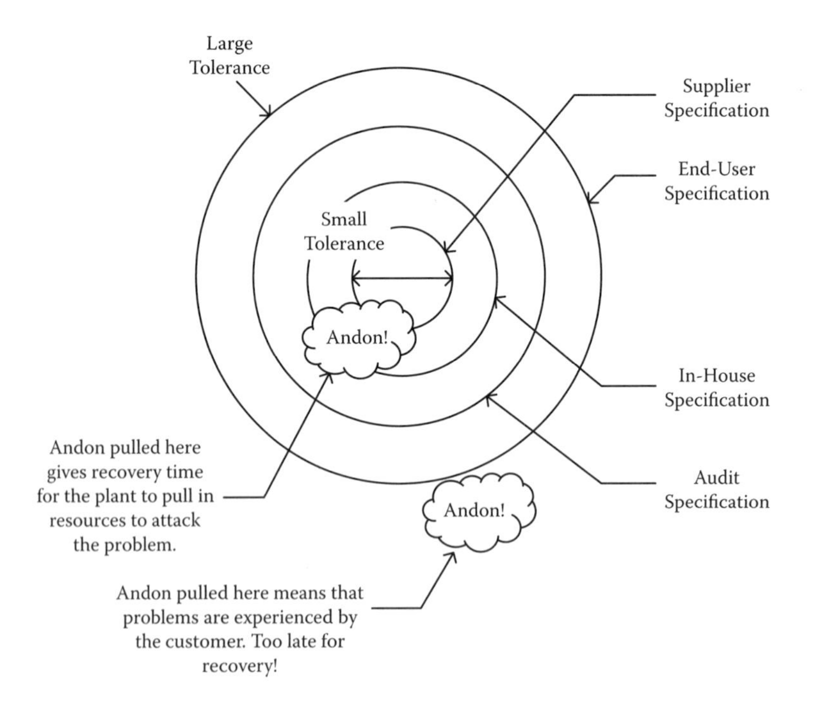

FIGURE 10.20
Early detection, early resolution.

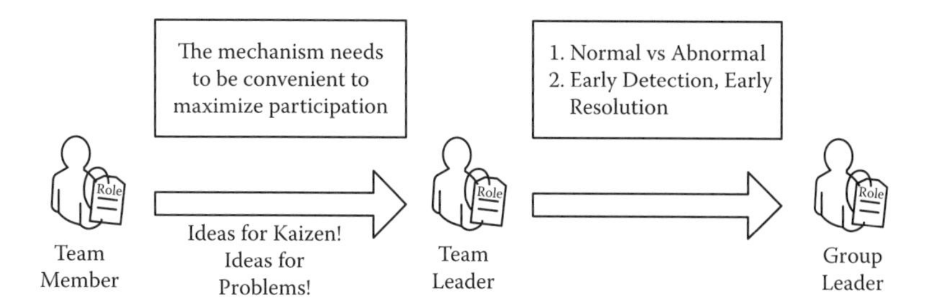

FIGURE 10.21
Coordinating and filtering ideas for kaizen and problem solving.

system before the operation slows or shuts down (i.e., early detection). Toyota believes that problems that are detected early can be solved easier than problems that are allowed to grow (i.e., early resolution).

Andon thinking (i.e., highlighting the problem thinking) is a way to encourage prevention rather than reaction for soliciting concerns from team members.

10.6 SUMMARY

The purpose of this chapter was to discuss the reversibility of systems as it relates to TPS. When energy can be converted in two directions—forward and backward—the efficiency of that system can be measured in terms of reversibility. The more reversible a system behaves, the more efficient it is at converting energy from one form to another. A reversible system is one that idealizes a perfect system that does not produce entropy or waste. The reversibility of systems is an important tool in system analysis because a totally reversible system represents the best possible conditions that a system can run without entropy.

In organizations, reversibility can best be described as the way labor and management work together or dominate one another (reversibility) in meeting organizational outcomes.

Since TPS is a system that tries to increase the rights of workers, union theory was used to analyze the ways TPS behaves as a form of organized labor. While most HRM theories only describe the positive or cooperative interactions between labor and management, union theory demonstrates a more realistic picture by describing the struggles that each side goes though in meeting organizational outcomes.

TPS shares characteristics of organized labor, mainly how workers are treated as a monopoly at Toyota, meaning they receive high wages and exercise greater control of the workplace. TPS satisfies these two union functions so that employees can be open to participating in kaizen. The way that management tries to increase their rights in the relationship is strongly connected to the way problems or ideas enter into the organization. Not all ideas can enter into TPS, and not all are allowed. While this may sound aggrandizing or limiting, Toyota views this as targeted creativity. The best way to accomplish targeted creativity is to use a division of labor (namely, the team leader) to filter good and bad ideas by coordinating issues. Because issues are trapped, both management and labor can reach consensus quicker because the problem is seen by both parties when it occurs. This low-filtering approach raises management's rights in the workplace and encourages only a specific set of ideas from ever entering the system.

REFERENCES

Adler, P. (1996) Teams at NUMMI, Teamwork in the automotive industry, draft chapter for *Teamwork in the Automotive Industry*, edited by Jean-Pierre Durand, Juan José Castillo, and Paul Stewart (https://msbfile03.usc.edu/digitalmeasures/padler/).

Akerlof, G. (1982) Labor contracts as partial gift exchange. *Quarterly Journal of Economics*, Vol. 97, 543–569.

Akerlof, G. (1990) The fair wage effort hypothesis and unemployment. *Quarterly Journal of Economics*, Vol. 105, 255–284.

Ariovich, L. (2007) Organizing the organized. Structural change, organizing strategies, and member participation in a union local of service workers, dissertation, Northwestern University, Evanston, IL.

Babson, S. (1993) Lean or mean: The MIT model and production at Mazda. *Labor Studies Journal*, Vol. 18, No. 2, Summer, 3–24.

Banks, A. (1984) Pensions, participation and labor education. *Labor Studies Journal*, Winter, Vol. 8, No. 3, 244–261.

Bhatti, K. and Qureshi, T. (2007) Impact of employee participation on job satisfaction, employee commitment and employee productivity. *International Review of Business Research Papers*, Vol. 3, No. 2, June, 54–68.

Berggren, C. (1992) *Alternatives to Lean Production: Work Organization in the Swedish Auto Industry*, ILR Press, Ithaca, NY ILR Press.

Cengel, Y. and Boles, M. (1989) *Thermodynamics: An Engineering Approach*, McGraw-Hill, New York, NY.

Cole, R. (1984) Some principles concerning union involvement in quality circles and other employee involvement programs. *Labor Studies Journal*, Winter, Vol. 8, No. 3.

Commons, J. (1905) *Trade Unionism and Labor Problems*, Ginn, Boston.

Conner, M. and Armitage, C. (1998) Extending the theory of planned behavior: A review for further research. *Journal of Applied Social Psychology*, Vol. 28, No. 15, 1429–1464.

Eaton, A. (1995) New production techniques: Employee involvement and unions, *Labor Studies Journal*, Fall, Vol. 20, No. 3, 19–41.

Freeman, R., and Medoff, J. (1984) *What Do Unions Do*? Basic Books, New York.

Fukami, C. V. and Larson, E. W. (1984) Commitment to company and union: Parallel models. *Journal of Applied Psychology*, Vol. 69, No. 3, 367–371.

Fullagar, C. and Barling, J. (1987). Toward a model of union commitment. *Advances in Industrial and Labour Relations*, Vol. 4, 43–78.

George, T. and Hancer, M. (2003) The impact of selected organizational factors on psychological empowerment of non-supervisory employees in full-service restaurants. *Journal of Foodservice Business Research*, Vol. 6, No. 2, 35–47.

Gintis, H. (1976) The nature of labor exchange and the theory of capitalist production. *Review of Radical Political Economics*, Vol. 8, No. 2, 36–54.

Gordon, S. (1987) *The State and Labor in Modern Japan*, University of California Press, Berkeley.

Grenier, G. (1988) Quality circles in a corporate antiunion strategy: A case study. *Labor Studies Journal*, Summer, Vol. 13, No. 2, 5–28.

Hall, A. (2006). *Introduction to Lean Sustainable Quality Systems Design: An Integrated Approach From the Viewpoints of Dynamic Scientific Inquiry Learning & Toyota's Lean System Principles and Practices*. Published by Arlie Hall, Lexington, KY. ISBN 0-9768765-0-7.

Hoell, R. (2004) How employee involvement affects union commitment. *Journal of Labor Research*, Vol. 25, No. 2, 267–278, April.

Holusha, J. (1985) Kentucky site picked by Toyota, a union role still at issue. *New York Times*, December 12, p. D9.

Ichniowski, C., Shaw, K., and Prennushi, G. (1997) The effects of human resource management practices on productivity: A study of steel finishing lines. *American Economic Review*, Vol. 87, No. 3, 291–313.

Jenkins, H. (2009) The meaning of nummi. *Wall Street Journal*, October 7, p. 19.

Kaufman, B. (2007) What do unions do? Evaluation and commentary, in *What Do Unions Do? A Twenty Year Perspective*, ed. Bennett, J. and Kaufman, B., Transaction Publishers, New Brunswick, NJ. pp. 520–562.

Kaufman, B. (2010) Institutional economics and the theory of what unions do. Working Paper 201002-2, February. http://aysps.gsu.edu/usery/Papers.html.

Kochan, T., Katz, H. C., and McKersie, R. C. (1986) *The Transformation of American Industrial Relations*, Basic Books, New York.

Kuruvilla, S., Gallagher, D., Fiorito, J., and Wakabayashi, M. (1990) Union participation in Japan: Do Western theories apply? *Industrial and Labor Relations Review*, Vol. 43, 374–389.

Kuruvilla, S. and Sverke, M. (1993) Two dimensions of union commitment based on the theory of reasoned action: Cross cultural comparisons. *Research and Practice in Human Resource Management*, Vol. 1, 1–16.

Lay, J. (1963) *Thermodynamics*, Merrill Books, Columbus, OH.

Liker, J. and Hoseus, M. (2010) Human resource development in Toyota culture. *International Journal of Human Resources Development and Management*, Vol. 10, No. 1, 34.

Makoto, K. (1996) *Portraits of the Japanese Workplace: Labor Movements, Workers and Managers*, Westview Press, Boulder, CO.

Mann, D. (2005) *Creating a Lean culture: Tools to sustain Lean conversions*, CRC Press, Boca Raton, FL.

Mathieu, J. E. and Zajac, D. (1990) A review and meta-analysis of the antecedents, correlates, and consequences of organizational commitment. *Psychological Bulletin*, Vol. 108, 171–194.

Meyer, J.P. & Allen, N.J. (1997). *Commitment in the Workplace: Theory, Research, and Application*. Sage Publications,Thousand Oaks, CA.

Meyer, S. (1981) *The Five Dollar Day: Labor Management and Social Control in the Ford Motor Company 1908–1921*, State University of New York Press, Albany, NY.

Moore, J. (1983) *Japanese Worker and the Struggle for Power*, University of Wisconsin Press, Madison.

New United Motors Manufacturing (NUMMI) (2001) Agreement between New United Motor Manufacturing, Inc., and the UAW, August 4th, labor agreement, unpublished.

Parkes, K. and Razavi, T. (2004) Personality and attitudinal variables as predictors of voluntary union membership. *Personality and Individual Differences*, Vol. 37, 333–347.

Raff, D. (1988) Wage determination theory and the five dollar day at Ford. *Journal of Economic History*, Vol. 48, No. 2, 378–399.

Saturn (2000) Memorandum of agreement, Saturn and the UAW, labor agreement, unpublished.

Steel, R. (2002) Turnover theory at the empirical interface: Problems of fit and function. *Academy of Management Review*, Vol. 27, No. 3, 346–360.

Tannenbaum, A. and Kahn, R. (1958) *Participation in Union Locals*. Harper and Row, Evanston, IL.

Turner, C. (1995) *Japanese Workers in Protest: An Ethnography of Consciousness and Experience*, University of California Press, London.

Weathers, C., and North, S. (2010) Overtime activists take on corporate titans: Toyota, McDonald's and Japan's work hour controversy, *Pacific Affairs*, Vol. 82, No. 4.

Webster, E. (2008) Recasting labor studies in the twenty-first century. *Labor Studies Journal*, Vol. 33, No. 3, 249–254.

Womack,J., Jones, D., and Roos, D. (1990). *The Machine that Changed the World: The Story of Lean Production*, HarperCollins, New York.

Yanarella, E. and Green, W. (1994) The UAW and CAW confront Lean production at Saturn, CAMI, and the Japanese automobile transplants. *Labor Studies Journal*, Vol. 18, Winter, 52–75.

11

The System Property of Negative Entropy in TPS

11.1 SYSTEM PROPERTY: NEGATIVE ENTROPY

In previous chapters, entropy was introduced as a depreciated form of energy that is inherently present in closed systems. Entropy represents a state of energy that is in less of a natural or ready form mainly because conversion processes are not completely reversible (Bailly and Longo, 2009). Entropy is a state of disorder that describes how systems decrease in structure and in organization (Kim, 1975).

For a system to survive, new energy or negative entropy must be received from its environment (Becvar and Becvar, 1999). When a system receives useful energy, its ability to organize, structure, and carry out its essential functions become possible. Without useful inputs, systems tend to approach maximum disorder. All systems need useful energy inputs to replenish and counteract the forces of entropy. A system that cannot import new energy or evolve (i.e., negative entropy) is left helpless to its surroundings. Simply, a system cannot survive if it cannot take on negative entropy from its environment to become more organized and structured to complete its basic functions.

Negative entropy or negentropy describes the tendency away from disorder toward a state of order and balance (Figure 11.1) The organization of a system is held to be the opposite of entropy. Some authors prefer a more positive term, such as evolution, rather than negative entropy because evolution means increasing complexity and higher organization (Putt, 1978). Organizations obey the second law of thermodynamics and have to compensate for their high entropy by continuously feeding on negative entropy.

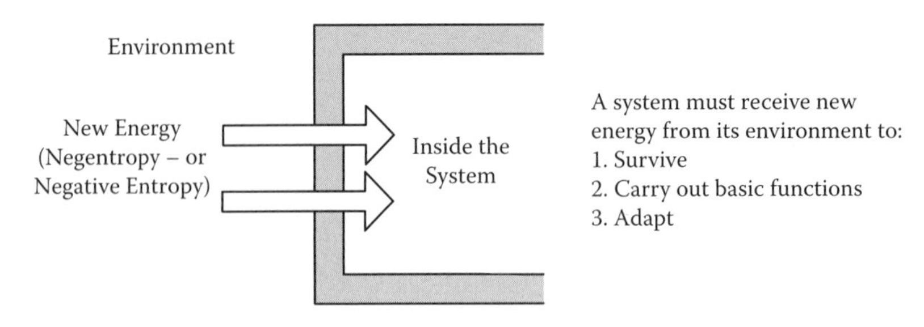

FIGURE 11.1
The purpose of negative entropy.

11.2 NEGATIVE ENTROPY AND THE TOYOTA PRODUCTION SYSTEM

In the Toyota Production System (TPS), negative entropy can best be described as employee learning (Figure 11.2). TPS cannot carry out its most important functions (problem solving and industrial engineering) if employees are not interested in pursuing kaizen. Even if employees are forced to do problem solving and industrial engineering, learning will be minimal because the dimensions of feeling and wanting are not present. Toyota believes that creativity is likely to manifest itself wherever one finds a new way of personally improving a situation. In this context, creativity is driven from personal interest rather than an external or formal manner of motivation. The fuel that drives creativity at Toyota is freedom to learn.

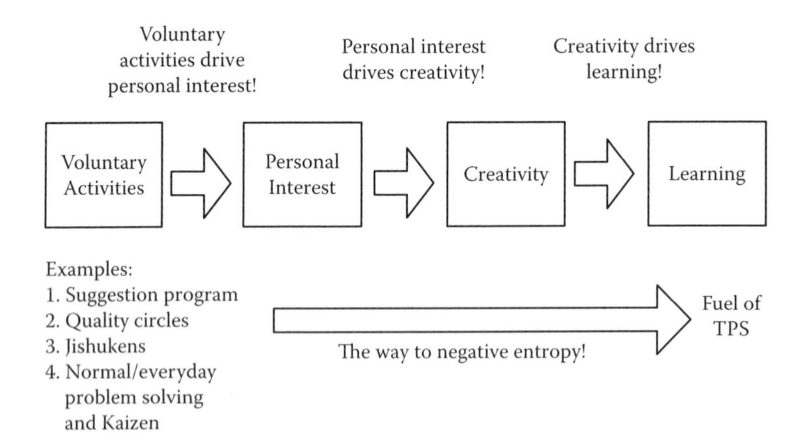

FIGURE 11.2
The negative entropy of TPS.

11.2.1 Exceptional Creativity, Mundane Creativity, and Lifelong Learning

When employees connect with TPS, they become a conduit of negative entropy for analyzing and improving the workplace. The challenge is establishing the way employees can use their creativity during kaizen. Researchers have shown that the best way to promote creativity is largely dependent on the level of autonomy in a worker's job (Amabile, 1997) and the level of support by their immediate supervisor to be creative (Sjin and Zhou, 2003). An organization can unleash creativity in the workplace when the constraints of conventional socialization can be weakened and there exists a challenging experience to increase a person's capacity to preserve in the face of obstacles (Simonton, 1994). Sternberg (1985) has shown that a worker is likely to demonstrate creativity when the following are evident:

Unconventionality. Demonstrates the ability to make up rules as he or she goes along; the person has a free spirit and is unorthodox and questions societal norms, truisms, and assumptions.

Imagination and integration of ideas. Can make connections and distinctions between ideas and things; has the ability to recognize similarities and differences; is able to put old information and theories together in a new way.

Flexibility and intuitiveness. Follows his or her gut feelings in making decisions after weighing the pros and cons; has the ability to change directions and use other procedures.

Drive for accomplishment. Motivated by goals; likes to be complimented on his or her work and is energetic.

Researchers categorize creativity into two different types: exceptional and mundane. Exceptional or extraordinary creativity is described as a rare form of creativity that requires long deliberation and considerable effort. Exceptional creativity is influential and tends to involve original or groundbreaking forms of expression (Barsalou and Prinz, 1998). Mundane creativity is more common, not groundbreaking, rare, or influential. Mundane creativity is performed with little effort or deliberation and is important for the individual but rarely important from a social perspective. Mundane creativity deals with combining concepts to function in everyday life (Turner, 1995). For exceptional creativity to exist, researchers

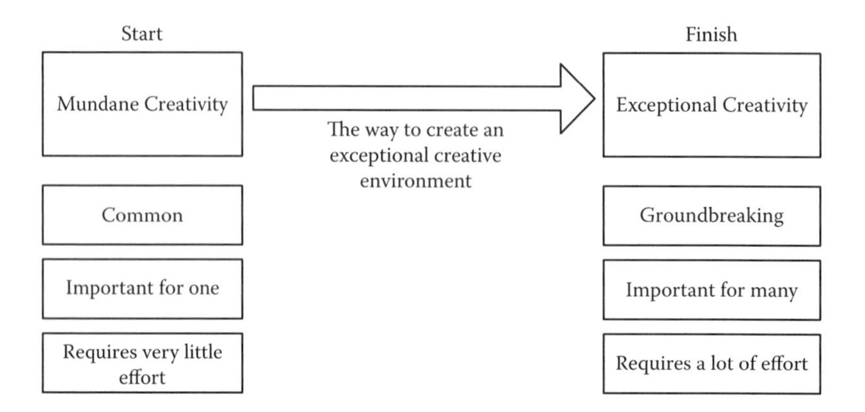

FIGURE 11.3
Traditional view on creativity and the path to exceptional creativity.

suggest that the work environment must at least foster mundane creativity (Figure 11.3). Consequently, when workers are not in a free state for small and creative ideas, the organization cannot expect to achieve influential and dramatic ideas on a larger scale.

Toyota's view of creativity is somewhat different. Toyota would agree that mundane creativity is a form of common creativity that should occur frequently, much like the concept of kaizen, for which workers are in a continuous state to find new ways to improve their work (Figure 11.4). However, Toyota would disagree that ordinary forms of creativity or workplace kaizen should not be shared, that it is important for one person to benefit. Toyota is constantly looking for ways to share ideas through *yokotan* (in Japanese, this means "across everywhere") and standardization. If

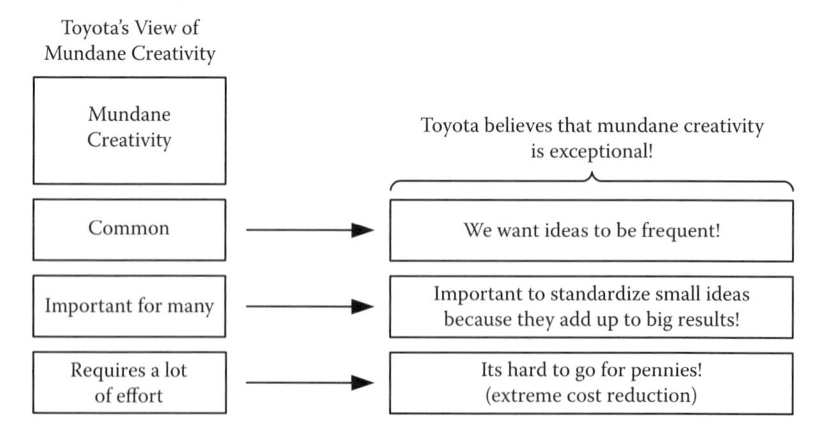

FIGURE 11.4
Toyota's view of mundane creativity.

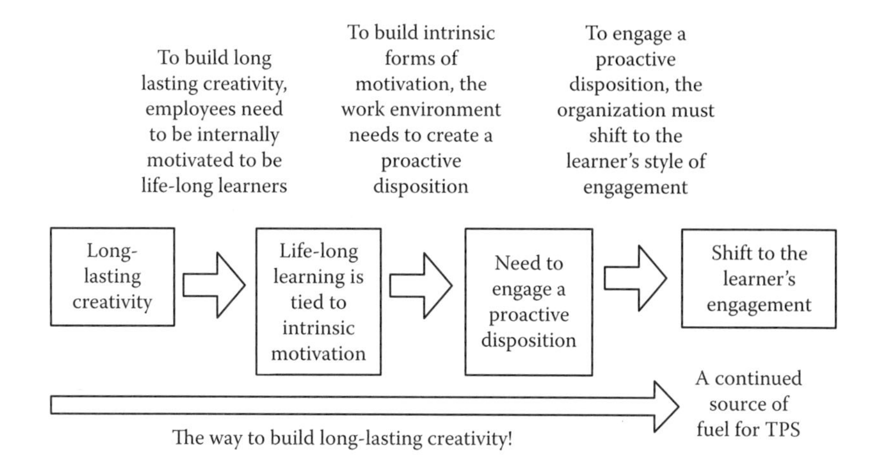

FIGURE 11.5
The way to long-lasting creativity (fuel) for TPS.

one small idea can be shared across the entire plant, Toyota would consider this to be exceptional, not mundane. There is also the belief that mundane creativity does not require much effort. Toyota believes that ordinary forms of creativity are extremely challenging because most processes have already gone through some form of refinement. A work environment that promotes a kaizen mindset would imply that workers must become extremely creative to find new ways to eliminate waste. In other words, searching for ways to eliminate pennies in an already-refined work environment is much more challenging than starting fresh in a work environment that has not completed any kaizen.

If an organization agrees with Toyota's view of creativity, it can be agreed that mundane or exceptional forms of creativity should be long lasting (Figure 11.5). Researchers suggested an environment that promotes long-lasting creativity is also one that promotes lifelong learning. Lifelong learning is a characteristic of an evolving system that constantly has to adapt due to uncertainty and changing surroundings. Individuals who are internally motivated are often associated with lifelong learners because they demonstrate a proactive disposition to create. Everyone is creative in some manner; however, long-lasting creativity is often centered around the principle of passion. Unlike cognitive dimensions of reasoning, the more feeling one associates with creative learning, the more likely one is to engage in a creative learning process that is long lasting and runs deep (Sternberg, 1985). To help this engagement, teachers are trained to shift from knowledge content to the engagement of the learner. The best teachers are those who can connect to each individual student's learning and communication style. Toyota believes that mundane creativity is the workhorse that accounts for the bulk of negative entropy for TPS, but for it to be long lasting, the company has to find ways to accommodate the learner.

11.3 ORGANIZATIONAL LEARNING AND COMMUNICATION: SOME BASIC SIMILARITIES WITH TPS

Toyota has been claimed as a learning organization because it has the ability to share information across groups and departments (Figure 11.6) (Adler and Cole, 1993). Toyota regularly practices concepts such as genchi genbutsu (also known as "go and see") to gather facts (Liker and Hoseus, 2008); standardization to share ideas; and a systematic problem-solving process that is helpful for keeping problems from returning (Spear, 2010). Possibly, the most well-known examples of Toyota's approach to learning

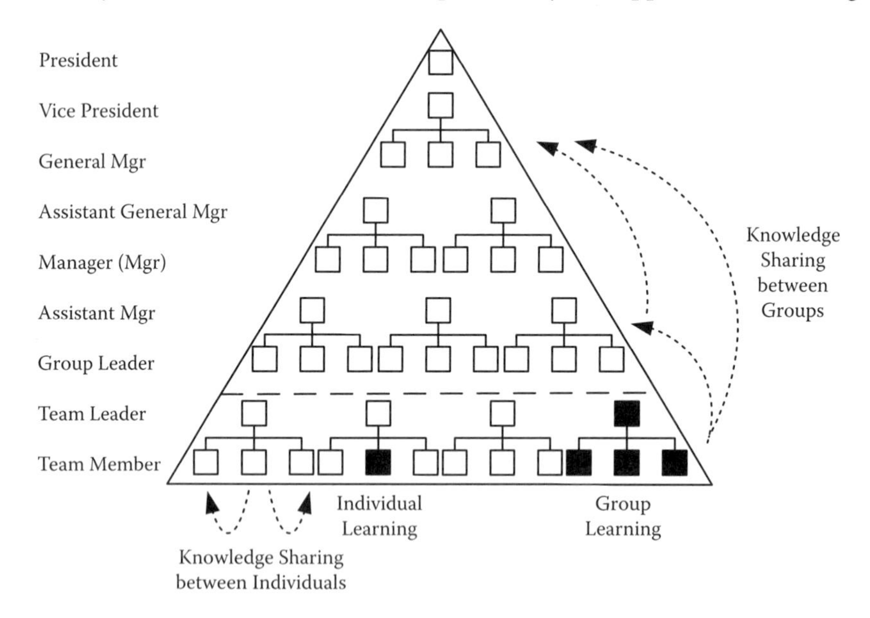

FIGURE 11.6
Adler and Cole's (1993) concept of organizational learning: sharing knowledge between groups and individuals.

at the workplace are their human resource development (HRD) practices. Few companies emulate Toyota's philosophy of promoting from within, which encourages in-house training, and its concept of a multifunctional worker, which for Toyota is at the heart of worker satisfaction and motivation (Hall, 2006).

11.4 LITERATURE REVIEW OF ORGANIZATIONAL LEARNING AND COMMUNICATION

Organizational learning (OL) is a diverse and multidisciplinary topic that has been described in terms of organizational psychology, sociology, anthropology (Czarniawska, 2001), political science, management science (Garrick and Clegg, 2000), organizational behavior (March and Simon, 1958), and systems engineering (Cohen and Levinthal, 1990). OL has been defined as a means of improving actions through better knowledge and understanding (Fiol and Lyles, 1985) and increasing organizational members' capacity to change the things they really care about (Karash, 1995). One of the most well-known definitions of OL is from Peter Senge's best-seller, *The Fifth Discipline*: Learning organizations are where people continuously expand their capacity to create the results they truly desire, where new and expansive patterns of thinking are nurtured and where people are learning to see the whole together.

The most appealing aspect of OL is that companies pursue it because they believe it will allow them to be more adaptive, flexible, and responsive to environmental changes and demand (Yang, 2003). Companies are increasingly facing turbulent and unpredictable markets and business circumstances (i.e., technological change, shorter product life cycles, etc.) for which becoming adaptive is appealing, which means if you can adapt, you can survive (Bennet and Bennet, 2008).

Adaptability is not the only reason for applying the concept of OL. OL increases a firm's competitive advantage, performance (Choe, 2004), and capability (Campbell and Cairns, 1994) by knowledge sharing (Nonaka and Takeuchi, 1995). OL also helps resource utilization (Smith et al., 1996) and the achievement of organizational goals by improving organizational strategy and implementation. One of the most fashionable reasons for applying organization learning is to promote innovation (McGrath,

2001) using organizational control (Kidd, 1998) and knowledge creation approaches (Choe, 2004).

The views of how organizations actually achieve OL vary. Several characteristics of a learning organization are illustrated in Figure 11.7. Initial ideas about OL showed that individuals must have a structured connection to the organization as the individual learns while trying to achieve the outcomes of the organization (Simon, 1969). This view stresses the importance of sharing and distributing know-how and information (Ong, 1982).

Other views of OL relate to how tacit information (which is knowledge that is difficult to articulate or put into words) is converted and decoded

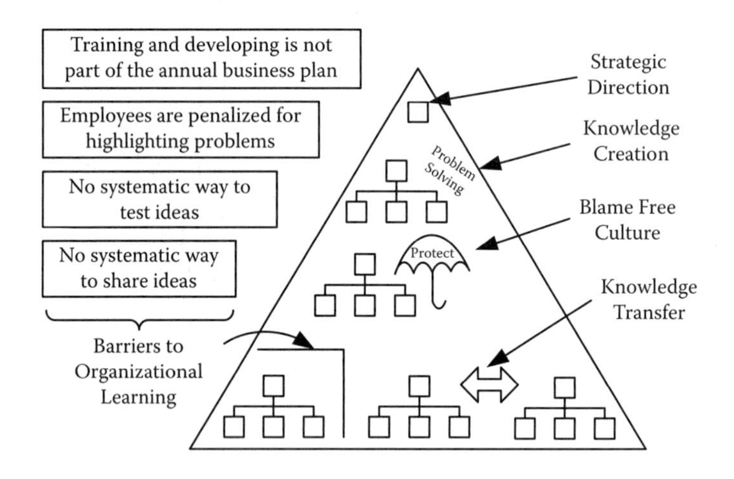

Strategic Direction	Learning is a deliberate aspect of the organization's business strategy. Formal approach to stimulating OL aspects into individual business units, divisions, and plants.
Knowledge Direction	The creation of new knowledge is seen as central to the organization. The organization finds ways of capturing new knowledge by promoting systems and techniques that document information in usable form.
Blame Free Culture	Learning is valued and encouraged which encourages experimentation without the fear of failure, humiliation, or blame. Doesn't imply that accountability does not exist, it only means that managers are going to promote ideas to be tested without blame of failure.
Knowledge Transfer	Knowledge and information can be easily learned and applied from other parts of the organization. Transferring knowledge allows successful ideas to be widely disseminated.

FIGURE 11.7
Structured views about organizational learning and some common barriers.

Tacit Knowledge	Explicit Knowledge
1. Personal	1. Social
2. Informal	2. Formal
3. Art form	3. Science
4. Apprenticeship based	4. Training based
5. Feeling and intuition	5. Methodological
6. Crude	6. Precise
7. Organic	7. Linear and mechanistic

FIGURE 11.8
Comparison between tacit and explicit knowledge.

into usable information known as explicit knowledge (Figure 11.8) (Nonaka and Takeuchi, 1995). Tacit knowledge, originally describe by Polanyi (1967), is initially knowing what decision to make or how to do something that cannot be clearly coded or articulated. Various ideas have been generated to help explain how tacit and explicit knowledge can be shared. Nonaka and Takeuchi (1995) gave several channels for how knowledge is transferred, such as socialization, externalization, combination, and internalization. Each of these mechanisms provides an interesting view of how knowledge is shared in various settings, such as learning in groups, by individuals, or by habitualization.

One of the most productive views of applying OL has been through the use of the human resources (HR) function. OL encapsulates a number of concerns and developments in HR management for bringing out the best in organizations and individuals (Jones and Hendry, 1994). The HR department is instrumental in promoting trust among employees (Denton, 1998; Sako, 1992) and for relieving defensive mechanisms that exist in organizations (Argyris and Schon, 1996). The work of Argyris and Schon showed that individuals are programmed with model 1 theories in use, which means that all individuals have the wish to be in control. This means that when individuals are faced with threatening or embarrassing issues, they act and reason in defensive ways. The problem is more amplified when trust does not exist. To overcome defensive behavior, organizations must create an environment of open exchange of valid information in which ideas can be tested publically (Antonacopoulou, 2006). For this reason, HR is typically charged with creating programs and incentives to encourage workplace learning and to find ways to strengthen the ways people work together (Dodgson, 1993).

Unfortunately, even with the best HR programs, it is still difficult to know how exactly to achieve OL. Organizations are extremely complex systems, consisting of personalities, small groups, intergroups, norms, values, and attitudes (Argyris, 1964). Organizations are open systems; they are influenced by the environment and are dynamic, which means output does not necessarily change with any change to the input. There is also no fixed constant relationship between output and input. Consequently, the future of an organization cannot be predicted by knowing its current state. Therefore, there is no one-size-fits-all best approach to OL (Reynolds and Vince, 2007).

11.5 ORGANIZATIONAL LEARNING THEORY

A more holistic approach in understanding OL is the use of theoretical models to describe various relations among concepts within a set of boundary assumptions and constraints. OL theory is largely based on frameworks derived from other disciplines and fields of research. OL theory is deeply rooted in psychology, education, cognition, and communication. Most OL theory is based on how individuals learn and communicate. A common area of study is the use of theoretical frameworks to analyze an individual's particular disposition toward learning or communication. Since most scientists agree that individual learning is the prerequisite to OL, research continues to center around the needs of the individual (Argyris, 1964; Bandura, 1977; Senge, 1990; Antonacopoulou, 2006; Gully and Chen, 2010).

Two of the most common frameworks for understanding how individuals learn and communicate are Kolb's learning inventory list (Kolb, 1984) and Merrill and Reid's communication style model (Merrill and Reid, 1991). These two frameworks combined offer an interesting insight into understanding a particular individual's preference in learning and communication. While no single preference is considered to be superior over another, the idea is that knowledge transfer can occur easier and more effectively if one is aware of another's style. Flexibility has been shown to distinguish the success of a manager in communication and learning with peers, supervisors, and subordinates. The use of models can help ensure that all styles of learning or communication have been addressed. Similarly, all the ways to communicate or learn have been presented to address each student's particular style.

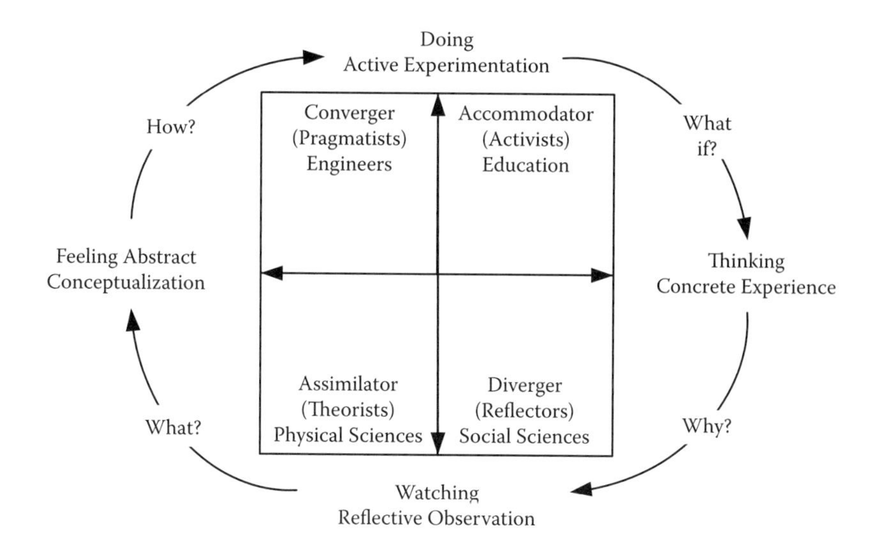

FIGURE 11.9
Adapted from Kolb's individual learning style grid. (Adapted from Kolb, D., 1984, *Experiential Learning*, Prentice Hall, Englewood Cliffs, NJ.)

Kolb analyzed learning styles using a learning style inventory, which measures the relative emphasis an individual learner attaches to concrete experience (thinking), reflective observation (watching), abstract conceptualization (feeling), and active experimentation (doing) (Figure 11.9) (Kolb, 1984). Kolb further classified these styles of learning as diverging, assimilating, converging, and accommodating.

Diverging (feeling and watching): Divergers are individuals who are able to look at a situation from various perspectives.

1. They watch and gather information.
2. They are open minded and sensitive and listen actively.
3. They are good in situations that require various ideas (i.e., brainstorming).
4. They have broad interests and are imaginative and emotional.
5. They prefer to work in groups.

Assimilating (watching and thinking): Assimilators are logical and concise thinkers:

1. They are excellent at organizing and processing information.
2. They are less focused on people.

3. They prefer theory rather than practical ideas.
4. They are analytical.

Converging (doing and thinking): People with a converging learning style can solve problems and will use their learning to find solutions to practical issues:

1. They are practical and linear thinkers.
2. They learn by trial and error.
3. They prefer technical aspects and problems.
4. They have good problem-solving skills and speed to the solution.
5. They prefer not to work in groups.

Accommodating (doing and feeling): This learning style relies on intuition rather than logic:

1. Their learning is hands on and practical.
2. They are attracted to new challenges.
3. They are excellent at executing plans.
4. They rely on gut instinct.
5. They are not big on analysis.
6. They tend to be social.

Merrill and Reid (1991) analyzed communication styles using a communication style inventory to measure the degree of assertiveness and responsiveness of an individual. Assertiveness and responsiveness are two dimensions of an individual's social style (Figure 11.10), meaning their tendency to react, associate, and adapt to others in communication (Wheeless and Reichel, 1990). Assertiveness is defined as the "capacity to make requests, actively disagree, express positive or negative personal rights and feelings, initiate, maintain or disengage from conversations, and stand up for oneself without attacking another" (McCroskey and

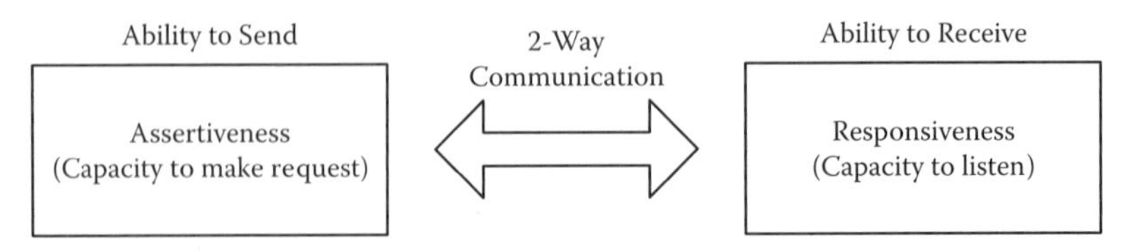

FIGURE 11.10
Two major dimensions of social style as it relates to communication.

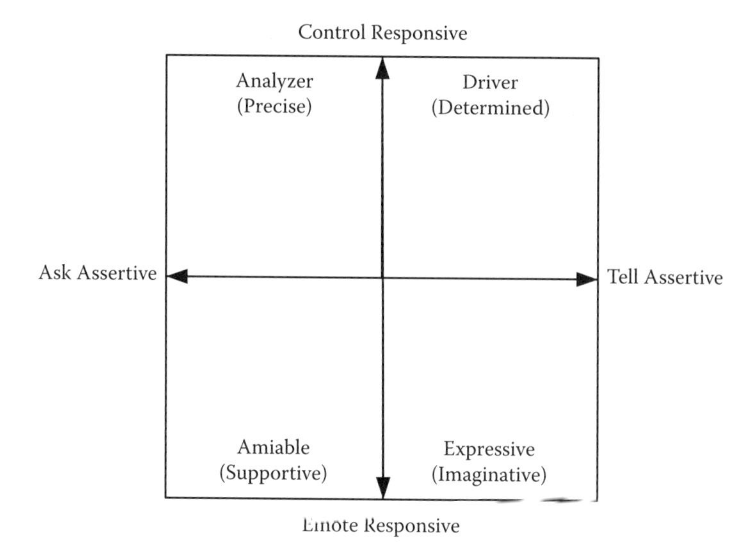

FIGURE 11.11
Adapted from Merrill and Reid's (1981) individual communication style grid (Kolb, 1984). (Adapted from Merrill, D. and Reid, R., 1981, *Personal Styles and Effective Performance*, Chilton, Radnor, PA; Kolb D, 1984, *Experiential Learning*, Prentice Hall, Englewood Cliffs, NJ.)

Richmond, 1996). Responsiveness is defined as the "capacity to be sensitive to the communication of others, to be a good listener, to make others comfortable in communicating, and to recognize the needs and desires of others" (McCroskey and Richmond, 1996).

Merrill and Reid (1991) identified four primary communication styles: amiable, analytical, driver, and expressive (Figure 11.11). As stated previously, each person tends to employ one of these dominant social styles as the way he or she gets along with people whose styles differ from their own.

Analyzer (ask assertive ± control responsive): An analyzer is one who is generally industrious, methodological, and well organized.

1. Is a precise, systematic, and deliberate communicator
2. Gathers data before making decisions
3. Strengths: precise and systematic
4. Weaknesses: exacting and inflexible

Driver (control responsive ± tell assertive): This communication style describes an individual who gets to the point quickly and expresses him- or herself openly:

1. Highly task oriented (valued for ability to get things done)
2. Determined, decisive, quick to make decisions
3. Results oriented and competitive
4. Strengths: determined and objective
5. Weaknesses: dominating and insensitive

Expressive (tell assertive ± emote responsive): This individual demonstrates numerous desirable social characteristics and usually is talkative and a good persuader and motivator:

1. Good at motivating others
2. Willing to take risk
3. Often associated with charismatic leadership
4. Can act decisively and view a situation from the big picture
5. Strengths: enthusiastic and imaginative
6. Weaknesses: undisciplined and unrealistic

Amiable (ask assertive ± emote responsive): Amiable communicators are excellent at resolving interpersonal conflicts and dealing with problems that have an impact on a group on a significant level:

1. Demonstrates respect when working with others
2. Capable of being sympathetic and understanding others
3. Tends to trust people and to be trusted by others
4. Strengths: supportive and easygoing
5. Weaknesses: conforming and permissive

11.6 CONNECTING THE KAIZEN MIND

Good teachers are those who can connect to each student's interest to put the student into a proactive disposition to desire to be creative and learn. Since each student has a unique style of learning and communicating, teachers must constantly adjust how they present and organize information to make learning effective. For example, science students over time become more analytical and less creative, while art students become more creative and less analytical. When teachers specialize in certain learning

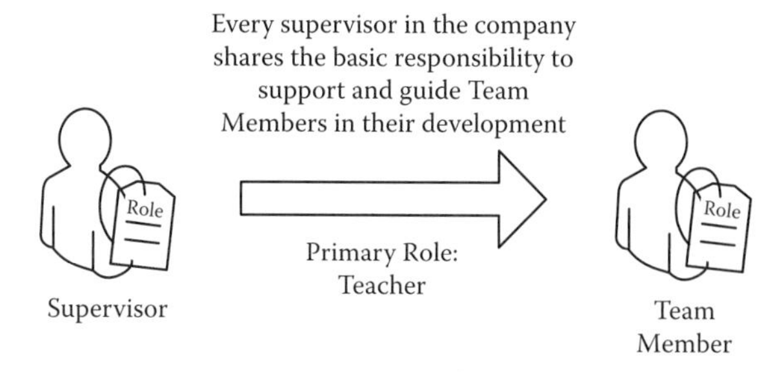

FIGURE 11.12
The supervisor has the role of teaching and coaching on the job.

or communication styles, knowledge transfer is less effective for diverse groups. To this end, teachers must find ways to be adaptive to the needs of their students.

According to education theory, a teacher can be more adaptive if the teacher uses a curriculum and delivery style that touches on all the bases of learning, such as experiencing, reflecting, thinking, and acting (Kolb, 1984). Kolb viewed the learning cycle to start with immediate or concrete experiences. These tangible experiences lead to observations and reflections. Reflections are assimilated (absorbed and translated) into abstract concepts with implications for action, which the person can actively test and experiment with, in turn enabling the creation of new experiences (Kolb, 1984).

Toyota believes the best way to get team members to touch on all the bases of learning is when their supervisor assigns them work on the job. Every supervisor in the company shares the basic responsibility to support and guide his or her team members in their development (Figure 11.12). Toyota believes that the supervisor is the best person to connect team members to kaizen because the supervisor has the authority to delegate decision making to the most appropriate person. Every supervisor has the role to let team members take responsibility for their ideas and to see them realized on the job (Figure 11.13). Toyota views that the immediate supervisor is the best conduit to connect team members to kaizen because the supervisor is the one who ultimately has the role to involve employees in decision making.

One of the most common ways that supervisors are trained to teach and coach team members in the workplace is through delegation. Toyota expects supervisors (i.e., management) at all levels to use delegation as a

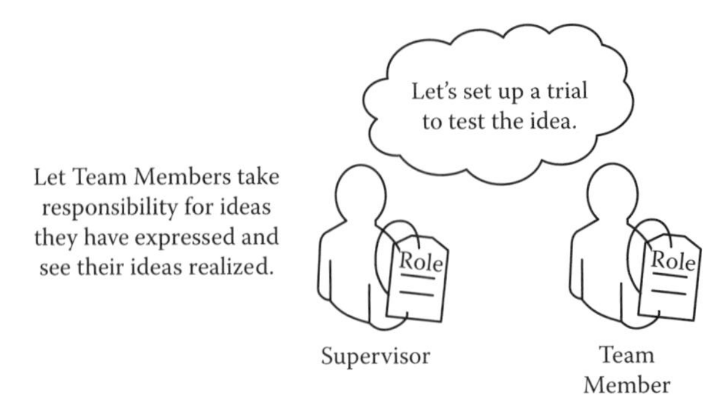

FIGURE 11.13
The supervisor has the role of letting team members realize their ideas.

way to accomplish the results of the unit while obtaining good performance and utilization of each member of the unit. Each supervisor cannot be expected to do all the work or produce everything the unit produces him- or herself. Toyota views that the most important thing in delegation is to make sure the team members feel that the job is theirs rather than just doing the job for the supervisor.

The way delegation becomes an excellent approach to teaching in the workplace is when subordinates fully understand their supervisor's approach and way of thinking (Figure 11.14), the goal, the priorities, and the results for the situation. Delegation is not giving the job away and leaving team members. Delegation is assigning work for a team member to practice the same type of thinking as the team member's immediate supervisor. The best way to utilize team members in sharing their manager's approach is to give each team member the whole job. Again, if

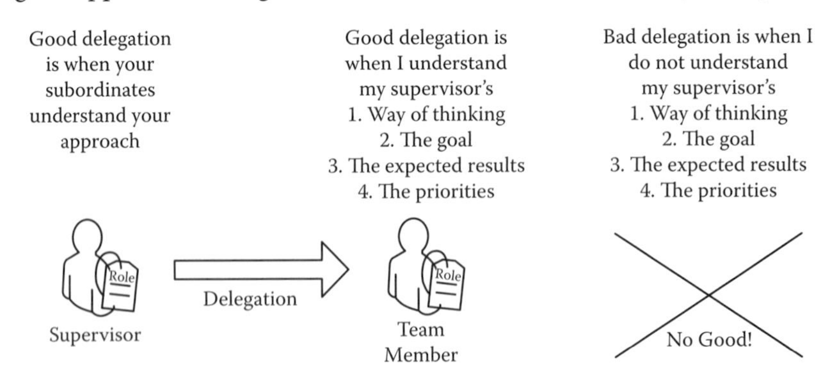

FIGURE 11.14
Good delegation is knowing your supervisor's way of thinking.

subordinates feel ownership of the whole job, they will be in a positive disposition to voluntarily take initiative.

This type of delegation can only work if there is daily communication between the supervisor and the subordinate. Supervisors must be in contact with their subordinates to listen, to speak openly about the issues, and to pass along information that could change the job itself. Communication is critical in delegation so that the supervisor and the subordinate are not surprised and feel that they are being left out of decisions that could affect either one of them. Daily communication is good when subordinates feel free to discuss issues with their immediate supervisor and are not worried that the job could be taken away or they could be blamed for raising issues. From a supervisor's perspective, good communication is timely reporting. When subordinates raise issues early, management can still have time to help. The speed and quality of management can be increased when management has a grasp of what is really going on. Supervisors can only assist if they have adequate information about the situation.

Delegation is also good for helping team members learn new things. Toyota would like for supervisors to establish the direction with their subordinates and talk through how they will be evaluated. Good delegation is asking subordinates to try for a little more than they feel they can do at the time. When employees are challenged, the activity not only increases in importance but also gives employees a sense of accomplishment. Supervisors should let their subordinates think through the actual methods on their own and step back and observe. Supervisors should coach and help them learn but not do the work. An appropriate way to delegate is to watch the results and the way the job is performed. Only then can a supervisor truly know when to intervene. Supervisors should be prepared to recover from small failures and seize teachable opportunities with their subordinates even if the subordinates do fail (Figure 11.15). A good supervisor will help his or her subordinate as much as possible to overcome difficulties and not see the subordinate as the problem when the subordinate fails or needs help.

Delegation done properly is one of the best ways to develop subordinates learning on the job. When jobs are fully delegated, a supervisor can create time to focus on intermediate and long-range planning. When subordinates are able to maintain the normal situation, a supervisor can step back and understand the unit's capabilities, prepare for future challenges, and anticipate problems from a larger perspective.

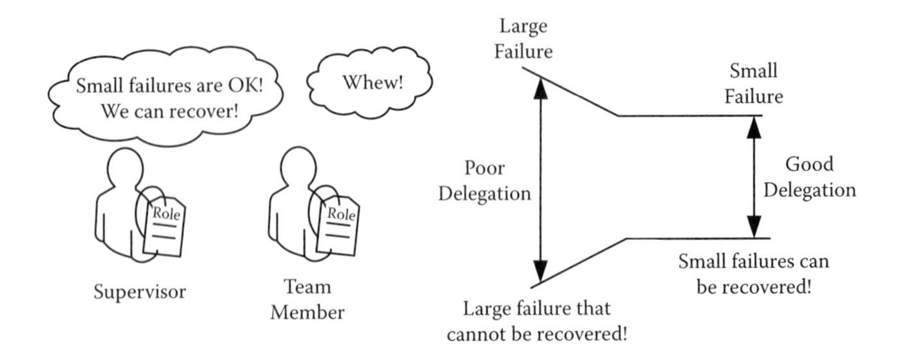

FIGURE 11.15
Good delegation is allowing small failures: good opportunities for learning.

11.7 THE PREFERRED LEARNING AND COMMUNICATION STYLE FOR THE KAIZEN MIND

Not only do teachers influence students and the impact of learning, but also students influence teachers and the outcome. This chapter discussed what makes an effective teacher, but it should also be understood what makes an effective student. Humans have the capacity to develop relatively stable behavior patterns, specifically how they learn and develop (Lashbrook and Lashbrook, 1979; Snavely and Waiters, 1983). Norton (1983) suggested that the style of communication is one of the most enduring patterns of human interaction. The way one verbally, nonverbally, and paraverbally interacts to signal how literal meaning should be taken, interpreted, filtered, or understood is largely fixed and does not change that much over time (Norton, 1983). If this is true, which communication and learning style is most effective for the kaizen mind?

The best way to describe the learning style for the kaizen mind is Kolb's diverger. Divergent-type learning is the primary style for teaching and training employees at Toyota. Divergent learners rely on concrete experience and reflective observation. Divergent learners have a strong imagination and can view concrete situations from many perspectives. Divergent learners have broad cultural interests and are generally interested in people. They are open minded, likely to help people, and enjoy trying different things. Divergent learners tend to be counselors, seek positions in HR, and are good at separating from the immediacy of events to reflect and rethink though previous experiences to relearn and organize thoughts.

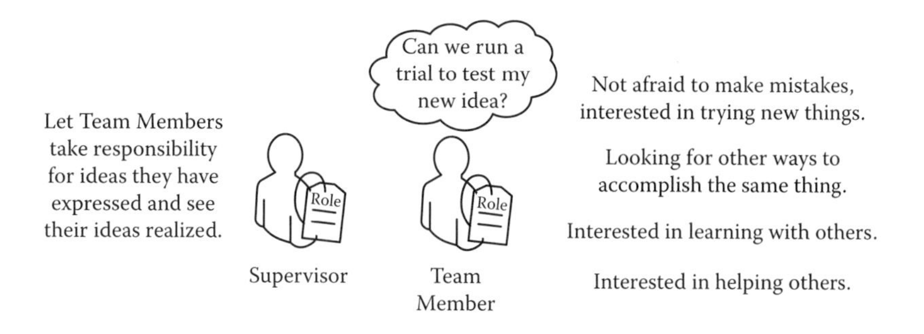

FIGURE 11.16
The preferred learning style for kaizen mind: the diverger.

The kaizen mind resembles divergent learning because TPS encourages employees to be highly open to trying different things. While most outsiders would view that another style would be more fitting, such as experimentation, reasoning, or thinking critically, openness to new ways of thinking and knowledge is the heart of kaizen and most importantly the path to creativity. To correctly apply the concept of kaizen (small improvements at little to no cost), an employee must be extremely creative. Divergent learners have an innate ability to think about a problem from numerous perspectives, which makes them ideal for generating a variety of solutions. Since divergent learners are not afraid to experiment or find new ways of doing things, they are ideal for fueling TPS. Divergent learners are natural at kaizen because they are open, flexible, and up for trying anything (Figure 11.16).

The preferred kaizen mind for communication can best be described as that of someone who is ask assertive and emote responsive, someone who speaks by raising questions, and someone who takes time to listen to others. The best communication style for the kaizen mind is that of someone who is amiable. Amiable communicators are good natured, and friendly; like to act one on one; and ask others what they think. They tend to be gentle, willing, eager, and tolerant. This type of individual communication style is suited for building an environment in which employees (both supervision and team members) treat each other with dignity and trust.

Toyota prefers amiable communication styles over expressive, driver, and analyzer styles, primarily because team members should feel comfortable voicing their opinions, ideas, suggestions, and concerns. Toyota believes that communication is fundamental to stable employee relations and for building trust at the workplace. One of the most essential areas

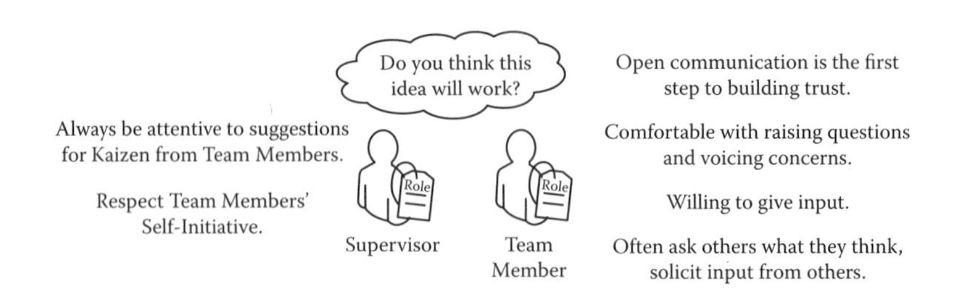

FIGURE 11.17
The preferred communication style for the kaizen mind: amiable.

of trust in a Lean environment is between the supervisor and the subordinate. Toyota utilizes open communication as a tool to help supervisors promote an environment in which team members feel comfortable asking questions. For this reason, Toyota prefers an amiable communication style to better understand a team member's point of view. Ideally, team members should know what is going on and should be asked what they are thinking. It is believed that Toyota prefers amiable communication styles so that supervisors can listen better to what team members have to say. Toyota believes that only through sincere communication can both parties understand one another and build mutual trust.

Divergent learners and amiable communicators promote the kaizen mind because they generally want to share the spirit of challenge with others. Divergent learners are interested in learning with others and helping others to learn. Amiable communicators prefer to ask others what they think and are comfortable giving their input (Figure 11.17). Both of these styles demonstrate the ideal kaizen mind because they thrive in group settings. Working in teams reinforces diversity in learning (the ability to generate a variety of different ideas) and an open climate in which everyone feels free to participate and raise issues. Coupled with Toyota's systematic problem-solving and industrial engineering identity, divergent learners and amiable communicators provide a strong complement to kaizen.

11.8 SUMMARY

The purpose of this chapter was to explain the fuel that drives TPS. For a system to live and carry out its most essential functions, it must have a way to import new energy or, in the thermodynamic sense, receive negative entropy. Negative entropy is the way the system brings order to itself so it can grow and, most importantly, adapt to its environment. In TPS, negative entropy is employee learning. Learning is the lifeline to TPS because it allows the system to refine itself over time. Since the system should be accessible to everyone, that system should be effective when it can solicit input and encourage learning from all those who use it. Even if employees are forced to do kaizen, they will not be in a proactive disposition to learn or do it continuously so that it is long lasting. The trick for generating negative entropy in TPS is to create a condition that allows creativity to thrive.

One approach that Toyota employs to encourage an environment of OL is the concept of delegation. TPS is able to draw negative entropy (i.e., ideas and suggestions) from its employees by delegating decision making to the most appropriate level. Toyota believes that each supervisor can encourage workplace learning by using delegation to involve employees in TPS. When employees have ownership of the entire job, they are in a much better position to be creative. Employees are also more willing and able to learn when delegation is used as a way to let employees fail without risk of losing their job. Supervisors can share their way of thinking (a complete learning cycle, often described as Plan-Do-Check-Act [PDCA]) to make learning more effective.

Last, this work presented the idea that students are just as responsible as teachers when implementing kaizen. The perfect kaizen mind would be described as that of a divergent learner who communicates with an amiable style. To apply kaizen, employees in groups need to give their ideas to improve the workplace. Divergent learners are excellent examples of persons who are open minded and flexible. They are extremely creative and imaginative. Amiable communicators seek trust and respect in the workplace. They generally want to elicit the input from others and work together to achieve common goals. Toyota's learning and communication styles are one of its most effective weapons for kaizen.

REFERENCES

Adler, P. and Cole, R. (1993) Designed for learning: A tale of two auto plants. *Sloan Management Review*, 34(3), 85–94.

Amabile, T. (1997) Motivating creativity in organizations: On doing what you love and loving what you do. *California Management Review*, 40, 39–58.

Antonacopoulou, E. (2006) The relationship between individual and organizational learning: New evidence from managerial learning practices. *Management Learning*, 37(4), 455–473.

Argyris, C. (1964) *Integrating the Individual and the Organization*, Wiley, New York.

Argyris, C. and Schon, D. (1996) *Organizational Learning II, Theory, Method and Practice*, Addison-Wesley, Reading, MA.

Bailly, F. and Longo, G. (2009) Biological and anti-entropy. *Journal of Biological Systems*, 17(1), 63–96.

Bandura, A. (1977) *Social Learning Theory*, Prentice Hall, Englewood Cliffs, NJ.

Barsalou, L. and Prinz, J.(1997). Mundane creativity in perceptual symbol systems. In ed. T.B. Ward, S.M. Smith, & J. Vaid, *Creative Thought: An Investigation of Conceptual Structures And Processes*. American Psychological Association, Washington, DC. pp. 267–307.

Becvar, D. and Becvar, R. (2003). *Family Therapy: A Systemic Integration* (5th ed.). Allyn and Bacon, Boston, MA.

Bennet, D. and Bennet, A. (2008) The depth of knowledge: surface, shallow or deep? *VINE: The Journal of Information and Knowledge Management Systems*, Bradford, 38(1) 405–420.

Campbell, T. and Cairns, H. (1994) Definition development and measurement in the learning organization. Moving from buzzwords to behaviors. Working paper presented at the 1994 ECLO conference, La Hulpe, Belgium.

Choe, J. (2004) *The Relationship among Management Accounting Information, Organizational Learning and Production Performance*, Kyungpook National University, Daegu, South Korea.

Cohen, W. and Levinthal, D. (1990) Absorptive capacity: A new perspective on learning and innovation. *Administrative Science Quarterly*, 35(1), 128–152.

Czarniawska, B. (2001) Anthropology and organizational learning, in *Handbook of Organizational Learning and Knowledge*, ed. Dierkes, M., Berthon, A., Child, J., and Nonaka, I., Oxford University Press, London. pp. 118–136.

Denton, J. (1998) *Organizational Learning and Effectiveness*, Routledge, New York.

Dodgson, M. (1993) Organizational learning: A review of some literatures. *Organizational Studies*, 14(3), 375–394.

Fiol, M. and Lyles, M. (1985) Organizational learning. *Academy of Management Review*, 10(4)803–813.

Garrick, J. and Clegg, S. (2000) Knowledge work and the new demands of learning. *Journal of Knowledge Management*, 4(4), 279–286.

Gully, S. and Chen, G. (2010) Individual differences, attribute-treatment interactions, and training outcomes, in *Learning Training and Development in Organizations*, ed. Kozlowski, W. and Salas, E. Routledge, New York.

Hall, A. (2006). *Introduction to Lean Sustainable Quality Systems Design: An Integrated Approach From the Viewpoints of Dynamic Scientific Inquiry Learning & Toyota's Lean System Principles and Practices*. Published by Arlie Hall, Lexington, KY. ISBN 0-9768765-0-7.

Jones, A. and Hendry, C. (1994) The learning organization: Adult learning and organizational transformation. *British Journal of Management*, 5, 153–162.

Karash, R. (1995) *Mental Models and Systems Thinking: Going Deeper into Systematic Issues*, The Systems Thinker, Pegasus Communications, Cambridge, MA.

Kidd, J. (1998) Knowledge creation in Japanese manufacturing companies in Italy. *Management Learning*, 29(2), 131–146.

Kim, J. (1975) Feedback in social sciences: Toward a reconceptualization of morphogenesis, in *General Systems Theory and Human Communication*, ed. Ruben, B. and Kim, J. Hayden, Rochelle Park, NJ.

Kolb, D. (1984) *Experiential Learning*, Prentice Hall, Englewood Cliffs, NJ.

Lashbrook, W. and Lashbrook, V. (1979). *The Statistical Adequacy of the Social Style Profile.* Wilson Learning, Eden Prairie, MN.

Liker, J. and Hoseus, M. (2008) *Toyota Culture—The Heart and Soul of the Toyota Way*, McGraw-Hill, New York.

March, J. and Simon, H. (1958) *Organizations*, 2nd ed., Blackwell, Oxford, UK.

McCroskey, J. and Richmond, V. (1996). *Fundamentals of Human Communication: An Interpersonal Perspective.* Waveland Press, Prospect Heights, IL.

McGrath, R. G. (2001) Exploratory learning, innovative capacity, and managerial oversight. *Academy of Management Journal*, 44(1) 118–131.

Merrill, D. and Reid, R. (1981) *Personal Styles and Effective Performance*, Chilton, Radnor, PA.

Merrill, D. and Reid, R. (1991) *Personal Styles and Effective Performance,* © Tramcom Corporation; (1999) CRC Press, Boca Raton, FL.

Nonaka, I. and Takeuchi, H. (1995) *The Knowledge-Creating Company,* Oxford University Press, New York, NY.

Norton, R. (1983) *Communicator Style: Theories, Applications and Measures*, Sage Publications, New York, NY.

Ong, W .(1982) *Orality and Literacy: The Technologizing of the Word*, Routledge, London.

Polanyi, M. (1967) *The Tacit Dimension,* Anchor Books, New York.

Putt, A. (1978) *General Systems Theory Applied to Nursing*, Little, Brown, Boston, MA.

Reynolds, M. and Vince, R. (2007) *The Handbook of Experiential Learning and Management Education*, Oxford University Press, New York.

Sako, M. (1992) *Prices, Quality and Trust: How Japanese and British Companies Manage Buyer-Supplier Relations*, Cambridge University Press, Cambridge.

Senge, P. (1990) *The Fifth Discipline*, Doubleday/Currency, New York, NY.

Shin, S. and Zhou, J. (2003) Transformational leadership, conservation, and creativity: Evidence from Korea. *Academy Management Journal* 46(6) 703–714.

Simon, H. (1969) *Sciences of the Artificial*, MIT Press, Cambridge, MA.

Simonton, D. (1994) *Greatness: Who Makes History and Why*, Guildford Press, New York, NY.

Smith, A., Vasudevan, P., and Tanniru, R. (1996) Organizational learning and resource based theory: An integrative model. *Journal of Organizational Change Management*, 9(6), 41–53.

Snavely, W. (1981). The imact of social style upon person perception in primary relationships. *Communication Quarterly*, 29, 132–143.

Spear, S. (2010) What to learn from Toyota for those who haven't already—Improvement and innovation needed now more than ever. *Leadership and Innovation*, June 13.

Sternberg, R. (1985) Implicit theories of intelligence, creativity, and wisdom, *Journal of Personality and Social Psychology*, 49(3), 607–627.

Turner, S. (1995). Margaret Boden, the creative mind. *Artificial Intelligence*, 79, 145–159.

Wheeless, L. and Reichel, L. (1990) A reinforcement model of the relationships of supervisors general communication styles and conflict management styles to task attraction, *Communication Quarterly*, 38, 372–378

Yang, J. (2003) *Qualitative Knowledge Capturing and Organizational Learning: Cases in Taiwan Hotels*, National Kaohsiung Hospitality College. Taiwan.

12

The System Property of Requisite Variety in TPS

12.1 SYSTEM PROPERTY: REQUISITE VARIETY

According to living systems theory, an organism cannot survive if it is not as complex as its environment. When the variety or complexity of the environment exceeds the capacity of a system (natural or artificial), the environment will dominate and ultimately destroy that system. A law that relates to the degree of adaptation of a system and its capacity for adaptation is often referred to as Ashby's law of requisite variety (Figure 12.1). Simply stated, the premise that "only variety can destroy variety" relates to how well a controller or regulator of a system can take in as many distinct situations as the system would allow (Ashby, 1956; Gigch, 1978). The performance of regulators is described by the degree of adaptability and the capacity for adaptability. A regulator that has a high degree of adaptability can consistently, accurately, and quickly control its internal parameters to meet external conditions. A regulator that has a high capacity for adaptability demonstrates flexibility over a wide range of environmental conditions. The best regulators are those that can fully represent all the variations of a system accurately and quickly (Conant and Ashby, 1970).

Ashby's law of requisite variety provides a reference point for understanding how systems behave and how they are likely to be managed (Ashby, 1956). The law describes how a system should cope with complexity. Complexity depends on the number and types of permutations or combinations of variables that can exist within a system (Gigch, 1978). Simply, anything that can be different in a system is complexity (Ashby, 1956). A system is likely to maximize its internal variety (or diversity) to prevent any foreseeable or unforeseeable contingencies (Heylighen, 1992). A system without requisite variety is likely to fail or be considered

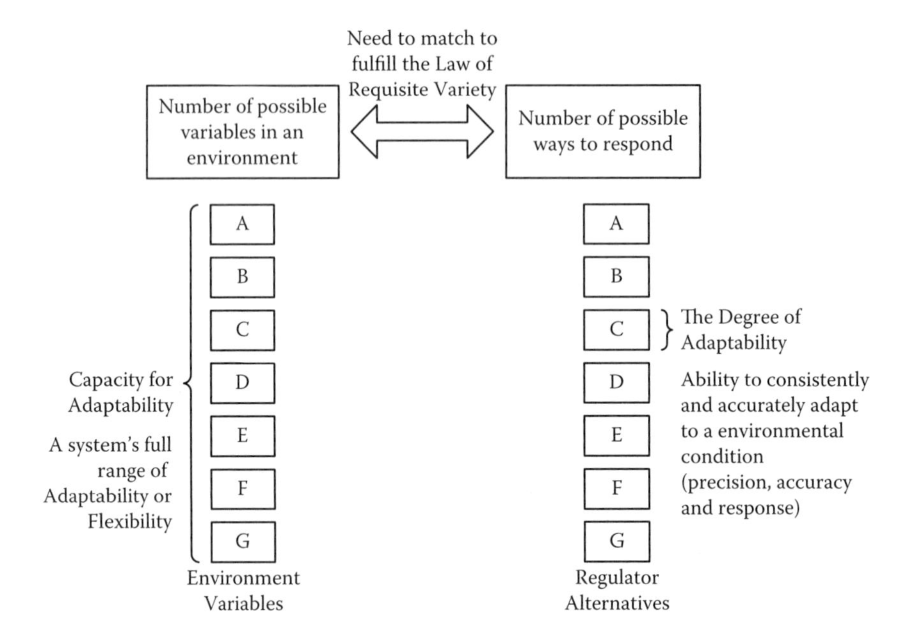

FIGURE 12.1
The law of requisite variety: for a system to live it must be precise and flexible.

unviable whenever it encounters an unexpected scenario (Gigch, 1978). Ashby's law is fundamental to systems theory because it describes how a system should be designed to deal with uncertainty.

12.2 A THEORETICAL RESPONSE TO COMPLEXITY AND VARIETY

The classic response to increased uncertainty for variety is to move a system's structure from mechanistic to organic (Figure 12.2) (Burns and Stalker, 1961) or from an undifferentiated structure to a differentiated structure (Lawrence and Lorsch, 1967). Mechanistic structures are characterized by extensive rules, procedures, and structures (Mintzberg, 1979). Mechanistic structures exhibit relatively few possible states, which requires a fewer number of decisions to be made in advance. On the other hand, organic structures are not confined by rules or regulations. Organic structures or systems flourish in ad hoc or impromptu decision-making environments, mainly because they can take on and assume more possible states (Scala et al., 2006). Katz and Kahn (1978) suggested that a shift from

	Structure	Complexity	Decision Making
Mechanistic Structure	Rigid rules; differentiated structure (clear boundaries)	Only able to handle a few number of states	Fewer decisions to make in advance (algorithm)
Organic Structure	Fewer rules, if any; undifferentiated structure (boundaries are not clear)	Able to handle a larger number of undefined states	Greater flexibility in decision making (ad hoc)

FIGURE 12.2
Comparison between mechanistic and organic structures.

a rigid bureaucracy toward a more informal and flexible structure fosters effectiveness in a more turbulent environment.

Researchers suggested that the response to dealing with variation and complexity should depend on the situation or environment. Because complexity can change over time and its distribution on the system, a more systematic response to complexity (case by case) is required. The work of Love and Cooper (2007) proposed that there should be design guidelines to better understand the locus of control and prevention of complexity in systems.

1. Identify relative distributions of complexity and variety and their impact on the total system
2. Identify all constituencies of ownership, power, and influence that largely have an impact on complexity
3. Identify the types of benefits that constituencies are likely to gain from the system over time due to their complexity
4. Identify the types of relative transaction cost when constituencies change
5. Identify how the configuration of a system changes over time

Overall, the variety and complexity of a system are more difficult to control when there are multiple constituencies of ownership or control or when the distribution of value changes in the system (Love and Cooper, 2007).

The best way to handle variety is by decreasing the complexity of the system itself (Figure 12.3) (Umpleby, 2007). When the complexity of a system decreases, the organization can assume a proactive stance because there are fewer disturbances and fewer possible outcomes to consider. Variety reduction is a preferred solution because variety-handling activities often

Ideal Situation	Last Resort When Dealing with Complexity
Decrease system complexity	Increase complexity of regulator
Reduce complexity of the system	Increase the number of responses by the regulator to counter the effects of complexity

FIGURE 12.3
Two ways to handle variety and complexity.

generate additional undesirable variety. The longer a system is in an abnormal mode, the greater the likelihood the system will have to generate more alternatives (Duimering et al., 1993).

The other way to handle variety is to increase the complexity of the regulator. Increasing the complexity of the regulator (i.e., destroying variety with variety) is one approach used by a system theorist to have control over a system's surroundings (Scala et al., 2006). While this strategy is not preferred because it increases the variety of responses made by the system, it gives a system a last-resort approach when dealing with variety.

12.3 TOYOTA PRODUCTION SYSTEM AND THE LAW OF REQUISITE VARIETY

One of the best ways to illustrate the complexity of the Toyota Production System (TPS) as it relates to the law of requisite variety is through the concept of production leveling. Production leveling is the ability of an organization to manage all of its resources effectively to achieve its intended purpose. Labor, materials, and equipment are all resources that have to be managed to meet customer demand. Because an organization and the environment are constantly changing, it is rare that internal capabilities match what is required from the customer. The law of requisite variety is useful because it describes the ability of TPS to regulate all of the organization's management systems (Figure 12.4). All organizations have a way to balance internal resources to match external requirements; however, their success is largely dependent on the extent the management system can digest complexity and variety in the system.

Unfortunately, most companies that attempt production leveling only focus on activities in manufacturing. Toyota views production leveling as

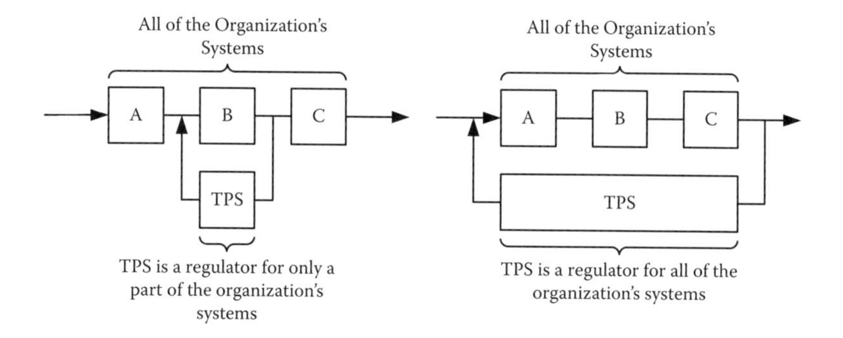

FIGURE 12.4
A comparison between a partial and total TPS regulator.

an interdependent activity that should be coordinated across the company. Toyota has extended the focus beyond manufacturing so that the interdependencies between departments can be properly managed to increase the benefits of its application and effectiveness as a means to manage the impact of demand fluctuations. When demand fluctuations affect manufacturing, coordination with other departments is essential to minimize the impacts of such variations.

Another major difficulty often underestimated or overlooked in emulating Toyota's approach to demand leveling is how management is integrated to support production leveling. Most outsiders treat TPS as a technical system that can exist without the human system (Hall, 2006). This perception of Toyota's practices is common, and convenient, as most organizations look for quick results without changing how the organization is led or managed. It is noteworthy that, at Toyota, managers are trained, encouraged, and rewarded for cooperating even when their own department may be impacted negatively. In many companies, especially those that rate managers only on how their department performs in isolation, one department manager is not inclined to help another unless there is a clear benefit to the former. Production leveling is not about how one department chooses to operate but how several departments chose to operate.

The law of requisite variety is essential in TPS because it considers a much larger picture of both the technical and the social structures of the organization (Figure 12.5). Ashby's law states that an effective regulator should be able to offer an alternative for each of the system's differential states. In this context, production leveling involves more than one department and cooperation from more than one manager. The social–technical

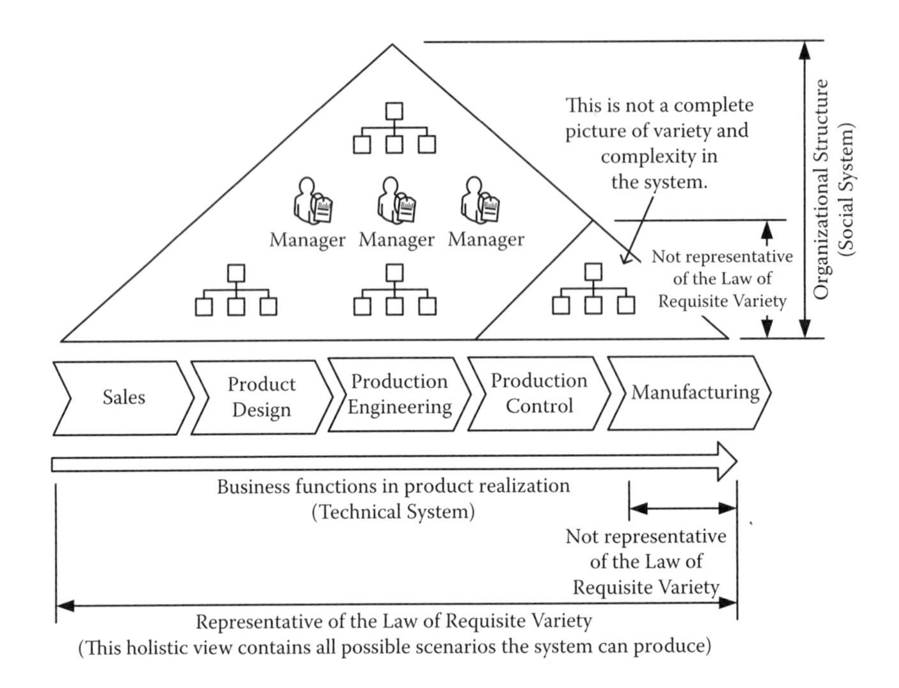

FIGURE 12.5
Requisite of variety for the social–technical structures of production leveling.

structure of production leveling is one of the best examples of how requisite variety is applied.

12.4 LITERATURE REVIEW: PRODUCTION LEVELING

The concept of production leveling is not new and has been studied and practiced by the industrial engineering community for well over 60 years (Baumol, 1951; Beckman, 1961; Elmaleh and Eilon, 1974). According to Maynard's *Industrial Engineering Handbook*, production leveling is the process by which, in mixed-model production lines, products are properly arranged, rather than manufactured in a random order, to minimize the variations in parts consumption and workload at the workstations (Figure 12.6) (Zandin, 2001). When performed properly, production leveling can maximize efficiency (Xiaobo et al., 1996); reduce inventory (Coleman and Vaghefi, 1994); increase the ability to make to stock (Swanson, 2008); eliminate spikes in production (Andel, 1999); prevent overburdened jobs (Rinehart, 1997); increase capacity (Yano and

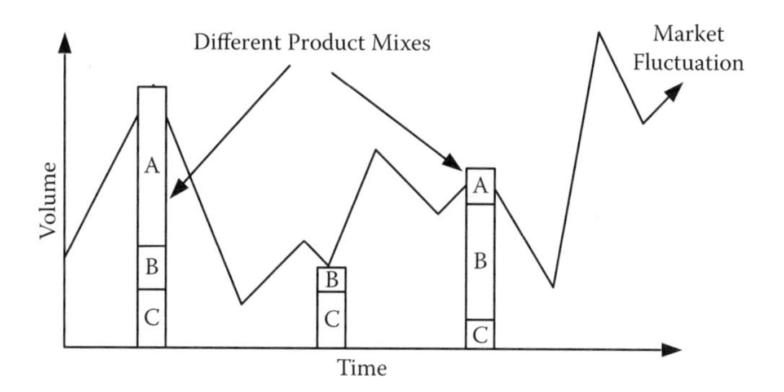

FIGURE 12.6
Example of variety and complexity from the market.

Rachamadugu, 1991); and smooth work demands throughout the supply chain (Monden, 1998). Production leveling is used to sequence various product models (or product variants) optimally during manufacturing so that the peaks and valleys in demand for products are smoothed out across the planning period to enable overall cost reductions and improved efficiencies (Figure 12.7).

Conventionally, production leveling is known as a method to sequence different products for mixed-model production to minimize the variations in parts consumption and level the workload at the workstations (Zandin, 2001). For example, research in this area has focused on developing heuristics (Okamura and Yamashima, 1979; Miltenburg and Sinnamon, 1992); algorithms (Monden, 1998); dynamic programming (Yano and Rachamadugu, 1991; Xiaobo et al., 1996); computational

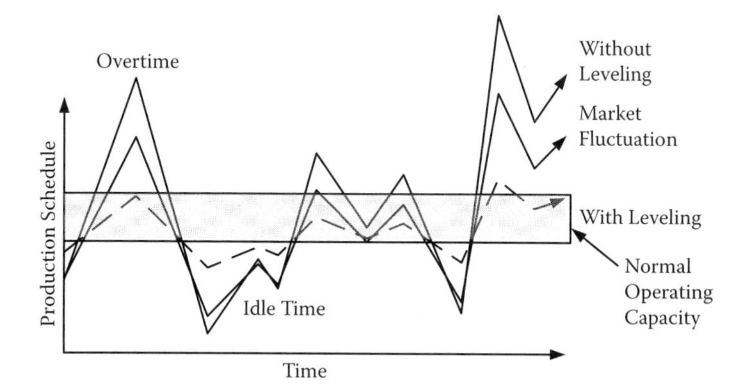

FIGURE 12.7
Example of smoothing effect from production leveling.

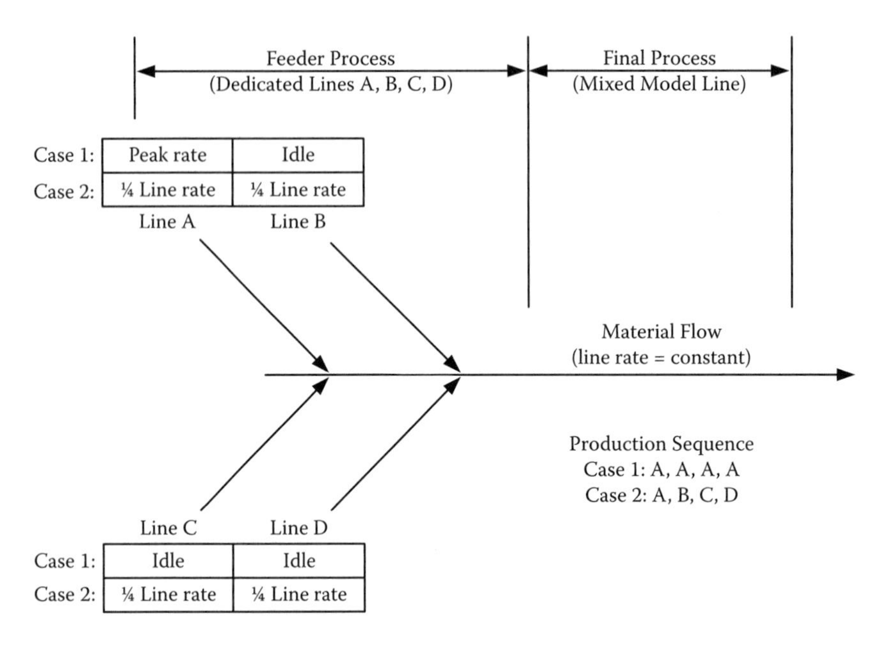

FIGURE 12.8
Production leveling: showing consistent use of resources in case 2.

analysis (Kovalyov et al., 2001); and simulations (Huttmeir et al., 2009). The main goal in all of this work is the determination of a model sequence (i.e., ABCDABE …) on a paced, mixed-model assembly line so production can be spread over a period. As shown in Figure 12.8, when production is not spread over the production period, parts of the plant speed up while other parts of the plant slow down (i.e., case 1). The end result is an uneven use of resources that is extremely costly. When products are spread out over the production period (i.e., case 2), the plant can more effectively use its resources evenly.

According to Ashby's law of requisite variety, production leveling should be applied in the following way:

1. A manufacturing system should contain less variety and complexity in its construction of products (i.e., decrease system complexity).
2. When the market fluctuates, manufacturing should be able to absorb variety and complexity through flexible operations (i.e., increased complexity in regulation).

An example of how Ashby's law applies to manufacturing systems is illustrated in Table 12.1. As stated previously, complexity reduction

TABLE 12.1

Activities to Reduce or Handle Complexity

Sources of Complexity	Complexity Reduction Activities (Preventive)	Complexity-Handling Activities (Reactive)
Diversification in products (shape, size, and function)	Standardize product configurations	Standardize manufacturing processes (i.e., production flow analysis, cellular manufacturing, product families, etc.)
Market fluctuation	Establish a product allocation allowance (limit the number and types of product sold)	Level product demand (increase flexibility in operations, schedule manufacturing evenly across the production period)
Varying operator skill levels	Rigorous selection criteria	Job instruction training (level skills for all workers for all jobs)
Defective parts delivered	Supplier qualification programs	Incoming parts monitoring
Erratic machine breakdowns	Preventive maintenance	Backup equipment, redundant system

Source: Adapted from Duimering, P. and Safayeni, F. (1991) A study of the organizational impact of the just-in time production system, *Proceedings of the International Conference on Just in Time Manufacturing Systems: Operational Planning and Control Issues* (Montreal), pp. 19–32; Gerwin, D. (1993) *Management Science*, Vol. 39, No. 4, 395–410; and Scala, J., Purdy, L., and Safayeni, F. (2006) *Journal of Manufacturing Technology Management*, Vol. 17, No. 1, 22–41.

activities offer a more proactive and preventive stance, while complexity-handling activities are mostly reactive. Both strategies are used today in dealing with variety and complexity in production systems.

12.5 THE TECHNICAL AND SOCIAL STRUCTURES OF PRODUCTION LEVELING

Toyota's approach to production leveling has been widely benchmarked as the foundation for flow and pull systems that allow the concept of just in time (JIT) to be achieved (Liker, 2004). The application of production leveling in Toyota's view requires a much broader view than the mere application of algorithms or heuristics to determine the optimal production sequence. Ohno described that once a company adopts the JIT philosophy, the entire production system has to be overhauled along with its management systems (Ohno, 1988). Departments generally tend to focus on their own interest, which makes activities such as production leveling difficult to achieve. For managers to cooperate, they need to overcome their individual focus and secure solutions that serve the company's larger interest (Shimikawa and Fujimoto, 2009). Figure 12.9 illustrates Toyota's reactive and preventive measures for achieving the technical and social features of production leveling.

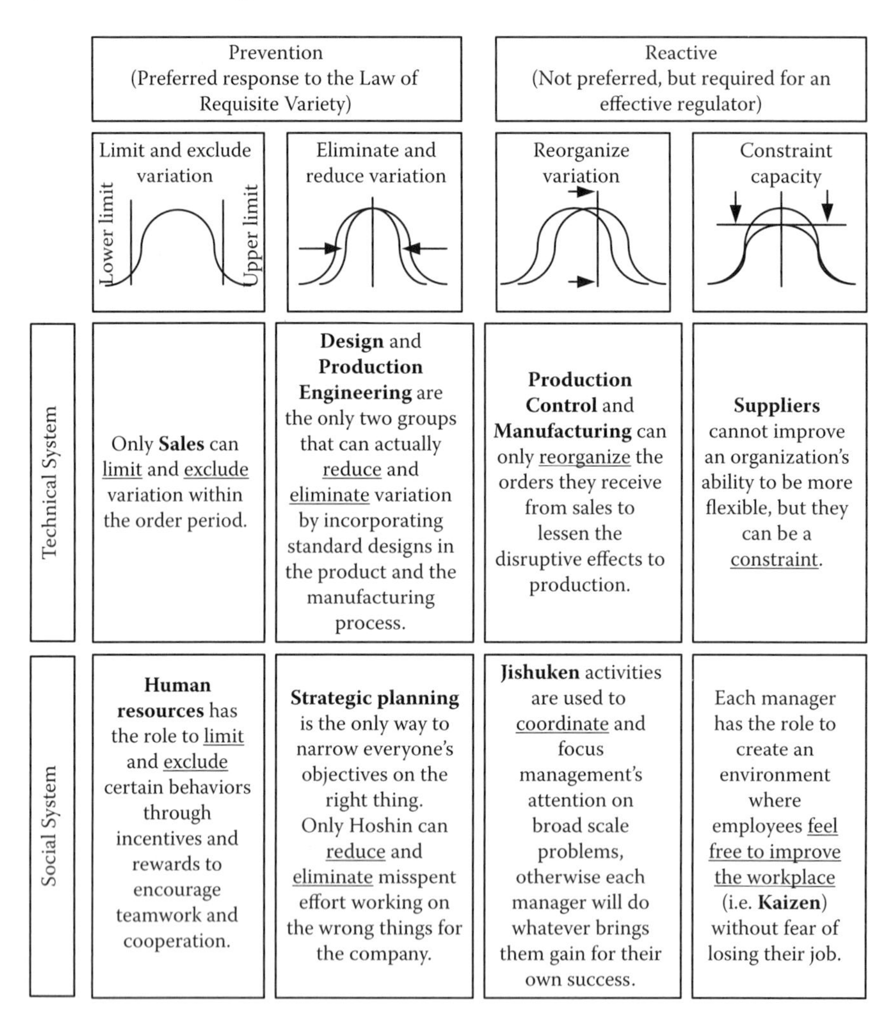

FIGURE 12.9
A social–technical perspective of production leveling as it relates to the law of requisite variety.

12.5.1 Technical System: Prevention

The technical system of production leveling relates to sales, design, and production engineering (PE) activities. These three departments are the only groups within Toyota that can prevent unnecessary complexity and variety from entering the system. Most research on production leveling does not emphasize these three groups. It is speculated that most organizations do not pursue variety reduction in these areas because they have an impact on the organization's policies and require a long-term perspective to show improvement.

Production leveling starts at the sales function mainly because orders can be excluded and adjusted according to an order cutoff period and the product allocation allowance. The order period serves to control the grouping of orders for the production build. Adjusting the order period allows market fluctuation to be excluded for later builds or combined with other orders when resources are low. The product allocation allowance describes the types and quantity of automobile models that can be produced and sold. Toyota manages the product allocation allowance, making demand more predictable and simplifying the complexity involved in performing production leveling (Figure 12.10). Both of these parameters are used by sales to avoid sets and combinations of orders that disrupt manufacturing.

The product design function has the goal of developing standard designs that allow common processes to be developed. Design is significant for product leveling because variation in manufacturing can be reduced by creating products that are less unique and require less specialized

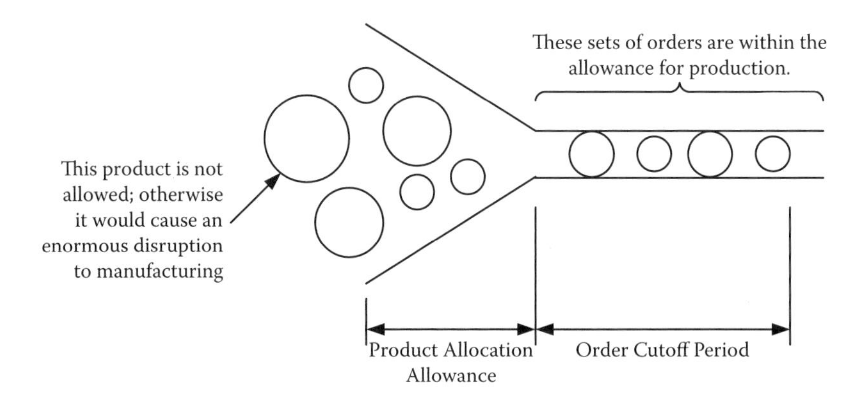

FIGURE 12.10
Complexity and variety elimination for sales.

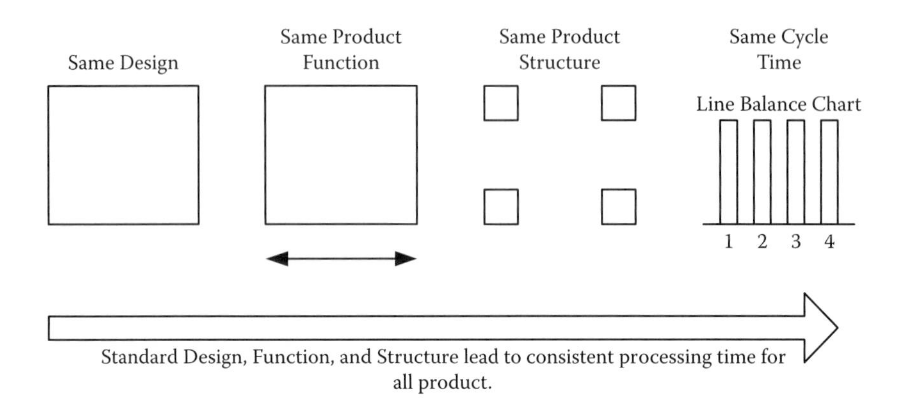

FIGURE 12.11
Complexity and variety reduction for design.

equipment (Figure 12.11). Toyota uses two criteria in the product design stage to improve the effectiveness of production leveling. The first criterion relates to the concept of standard designs for product functionality. Often, design has to differentiate the product and offer various combinations of features to maximize customer value. When design offers products with different functions, the goal is to use the same manufacturing process so that processing time can be held constant. Operations can be impacted negatively if each new design causes processing times to vary. Ultimately, product design variations are avoided using standard designs so that manufacturing can perform common operations consistently. Another critical factor in product design is the product structure or inner features of the product often used for material handling. The product structure offers opportunities for design to make the product more manufacturing friendly. If the product structure is not standardized, processing time may vary due to variations in changeover and the use of different jigs, tooling, and equipment.

The PE function defines and develops the manufacturing process; therefore, it will consider various production-leveling factors and plan for activities that lead to shorter processing times. Toyota employs four parameters in PE to reduce and eliminate production variation caused by market fluctuation: production layout, repeatable work sequences, elimination of uniqueness in equipment, and the establishment of common work methods.

Production layout aims to reduce processing time by arranging equipment and processes according to the flow of value-added operations.

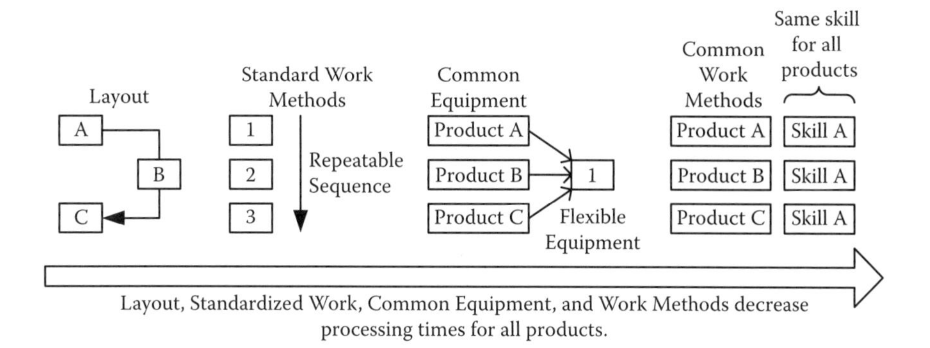

FIGURE 12.12
Complexity and variety reduction for production engineering.

Reductions in processing time are significant because they allow manufacturing to hold less inventory and run closer to actual orders. One of the main goals of PE is to develop work methods so that operations can be repeatable. Toyota employs standardization, standardized work, and work standards to differentiate normal conditions from abnormal conditions in the manufacturing process. This differentiation is a prerequisite for maintaining stable and predictable operations. Third, PE has the objective of eliminating uniqueness in equipment. Equipment versatility is a direct factor for maintaining constant processing times across different products. Last, production variation cannot be reduced if common work methods are not applied to several model types. Ideally, a change in the model does not appear different to the team member performing the job. In this context, PE has the role of making work methods common (Figure 12.12).

12.5.2 Technical System: Reactive

The regulation side of production leveling is the starting point for most outside practitioners when emulating Toyota. This is because most authors and researchers portray production leveling solely as a manufacturing activity. As stated previously, a fully functional regulator must be able to handle variety and complexity from the entire system. Unfortunately, the operations such as production control (PC) and manufacturing do not have an equal number of responses for each environmental condition. Most attempts to apply production leveling in PC and manufacturing fail because there are just too many alternatives to consider. Even if an organization does well at regulating resources, the organization is most likely to stay in an organic state in which ad hoc decision making becomes the norm. As one might expect, Toyota is trying to move from unknown to known, which is a mechanistic view of creating a system in which only a few alternatives have to be planned well in advance.

Toyota utilizes the PC function to lessen the disruptive effects of market fluctuation by reorganizing orders they receive from sales as the production plan is created. This reorganization is an attempt by PC to provide worker allocation opportunities for manufacturing. If manufacturing can receive a predictable schedule, downtime can be minimized, and the use of unnecessary resources can be avoided. When there are problems in meeting the production plan, PC and manufacturing will work together to make adjustments to prevent worst-case outcomes. Figure 12.13 shows the relationship and interdependence between PC and manufacturing.

The chart in Figure 12.13 shows that PC generates two schedules: a forecast and a actual schedule. A forecast is used for processes that have a long lead time; an actual schedule is used for processes that have a short lead time. Every month, PC analyzes the work content on a macro level to create a schedule that attempts to avoid major disruptions in manufacturing. PC has the goal to turn lines, cells, and portions of the plant "on" and "off" to level customer demand. Once a process is turned on, manufacturing has the role of balancing out the workload within the line or cell. In summary, PC levels the work across the plant, while manufacturing balances the work across the line or cell.

A technique that is used to level production is pattern production. Pattern production allows manufacturing to receive a predictable schedule that is a repeating sequence of orders throughout a period. The repetition in the schedule allows manufacturing to prepare how to utilize resources most effectively over a planning period. Without using pattern production, feeder processes would speed up and slow down, causing resources to be wasted. While not all possible scenarios can be avoided for manufacturing due to the complexity of option content in vehicles, PC has the goal to analyze work content and avoid combinations that can disrupt manufacturing (Figure 12.14).

Manufacturing's role in production leveling is to keep workstation loading as even as possible. This can only happen if PC can produce a leveled schedule that allows the line or cell to be loaded as evenly as possible. If PC issues a schedule that makes the line or cell speed up or slow down to support final assembly, workstation loading will be sporadic. In a way, PC

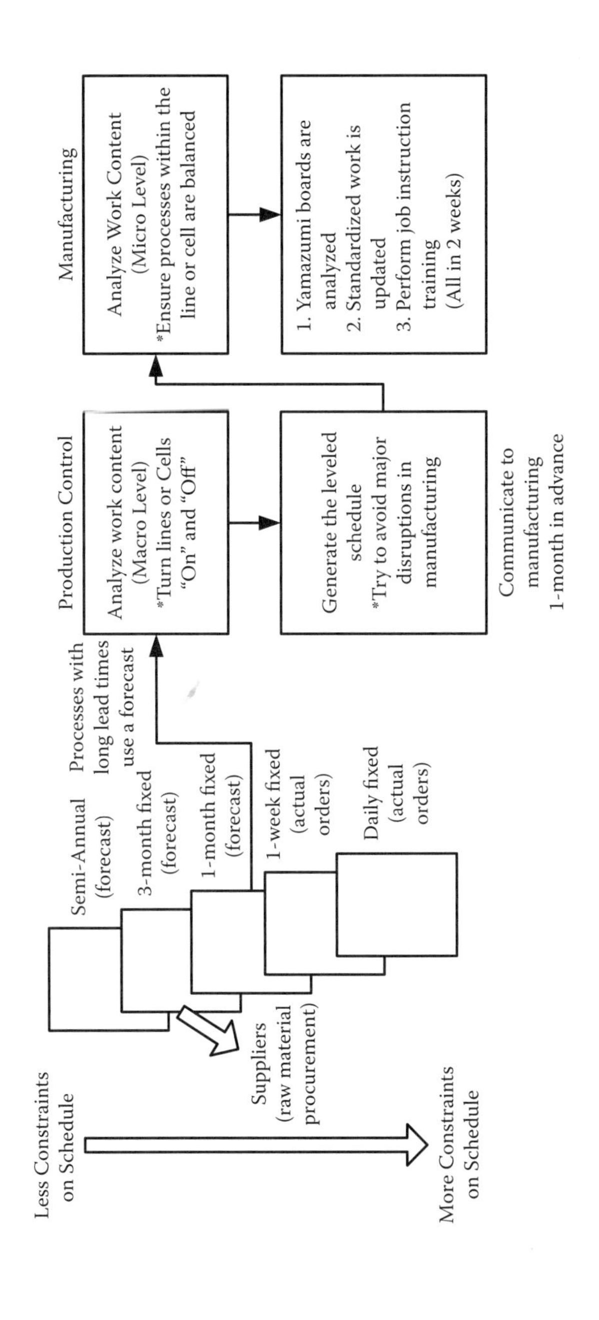

FIGURE 12.13
The regulator for production leveling: PC and manufacturing.

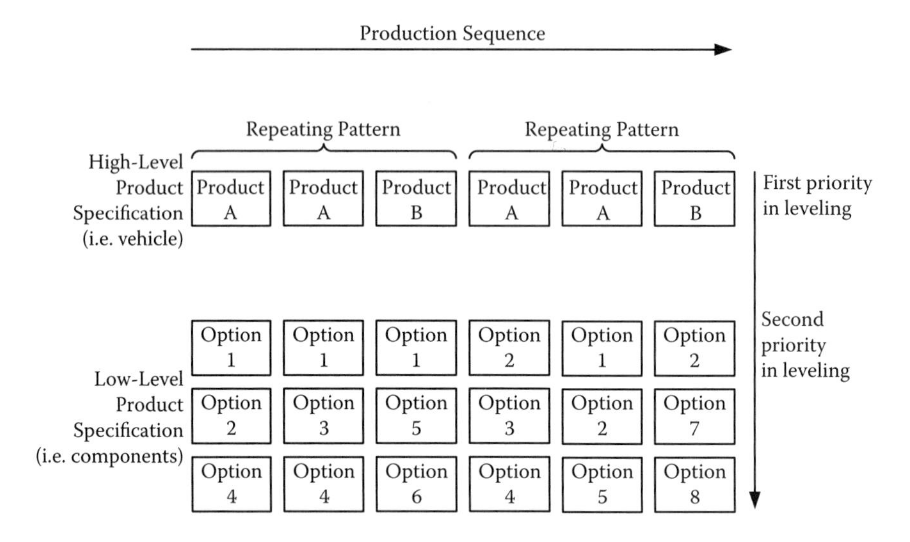

FIGURE 12.14
Pattern production and the priority for a mixed-model vehicle assembly line.

has the goal of loading feeder processes evenly (i.e., at the macro level), while manufacturing has the goal of loading workstations evenly (i.e., at the micro level).

A technique that manufacturing uses to regulate variety and complexity at the workstation level is yamazumi, more commonly known as line balancing (Figures 12.15 and 12.16). Yamazumi is a visual technique used

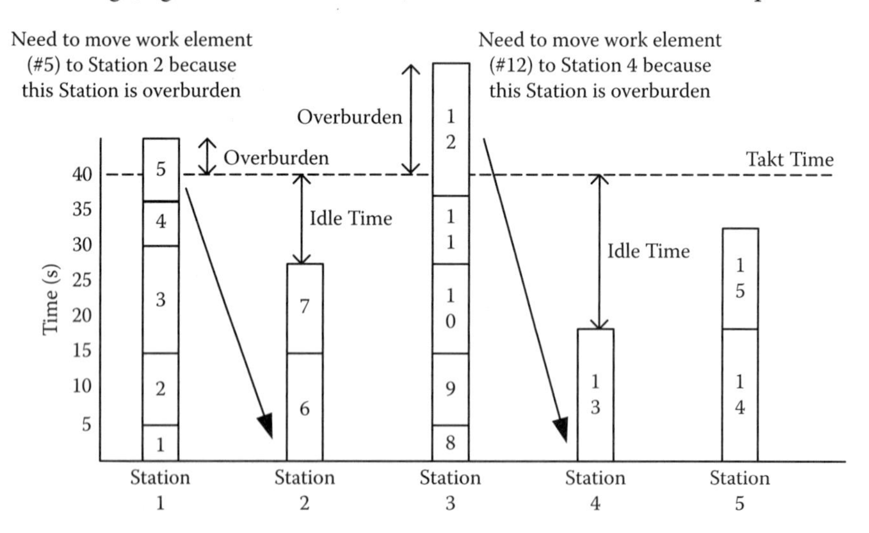

FIGURE 12.15
Yamazumi: unbalanced.

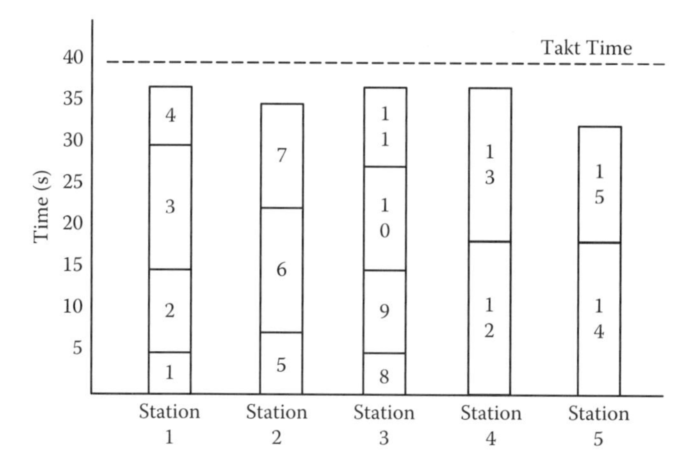

FIGURE 12.16
Yamazumi: balanced.

by manufacturing to allocate and distribute work at the line or cell level. Just as in conventional line balancing, major work elements of the job are identified and distributed to different areas of the line, cell, or station. The goal is to prevent overburdening and to reduce idle time.

12.5.3 Social System: Prevention

It is argued that Toyota's management system (i.e., social system) in production leveling is equally, if not more, important than its technical capabilities (Figure 12.17). The social aspects of production leveling relate to the company's ability to eliminate barriers that keep management from applying teamwork throughout the organization. Most Lean practitioners view that the technical system can operate independently of an organization's management system, which means that the management system can use any incentive system. Consider the task of implementing production leveling without asking each manager to understand the inner functions of other departments. Managers will have to be open and willing to spend significant time outside their department and be convinced that helping other managers benefits them and the company overall. The social factors of interdepartmental cooperation, employee relations, and management development are much more complex and varied than resolving how to establish a production sequence.

Human resources (HR) provides many vital elements in the achievement of production leveling, especially how managers are expected to

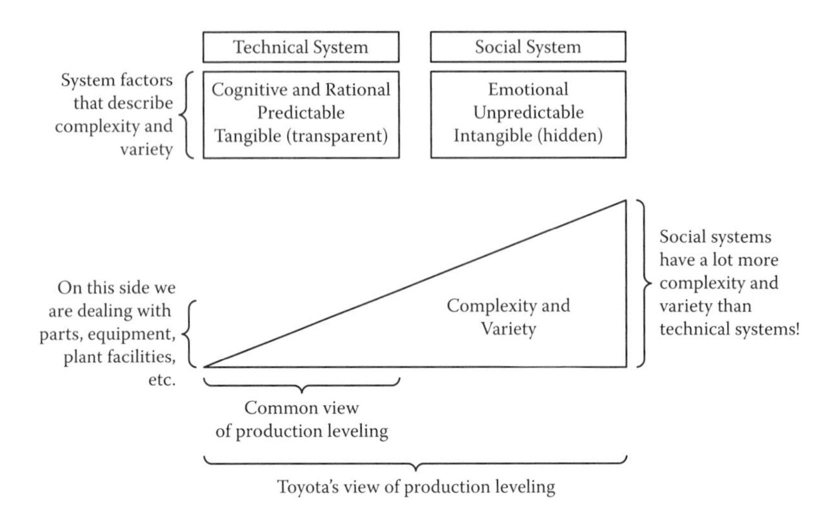

FIGURE 12.17
A comparison of complexity and variety between the technical and social systems of production leveling.

cooperate in achieving companywide targets and goals. HR is significant because production leveling introduces many unintended social mechanisms that can impede teamwork. For example, matching internal resources to external demand often causes resources to shift from one area to another. Managers naturally want to protect the interest of their unit, which creates a protective boundary over resources between units (Figure 12.18). Most managers are not inclined to give away resources unless it benefits them in the process. Because managers inherently will do whatever brings them gain and avoids problems for them, most managers are not motivated to cooperate unless it benefits them. The end result is managers who are less committed in achieving another department's target or goal. HR is essential in production leveling because it establishes the expectation for teamwork.

HR acts as both a preventive and a regulatory function in reducing and responding to variety and complexity as they relates to production leveling. HR has the role to limit and exclude certain behaviors from the workplace, such as managers who are not willing to cooperate or work in a team. The way that HR limits non-TPS behaviors and attitudes is through the appraisal performance process. HR evaluates managers on how well they solicit understanding from impacted groups, as well as how often they participate in another group's activities. Because production leveling will have an impact on how many departments operate, HR is necessary

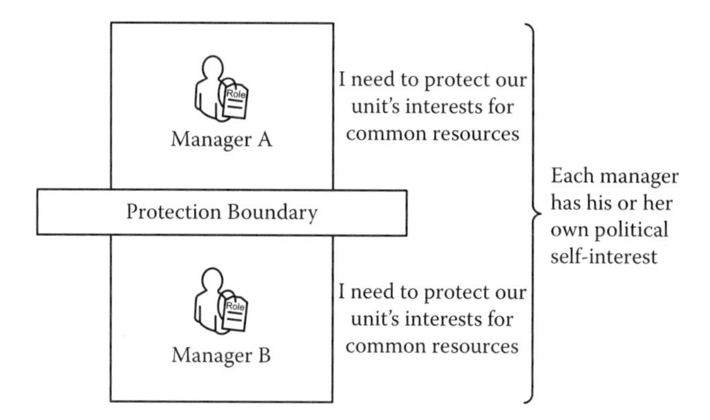

FIGURE 12.18
Managers typically generate their own political self-interest for common resources.

for setting the level of commitment for each manager. In most situations, managers will participate in production leveling but will have no intention of implementing such systems because individual incentives do not support it. Toyota achieves the right level of commitment because it rewards managers based on teamwork, specifically how the team performed and carried out each of the unit's functions. HR systems that do not consider evaluations based on teamwork and process are less able to limit and exclude undesirable behaviors when working toward production leveling.

Another way that TPS acts as a way to prevent unnecessary complexity and variety in the social system of production leveling is through hoshin kanri. Hoshin aligns the resources of the organization and identifies how each unit is expected to contribute. Hoshin encourages departments to share the same goals, establish strong lines of vertical and horizontal communication among departments, and coordinate what must be done from all areas of the company. Hoshin is a preventive measure against complexity and variety because it allows work groups to work together in advance to understand how each unit's operations affect one another. In this way, joint agreements can be established between departments to avoid unnecessary abnormality handling. While not all variety handling can be prevented, joint agreements provide departments a way to control abnormality between units before problems grow and expand. In the long run, hoshin provides a formal and mechanistic approach to plan in advance a few of the most obvious alternatives to dampen the effects of complexity and variety (Figure 12.19).

Hoshin is also critical for production leveling because it provides a constant tracking mechanism that can direct and aim management's attention

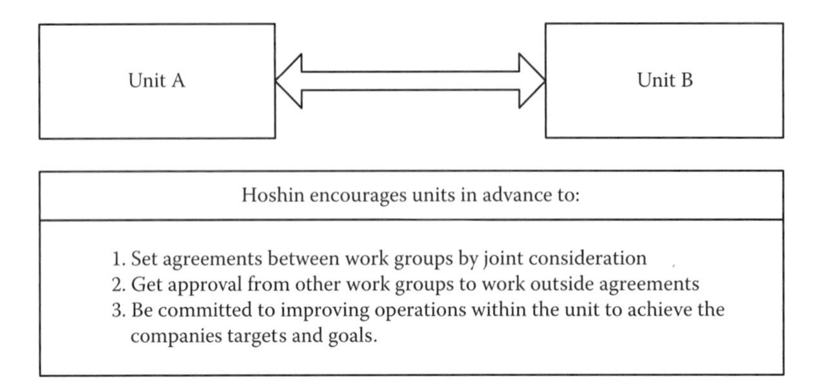

FIGURE 12.19
Annual business planning: preventive measure to reduce and eliminate variety and complexity.

to the changing conditions of the company. Due to the open nature of production leveling, it is sometimes difficult for managers to predict how to apply or adjust their leveling strategies. Without hoshin, departments outside manufacturing will not know how to contribute or know that they are contributing in the right way. Because of the complexity and variety in designing a technical system that is constantly changing due to the dynamic nature of the market, a manager's best intentions are not always enough to support another department's functions effectively. In a lot of ways, it is more important for managers to be wrong together rather than any one manager to be right by him- or herself. Hoshin provides a mechanism that forces managers to share the same objective by staying connected with each other's operations (Figure 12.20). This is instrumental because no department manager can solve all of his or her problems alone, and no department manager can improve the performance of the organization alone. Hoshin reinforces the idea that teamwork is required.

Finally, hoshin is used to reduce and eliminate variation in the most important stages of new model development, namely, design and PE. Hoshin provides a medium- to long-term perspective that is necessary for designs, structures, and equipment operations to be standardized. Standardized operations imply that cycle time variations in the manufacturing process can be lessened. Hoshin is critical for production leveling because it provides the proactive planning that allows many of the departments outside manufacturing to participate in removing unwanted cycle time variations associated with different products. Hoshin is an effective management tool because it forces managers to reflect on production

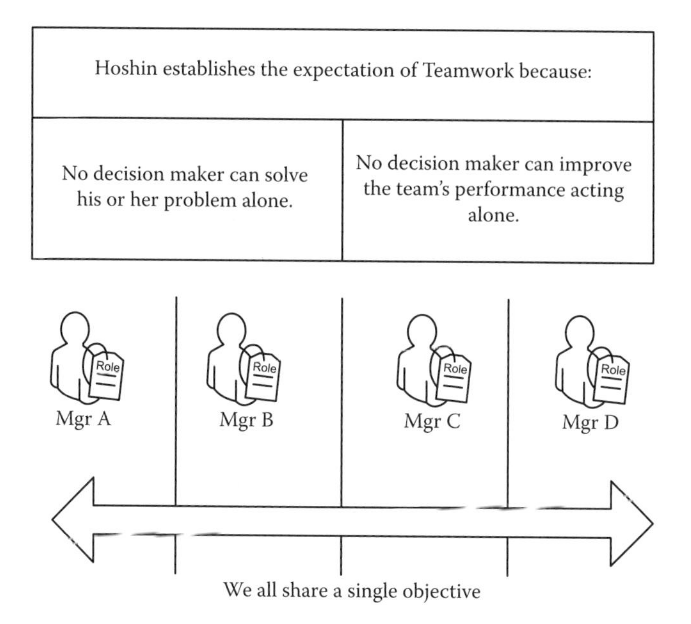

FIGURE 12.20
Hoshin kanri cuts across departments by defining each unit's role for companywide goals and targets.

activities that did not go well during the previous year. The reflection process in hoshin encourages managers to engage in problem solving to understand the factors that led to an unstable and unsuccessful system. Simply, hoshin is the basic planning tool that anticipates problems, tracks problems, and keeps problems from returning.

12.5.4 Social System: Reaction

The reactive and social approach to production leveling relates to activities that managers perform to regulate variety and complexity. In this context, managers must be able to respond to each social condition in production leveling. One of the ways managers are able to work out problems and issues that do not have a predetermined alternative to complexity and variety are jishukens (Figure 12.21). Jishukens are management-led problem-solving activities that attack broad-scale system-related issues. Jishukens coordinate and focus management's attention on problems that cannot be solved within the unit. When managers have difficulty coping with system-related issues, jishukens are a natural mechanism to call for help. While jishukens are structured, team-based activities, they tend

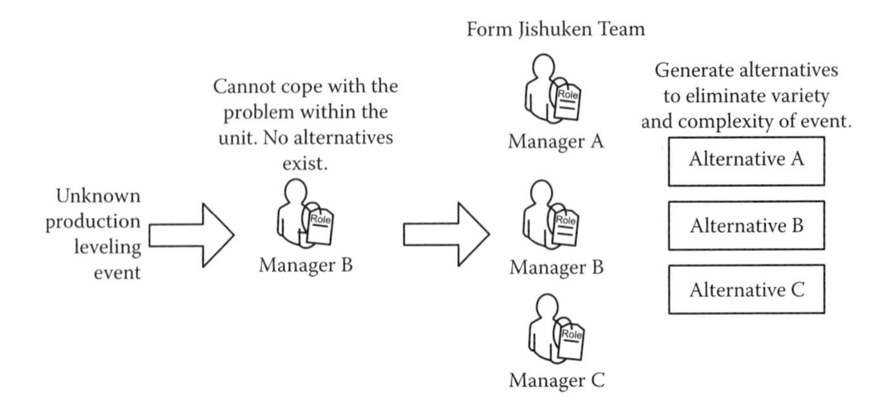

FIGURE 12.21
Variety handling using jishukens in production leveling.

to be organic in nature, mainly because they require a diverse group of managers to solve open-ended problems. Jishukens provide a medium for managers to sort out a variety of different alternatives to reduce the complexity of achieving production leveling.

Toyota's concept of kaizen also allows employees to work with trust in the workplace without fear of losing their job. This is significant in production leveling because departments will attempt to develop methods never tried before and will make mistakes while exploring various production-leveling alternatives. In traditional management systems, mistakes are generally negative. Employees often hesitate in informing management about problems or do not feel comfortable exposing problems. The sprit of kaizen provides employees a level of security that if they make a mistake trying to improve the workplace it will never be the cause of being fired. Thus, the development of production leveling should promote departments to experiment freely without fear that their job is ever in jeopardy.

12.6 SUMMARY

This chapter started with the idea that a system must have at least the same number of alternatives as the complexity of its environment for it to survive. When an environment creates a condition that cannot be prevented or regulated, that organism cannot survive. According to Ashby's law of requisite variety, variety can only be destroyed by variety, meaning a system must be able to have as much complexity as the environment for it to live.

The best example of complexity as it relates to TPS is production leveling. Production leveling is the activity that tries to match internal resources to external conditions. All organizations go through production leveling to some degree, but Toyota has been extremely successful at it because it considers a larger scale of factors that ultimately make up the complete variety and complexity of the system.

Technical and social factors were presented to show how Toyota responds proactively and reactively in achieving production leveling. Systems theory emphasizes that organizations that are more proactive in complexity reduction require less variety handling, which is the best choice for managing variety. This chapter illustrated various ways that Toyota's technical system is superior to most organizations by its efforts in the early stages of new model development, specifically sales, design, and PE. Manufacturing and PC can only reorganize variety, which is not much of a variety-handling alternative compared to the complete picture of complexity management. The social system of production leveling was also introduced because of the uncertainty and unpredictability of managers working in a system that causes natural tendencies to go against departmental cooperation and teamwork. Toyota's proactive stance in HR and hoshin limits and reduces variety in getting managers to cooperate to achieve companywide targets and goals. Last, regulatory measures were also presented as they relate to the social system of TPS, specifically how managers sort out problems when dealing with unexpected scenarios in production leveling and how employees should feel comfortable and confident in establishing kaizen in the workplace.

REFERENCES

Andel, T. (1999) Accentuate heijunka, eliminate junk. *Supply Chain Flow*, Vol. 54, No. 8, 77.

Ashby, W. (1956) *An Introduction to Cybernetics*, Vol. 2, Chapman and Hall, London.

Baumol, W. J. (1951) *Economic Dynamics*, Macmillan, New York.

Beckman, M. J. (1961) Production smoothing and inventory control. *Operations Research*, Vol. 9, 456.

Burns, T. and Stalker, G. (1961) *The Management of Innovation*, Tavistock, London.

Coleman, J. B. and Vaghefi, M. (1994) Heijunka: A key to the Toyota Production System. *Production and Inventory Management Journal*, Vol. 34, No. 4, 31–35.

Conant, R. and Ashby, W. (1970) Every good regulator of a system must be a model of that system. *International Journal of Systems Science*, Vol. 1, No. 2, 89–97.

Duimering, P. and Safayeni, F. (1991) A study of the organizational impact of the just-in time production system, in *Proceedings of the International Conference on Just in Time Manufacturing Systems: Operational Planning and Control Issues (Montreal)*, pp. 19–32.

Duimering, P., Safayeni, F., and Purdy, L. (1993) Integrated manufacturing: Redesign the organization before implementing flexible technology. *Sloan Management Review*, Vol. 34, No. 4, 47–56.

Elmaleh, J. and Eilon, S. (1974) A new approach to production smoothing. *International Journal of Production Research*, Vol. 12, No. 6, 673–681.

Gerwin, D. (1993) Manufacturing flexibility: A strategic perspective. *Management Science*, Vol. 39, No. 4, 395–410.

Gigch, J. (1978) *Applied General Systems Theory*, Harper & Row, New York, NY.

Hall, A. (2006) Introduction to Lean—sustainable quality systems design, published by Arlie Hall, EdD, Lexington, KY, ISBN 0-9768765-0-7.

Heylighen, F. (1992) Principles of systems and cybernetics: An evolutionary perspective, in *Cybernetics and Systems '92*, ed. Trappl R. Singapore: World Science, pp. 3–10.

Huttmeir, A., Treville, S., Ackere, A., Monnier, L., and Prenninger, J. (2009) Trading off between heijunka and just-in-time-sequence. *International Journal of Production Economics*, Vol. 118, No. 2, 501–507.

Katz, D. and Kahn, R. (1978) *The Social Psychology of Organizations*, 2nd ed., Wiley, New York.

Kovalyov, M., Kubiak, W., and Yeomans, J. (2001), A computational analysis of balanced JIT optimization algorithms. *INFOR*, Vol. 39, No. 3, 229–316.

Lawrence, P. and Lorsch, J. (1967) Differentiation and integration in complex organizations. *Administrative Science Quarterly*, Vol. 12, 1–47.

Liker, J. (2004) *The Toyota Way*, McGraw-Hill, New York.

Love, T. and Cooper, T. (2007) Complex built-environment design: Four extensions to Ashby. *Kybernetes*, Vol. 36, No. 9/10, 1422–1435.

Miltenburg, J. and Sinnamon, G. (1992) Algorithms for scheduling multi-level just-in-time production systems. *IIE Transactions*, Vol. 24, 121–130.

Mintzberg, H. (1979) *The Structure of Organizations*, Prentice Hall, Englewood Cliffs, NJ.

Monden, Y. (1998) *Toyota Production System—An Integrated Approach to Just in Time*, 3rd ed., Engineering and Management Press, Norcross, GA.

Ohno, T. (1988) *Toyota Production System: Beyond Large-Scale Production*, Taylor and Francis, Boca Raton, FL.

Okamura, K. and Yamashima, H. (1979) A heuristic algorithm for assembly line model-mix sequencing problem to minimize the risk of stopping the conveyor. *International Journal of Production Research*, Vol. 17, No. 3, 233–247.

Rinehart, J. (1997) After Lean production: Evolving employment practices in the world auto industry. *American Journal of Sociology*, Vol. 104, No. 4, 1212–1214.

Scala, J., Purdy, L., and Safayeni, F. (2006) Application of cybernetics to manufacturing flexibility: A systems perspective. *Journal of Manufacturing Technology Management*, Vol. 17, No. 1, 22–41.

Shimokawa, K., and Fujimoto, T. (Eds., 2009). *The Birth of Lean*. The Lean Enterprise Institute, Cambridge, MA.

Swanson, R. (2008) A generalized approach to demand buffering and production levelling for JIT make-to-stock applications. *Canadian Journal of Chemical Engineering*, Vol. 86, 859–868.

Umpleby, S .(2007) Ross Ashby's general theory of adaptive systems. *International Journal of General Systems*, Vol. 38, No. 2, 231–238.
Xiaobo, Z., Zhou, Z., and Asres, A. (1996) A note on Toyota's goal of sequencing mixed models on an assembly line. *Computers and Industrial Engineering*, Vol. 36, 57–65.
Yano, C. and Rachamadugu, R. (1991) Sequencing to minimize overload in assembly lines with product options. *Management Science*, Vol. 37, No. 5, 572.
Zandin, K. B. (2001) *Maynard's Industrial Engineering Handbook*, 5th ed., McGraw-Hill, New York.

13

The System Properties of Interrelationship and Interdependence in TPS

13.1 SYSTEM PROPERTIES: INTERRELATIONSHIP AND INTERDEPENDENCE

One of the objectives of systems is to ensure that all subsystems work together and contribute toward the system's objectives. Relationships within a system are structured depending on the type of problem and how the parts of the system are satisfied as a whole. The basic idea of interdependence in systems is that the goals of a system determine how components interact given a specific environment (Figure 13.1) (Deutsch, 1949).

A benefit of interdependence theory is that it proposes a taxonomy of situations based on an analysis of various outcomes; furthermore, it relates categories of situations to particular types of goals. Interdependence in systems moves the thinking from a linear notion of causality into a dynamic interaction. The concept of recursion in systems implies that there is an ongoing mutual influence and shared contribution of components in a system. Relationships in a system evolve to the fluctuations of the environment and the responses from the interactions within the system.

The interdependence of variables in a system describes how a change in one part of the system can affect the total system. Interdependence exists when the outcome of a component within a system is affected by the actions of another. There are two types of interdependence, positive and negative. Positive interdependence is when the action of an entity promotes the achievement of joint goals. Negative interdependence is when the actions of a component obstruct the achievement of another within the same system (Johnson and Johnson, 1989). In organizations, positive and negative interdependence is often compared to cooperation and competition.

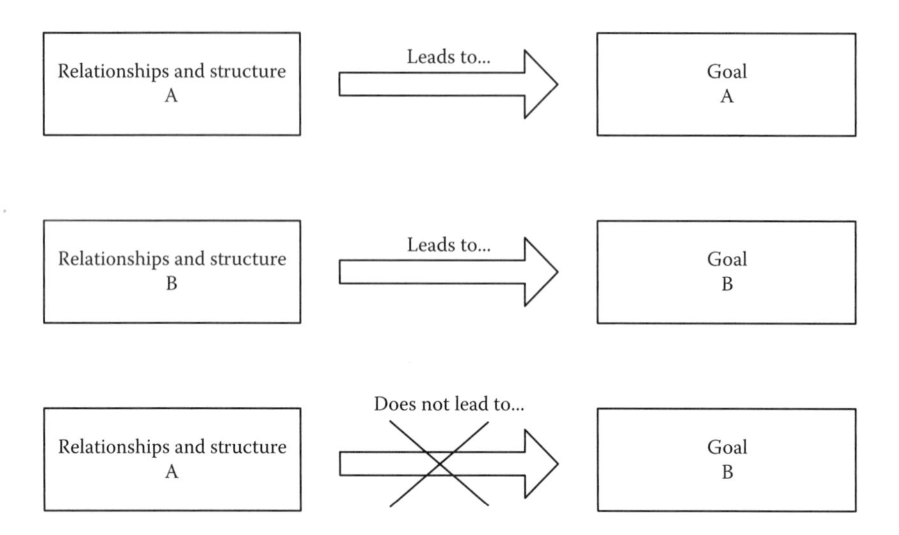

FIGURE 13.1
The goal of a system defines its structure.

Cooperation is when groups within an organization exchange information and resources to assist reaching common goals. Competition tends to obstruct parts of the organization by using tactics of coercion, threats, and deceptive communication to win at the expense of others (Deutsch, 1985).

The basis for the types of structures that exist in a system is largely dependent on how much a subsystem is interdependent (Nauta and Sanders, 2000). A system is dependent when the goal achievement of system A is affected by system B's actions. A system is independent when the goal achievement of system A is unaffected by system B's actions. When a system is independent, its parts can be treated in isolation because the behavior of one variable does not affect the total functioning of the system. However, according to experts, truly independent variables do not exist (Gigch, 1978). All systems have a relative wholeness, or else the system could not function as a unit (Putt, 1978).

Researchers have proposed that various forms of interdependence exist in systems (Figure 13.2). The weakest level of interdependence that can exist in systems is pooled interdependence. Pooled interdependence is when each part within a system can work independently of one another, but their indirect resources are shared (Thompson, 1967). Another interesting characteristic of pooled interdependence is that the total system is not affected by the order of action of lower subsystems' actions.

More complex forms of interdependence are sequential and reciprocal interdependence. Sequential interdependence describes interactions

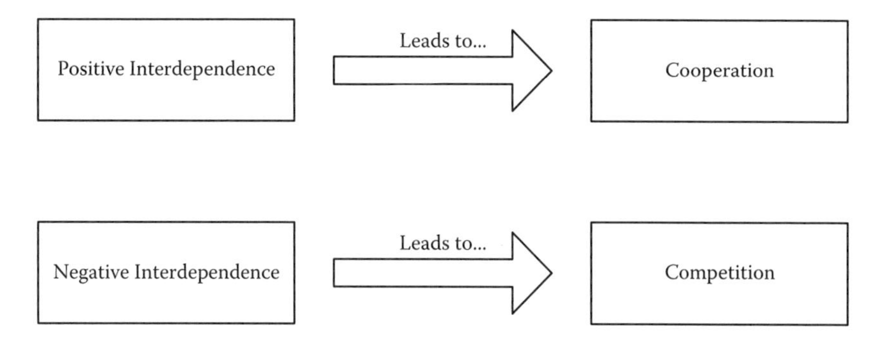

FIGURE 13.2
Structures in a system describe cooperative or competing behaviors of interdependence.

between components directly, meaning the input to one component is the output to another. Reciprocal interdependence refers to when the outputs of each component become the inputs for others. The key difference between sequential interdependence and reciprocal interdependence is that a system and another can influence each other simultaneously (Thompson, 1967).

A system is helpless when it or another part of the system cannot influence the system's goal achievement. In this context, the system is not inducible or does not have the ability to be influenced by another system. Inducibility provides the basis for channeling individual efforts of a subsystem to move toward a more cooperative (Deutsch, 1949) and interdependent form (Figure 13.3).

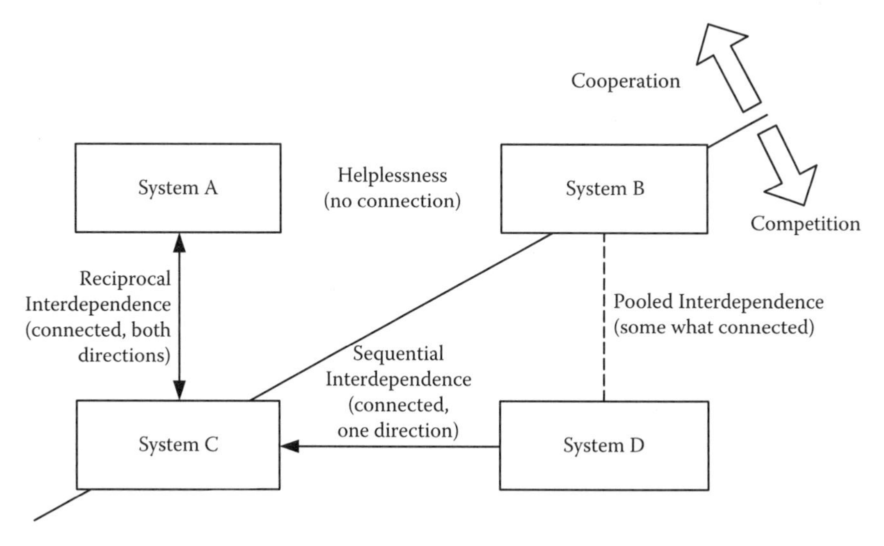

FIGURE 13.3
Types of interdependence in systems.

Research showed that cooperation promotes higher achievement and greater productivity than does competition in system effectiveness (Johnson and Johnson, 1989). Interdependent structures in systems that are reciprocal in nature behave in noncompetitive ways by meeting the goals of all subsystems involved (Nauta and Sanders, 2000). The positive interdependence of a system depends on the nature that components within a system can reach an agreement. In many situations, individual action by a component within a system is not enough, which leads to more inducible forms of cooperation. Usually, the effort required equals the alternative of failing when cooperation does not occur.

When a subsystem works toward competitive goals, individual acting components tend to engage in activities that are self-protecting. Self-protection is when a component of a subsystem operates as a self-contained, self-efficient organization, pursuing its own objectives. A variable that can influence the degree of self-protection in a system is the distribution of power. When there is an unequal distribution of power, the interdependence of the system will naturally serve the system's higher authority (Nauta and Sanders, 2000). In other cases, cooperation may be considered costly because of the amount of effort it takes to establish and maintain a cooperative system. If the individual action of a subsystem is feasible (i.e., pooled interdependence), cooperation is not a likely alternative to improve the system's performance.

13.2 INTERRELATIONSHIP AND INTERDEPENDENCE IN THE TOYOTA PRODUCTION SYSTEM

One of the most essential and basic examples of relationships and interdependence in the Toyota Production System (TPS) is the supply chain structure. No other example within TPS is more widespread or representational than Toyota's supply chain network. Suppliers make up 80% of the cost to produce a vehicle compared to internal operations, which explains why interrelationships and interdependence are significant in TPS.

Toyota's suppler development practices also demonstrate the minimum requirements that a structure must be organized for it to achieve its goals. Since no management system can completely control all of its interdependencies, supplier relationships represent the basic aspects of teamwork worth replicating. Not all interdependent structures can be effectively

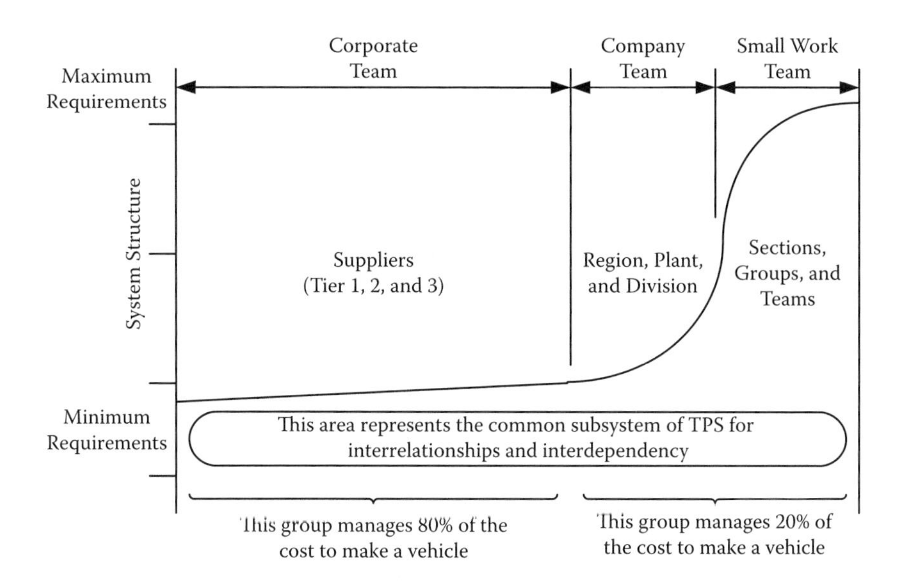

FIGURE 13.4
Interrelationship and interdependence in TPS.

maintained throughout the supply chain. Interdependence among suppliers is a choice for improved competitiveness rather than a cost to be minimized or avoided. The best way to understand the basic subsystem of interdependence within TPS is to study how Toyota forms relationships with its suppliers (Figure 13.4).

13.3 LITERATURE REVIEW: SUPPLIER DEVELOPMENT

Supplier development has long been used by companies as a way to compete globally (Motwani et al., 1999). In today's markets, properly managing the supply chain is increasingly important as success is tied to the capabilities and performance of suppliers (Carr et al., 2008). Suppliers have an impact on an organization's quality, technology, delivery, flexibility, and profits (Krause et al., 2000; Humphreys et al., 2004; Li et al., 2007; Carr et al., 2008). Supplier performance is often viewed as one of the leading contributors to enhance an organization's competitive advantage (Lemke et al., 2003).

Supplier development was first used by Leenders (1966) in a doctoral dissertation to describe efforts by manufacturers to increase the number

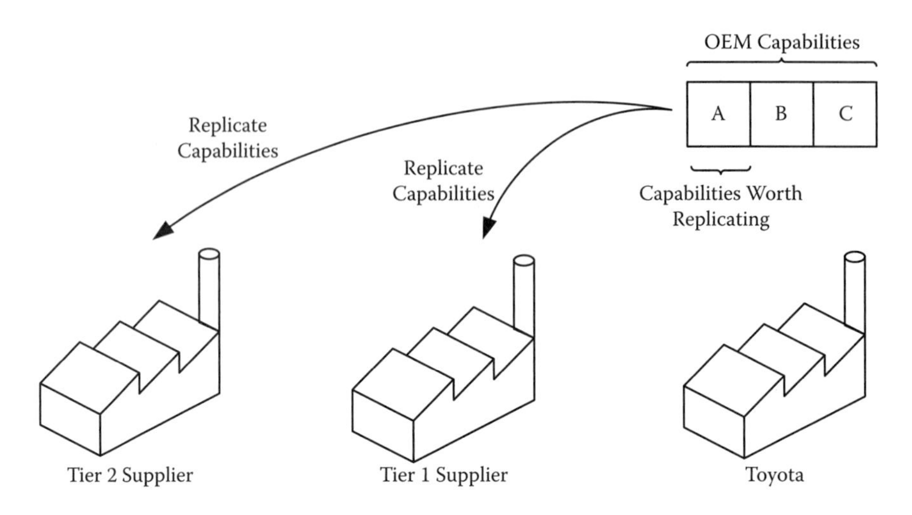

FIGURE 13.5
Supplier development purpose: to replicate OEM capabilities.

of viable suppliers and to improve suppliers' performance. Supplier development is a way to transfer and replicate an organization's in-house capability to improve a supplier's performance (Figure 13.5) (Sako, 2004; Knemeyer et al., 2008). The two most common replication approaches are system transfers (something that can be applied to the whole company) and process transfers (a specific practice that can be applied to a model line or product) (Figure 13.6). The overall approach to supplier development

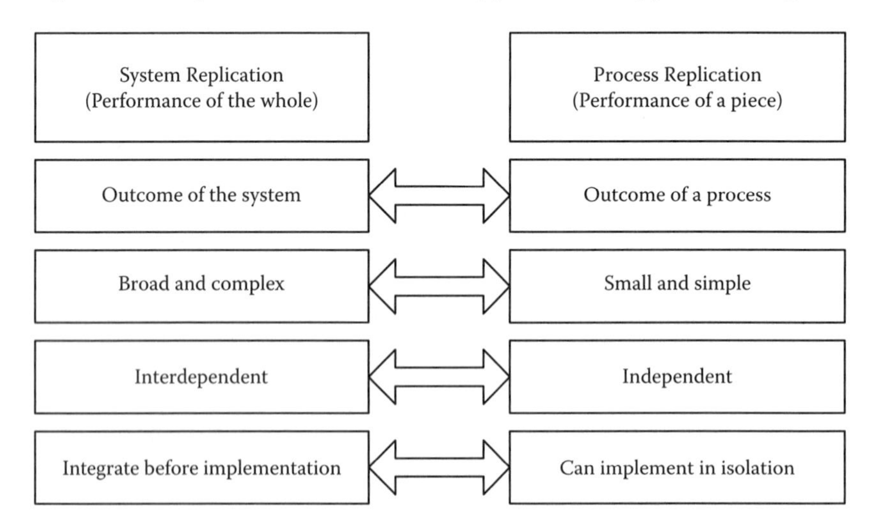

FIGURE 13.6
Two common views for supplier replication: system and process.

varies yet is often described as any effort of a buying firm to increase a supplier's performance (Krause and Ellram, 1997). In the most traditional terms, supplier development tries to qualify, integrate, and improve supplier capability (Lo et al., 2006).

Managing supplier relations has historically been the responsibility of an organization's purchasing department, sometimes referred to as procurement, sourcing, or materials management. Purchasing's basic function is to maintain a network of viable and capable suppliers (Krause and Ellram, 1997). Purchasing traditionally has supplier sourcing responsibilities (i.e., the searching for alternative sources); supplier integration responsibility (i.e., bringing the needed product in house); ongoing supplier quality assurance responsibilities (i.e., monitoring and assessing suppliers performance); and supplier development (i.e., replicating internal capabilities at suppliers) (Figure 13.7) (Wagner, 2006).

Currently, there exists a wide range of tools and strategies for establishing and assisting suppliers. While some focus is on the training aspects of supplier development, such as TQM (total quality management), TPM (total productive maintenance), SPC (statistical process control), and

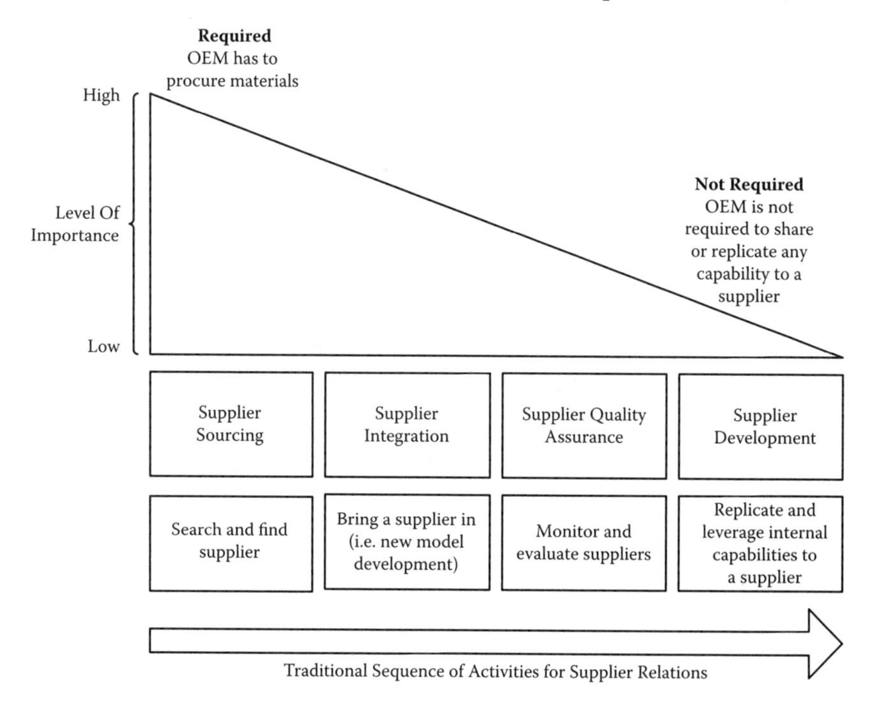

FIGURE 13.7
Purchasing functions and the traditional level of importance.

ERP (enterprise resource planning) systems (Wagner, 2006; Cagliano et al., 2006; Modi and Mabert, 2007; Govindan et al., 2010), most initiatives are centered on supplier performance. One of the most well-known supplier development initiatives is the International Organization for Standardization (ISO). Often a prerequisite for doing business in any parts of the world, ISO requires that suppliers incorporate quality into the production process from start to end. It requires a third-party assessment of both the products and the processes (Peach, 1992).

Engaging in long-term relationships with suppliers includes a wide range of benefits. Ellram (1991) indentified potential advantages of forming partnerships based on some of the early literature and observations of company experiences, namely, management advantages, technology advantages, and financial advantages. Partnering with suppliers or developing close relationships can enhance planning and information sharing, improve resource savings, and stabilize cost through long-term commitments and contracts (Knemeyer et al., 2008). Close ties with suppliers can also lessen risk associated with changing markets and volatile business conditions (Hartley and Choi, 1996). One of the primary advantages of building relationships with suppliers is that suppliers are more willing to share technology and offer their support, knowledge, and experience (Juettner, 2005).

But, not all suppliers are willing to open their books to the buying organization to help reduce costs (Ellram, 1996) or are willing to achieve common goals through mutual long-term efforts (Zaheer et al., 1998). The bottom line in supplier development is often a matter of collaboration (i.e., trust) or compliance (i.e., power). Most supplier development programs are accused of feel-good talk, yet performance measurement is often a higher criterion in supplier assessment in the short term (Chin et al., 2006; Sako, 2004).

An approach that has been used to offset trust and collaboration with suppliers has been the adoption of Japanese management practices popularized in the late 1980s. Japanese approaches tend to emphasize a more hands-on approach to supplier development, working side by side in cooperation, treating suppliers as equals (Hartley and Choi, 1996). It is speculated that Japan's unique organization and government structures are more likely to encourage a long-term relationship for both parties. However, in the United States the majority of suppliers are less dedicated or dependent on their customers for long-term survival as in Japan (Sako, 2004; MacDuffie and Helper, 1997). Supplier dependence is a strong factor/measure in the development of suppliers (Krause et al., 2007) and is often a function of sales provided to the vender (Crook and Combs, 2007). Suppliers who are more dependent on a buyer may be more willing to cooperate with the buyer (Laamanen, 2005), such as regarding change specifications, internal processes, or attending supplier training (Modi and Mabert, 2007).

13.4 A REVIEW OF TOYOTA'S SUPPLIER DEVELOPMENT PRACTICES

A more targeted approach to supplier development has been the adoption of Toyota's supplier development practices. An annual North American OEM (original equipment manufacturer) survey showed that Toyota was ranked number one for supplier relations for the seven-year period from 2003–2010 (Henke, 2010). The study further showed that Toyota was superior to the major U.S. and Japanese OEMs in supplier communication, OEM assistance, supplier profit, and partnering (Henke, 2010).

Toyota is recognized as a leader in supplier development, yet it was not until Toyota partnered with General Motors (GM) in 1984 that its supplier development relations were noticed on American soil. The United Auto Workers (UAW) not only had a strong presence in the big three but also had a long track record of constant clashes with American management practices (Newman and Rhee, 1990). UAW in many ways believed that workers pay the price of failure and management survives no matter what happens (Bieber, 1988). Toyota was in no position to export the concepts of TPS to suppliers if did not establish a solid base of support among its own people and operating groups. To make matters worse, at the time of the establishment of NUMMI (New United Motors Manufacturing), GM employees still had over 1,000 grievances and disputed firings on the books. Nonetheless, the NUMMI case is a historic event: Former GM employees were give the opportunity to be rehired and operate under a much different style of management. From Toyota's view, it would have been much easier to smooth out worker relations if they could revert to their traditional Japanese employment model. In Japan, workers are hired at the completion of high school, so there are no bad habits to unlearn or issues dealing with a tainted workforce (Newman and Rhee, 1990).

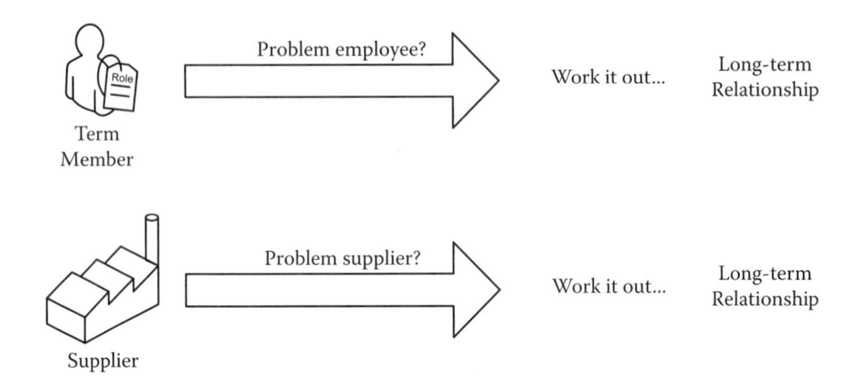

FIGURE 13.8
Toyota's supplier and employee relationship.

The NUMMI experience for Toyota could hardly be described as harmonious, both internally with their employees and for their suppliers (Figure 13.8). The common denominator was that both employees and suppliers were kept at "arms length," yet in Toyota's eyes a distant relationship was a natural creator of waste (Newman and Rhee, 1990). Toyota's purchasing values and buyer expectations represented an entirely new operating culture for American suppliers. Toyota's purchasing values are "to be the most admired auto manufacturer through respect for people and continuous improvement" (Toyota Motor Corporation (TMC), 2006a). Toyota's business partner approach meant that suppliers and dealers would work together through a long-term relationship to realize mutual growth based on trust (Toyota Motor Corporation (TMC), 2006b). Toyota's collaborative relationship with suppliers was not all that different compared to their own relationships with employees. Toyota believed that a work environment should be based on respect for employees, described as a monozukuri-based (value-adding) genchi-genbutsu (go-and-see) approach to work through relentless kaizen efforts and effective two-way communication. Toyota's supplier relationship has often been referred to as kieretsu or family-like (Dyer, 2000; Wu, 2003; Mollenkopf et al., 2010). While American practices were to blame suppliers when things went wrong, Toyota wanted to emphasize joint problem solving by discussing areas of mutual concern and offering suggestions for improvement.

Although Toyota's collaborative approach to supplier relations was refreshing for suppliers, it was not without difficulty. One of the most demanding aspects of supplier cooperation was getting suppliers to adopt the just-in-time (JIT) philosophy. JIT requires suppliers to make frequent deliveries in small lot sizes through synchronization and optimization of inventory levels. JIT also forces suppliers to be closer by exchanging production, technical, and logistic information (Cagliano et al., 2006). Most suppliers are willing to support JIT delivery using large end-of-line inventories but are not so eager to adopt JIT manufacturing if all inventory levels are optimized within the supplier's plant. Toyota's purchasing contract stipulates that suppliers must abide by all changes in delivery schedules or incur direct temporary suspension of scheduled shipments. Toyota will have no payment liability for undershipments, overshipments, nonconforming items, or delays. At the supplier's cost, shipments must be made in a timely manner as instructed by Toyota. If the supplier cannot ship, Toyota has the right to immediately acquire substitute or replacement items from one or more alternate sources (Toyota Motor Corporation (TMC), 2010). Interestingly, Toyota does excuse delays or failures that are beyond the control of suppliers, such as fires, floods, windstorms, explosions, riots, and acts of terrorism. However, suppliers are not excused from their obligations if their employees go on strike or participate in a slowdown or a lockout. Simply, Toyota believes that management is responsible for employee relations. While Toyota's delivery expectations may sound severe, JIT requires close coordination and a clear understanding of each other's interlocking network (Srinivasan, 2004).

In the 1980s, a number of internal and external groups grew in support of Toyota's supplier development network. While some groups were created from initiatives, some became permanent and gradually changed, allowing other departments or groups to absorb functions during maturity. One example is Toyota's Supplier Commodity Engineering (SCE) group. SCE began as the Supplier Improvement Committee (SIC) in 1988 and was a hybrid group composed of purchasing, quality, and production control (PC) to improve supplier quality. SCE was made a permanent corporate-level entity around 1996 after several successful initiatives by the SIC. Today, the SCE group acts much different from its original SIC predecessor. The SCE group can best be explained like a production engineering function for suppliers (Alloo, 2010). SCE evaluates suppliers' process engineering functions for major cost reductions associated with supplier tooling, equipment, fixtures, and technology. Since Toyota purchases most supplier tooling, it can be understood why Toyota would devote an entire corporate group to analyze cost savings with major purchases.

Other corporate groups that offer assistance for suppliers are the Operations Management Consulting Division (OMCD) and the Operations

Management Development Division (OMDD). OMCD was established in 1969 (originally named the Production Survey Office) as the guardian of TPS to implement TPS at Toyota plants and the plants of Toyota suppliers. OMCD is housed under the Production Control Division and today contains around fifty members (Shimokawa and Fujimoto, 2009). OMDD is the regional corporate office for North American operations. OMCD and OMDD jointly conduct improvement activities (more commonly known as jishukens) with suppliers on site as needed (Marksberry et al., 2010, 2011). Since OMCD and OMDD are decoupled from purchasing, suppliers are more likely to collaborate with Toyota's jishuken processes for improvement without fear of passing on price or cost reductions to purchasing (Sako, 2004).

There are also a number of external sources to assist Toyota suppliers. In 1989, an independent corporation named the Bluegrass Automotive Manufacturers Association (BAMA) became available to assist local Kentucky suppliers of parts to TMMK (Toyota Motor Manufacturing Kentucky). The purpose of BAMA is to share the Toyota philosophy, techniques, and experiences with the supply base, providing an opportunity for interaction between Toyota and BAMA members and among members themselves (http://www.bama-group.org). The BAMA forum provides Toyota suppliers lateral opportunities to learn from one another by performing shop floor improvement activities at each other's plant sites. Toyota acts as a consultant and an administrative facilitator in the development of the program, offering assistance when needed. The sharing of continuous improvement methods between members should lead to long-term survival and mutual prosperity among members (http://www.bama-group.org).

Another entity that supports supplier development is the Toyota Supplier Support Center (TSSC), renamed in 2009 as the Toyota Production System Support Center. TSSC was established by Toyota in 1992 to assist North American manufacturers in implementing their own version of TPS. Formed in response to increasing interest in TPS, the TSSC philosophy emphasizes customers first, people as intellectual assets, kaizen (continuous improvement), and shop floor first (Toyota Supplier Support Center (TSSC), 2010). While BAMA provides an opportunity for Toyota suppliers to share best practices, TSSC is a more dedicated resource for assisting suppliers on specific TPS applications otherwise not available in a forum format (Figure 13.9).

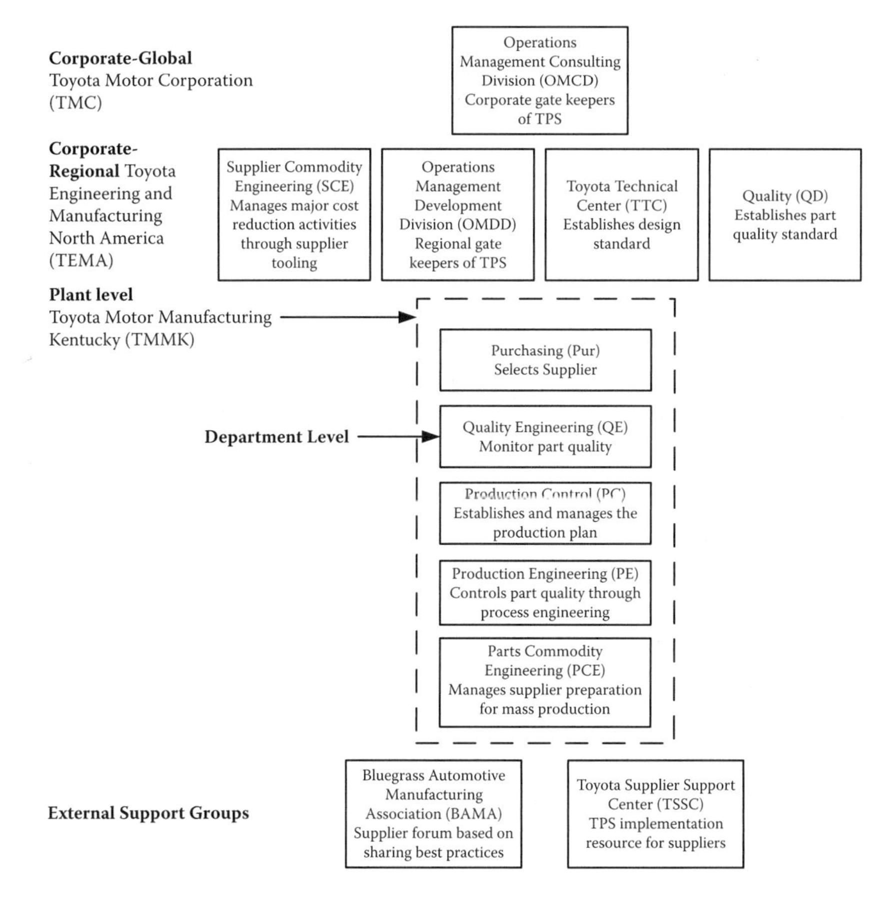

FIGURE 13.9
Representation of Toyota's supplier development structure.

In 2001, Fujio Cho, at the time president of Toyota Motor Corporation, released a collection of managerial values and beliefs named the "Toyota Way" to unify how every team member at Toyota is expected to approach work (Toyota Manufacturing Corporation (TMC), 2001). The Toyota Way is based on two pillars, continuous improvement and respect for people. The continuous improvement pillar is represented by three other values: challenge, kaizen, and genchi genbutsu. The other pillar is also represented by two other values: teamwork and respect for people. Toyota's two pillars are sometimes described as the hard and soft aspects of the Toyota Way. The Toyota Way is believed to be a more articulated explanation of Toyota's management system that summarizes both the human and the technical aspects of TPS.

One of the ways that Toyota tries to develop a partnership with suppliers is through teamwork. Toyota views teamwork as an essential element in developing strong and mutually beneficial supplier relations. Toyota believes that teamwork can best be achieved by setting expectations early and establishing clear roles using cross-functional project teams. One of the ways that suppliers benefit from the Toyota Way is through open and honest communications. Toyota expects its suppliers to feel comfortable communicating "bad news first" to identify problems promptly before they grow and expand. While no supplier wants to inform the customer when there are problems, Toyota recognizes how the supplier–customer relationship can influence how quickly suppliers notify when problems occur.

The respect for people value includes the concerns of suppliers as well as Toyota's own employees. One of the ways that Toyota respects its suppliers is through establishing systems and channels that encourage suppliers to voice their opinions and concerns throughout the development cycle. Toyota utilizes a Supplier Parts Tracking Team (SPTT) to track suppliers during preproduction. The SPTT is composed of purchasing, Parts Commodity Engineering (PCE), Production Engineering (PE), Quality Engineering (QE), Toyota Technical Center (TTC), Quality Development (QD), and Production Control (PC). Toyota's cross-functional team allows suppliers multiple channels to exchange information throughout the product development cycle. Toyota's development cycle is also divided into stages to allow suppliers to voice concerns early and to minimize risk to the program.

The hard side of the Toyota Way influences supplier relations more directly. While the soft side of the Toyota Way drives culture between supplier relations, the hard side drives productivity. One of the first values, challenge, represents how suppliers are to achieve cost reduction through benchmarking. Suppliers are expected to engage in cost reduction activities with Toyota in the initial stages of product development. For six weeks, Toyota evaluates suppliers' capability in achieving the best design for North America where suppliers are providing design alternatives. This activity is challenging for suppliers because sourcing has not been finalized, nor will it come soon in the next stage of product development. Toyota continues working with suppliers on evaluating design alternatives right up to tooling approval, which for the automotive industry is late at twenty weeks, when at forty weeks, mass production starts.

Kaizen for suppliers has been discussed as it relates to purchasing's approval process; however, in the development cycle it relates to how

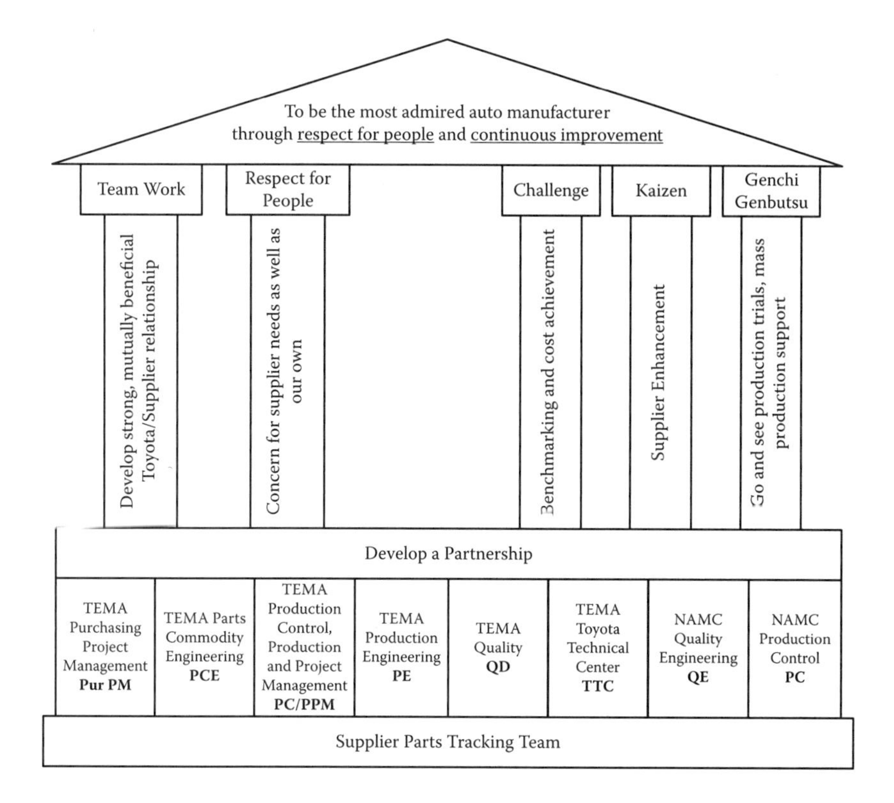

FIGURE 13.10
Representation of the Toyota Way for suppliers.

suppliers mitigate risk. Toyota refers to kaizen as supplier enhancement, which involves a set of tools to identify high-impact parts and design features that are critical in performance and function. Toyota provides a standardized approach for suppliers to mitigate risk in their products to improve their development process over time. The Toyota Way for suppliers is standardization, then kaizen.

The dominant value of the Toyota Way for suppliers is genchi genbutsu (Figure 13.10). Genchi genbutsu (which means "go and see") relates to Toyota's involvement in evaluating suppliers' production trials. Toyota makes planned visits to suppliers at various stages of the development process to confirm part quality and operations and manufacturing capability. Genchi genbutsu not only is important in product development to gauge supplier performance but also is a direct way to provide assistance to suppliers when there are problems. In all cases, suppliers and Toyota want to minimize risk to launch and provide a smooth transition into the start of production (SOP).

13.5 BUILDING POSITIVE AND INTERDEPENDENT STRUCTURES IN TPS

The beginning of this chapter stressed that interdependent structures are an important and essential property for systems to achieve their intended purpose. Without positive and interdependent relationships, a system is likely to compete for internal resources or engage in self-protecting behavior. From the outside, it would appear that an examination of Toyota's structures from a supplier development point of view would indicate that TPS is completely replicated throughout the supply chain. Specifically, all of the interacting components of TPS, such as the Toyota Way and the Toyota Business Practices, are required components to do business with Toyota. Currently, organizations are scrambling to emulate Toyota's supplier development practices by asking their suppliers to follow their version of TPS (Figure 13.11). There is the expectation that if suppliers adopt their *X* production system or *X* Way, they will be just as successful as Toyota.

While the Toyota Way does encourage positive supplier development relations, Toyota does not expect suppliers to adopt their managerial values, principles, or problem-solving technique. Toyota does not expect suppliers to install suggestion programs, establish quality circles, or adopt

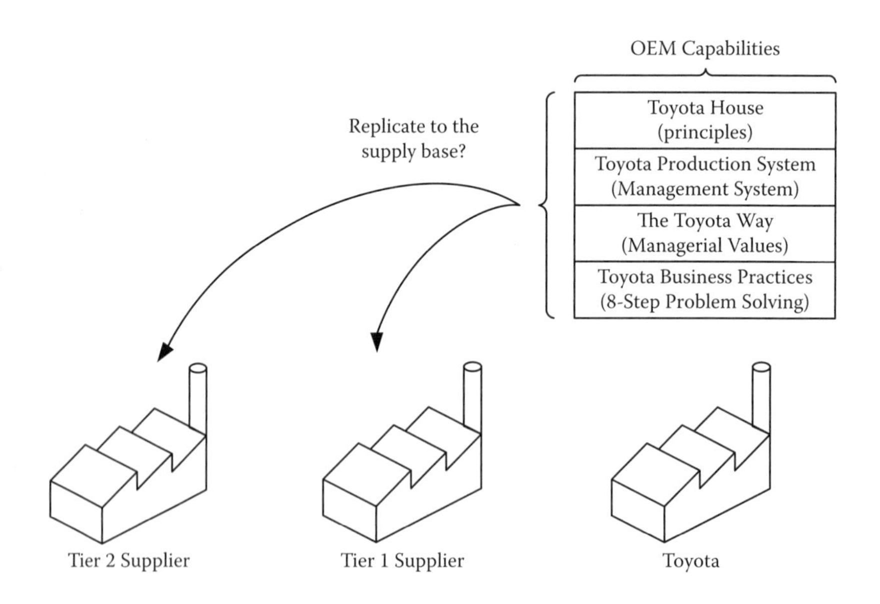

FIGURE 13.11
Traditional view of Toyota's approach to replicating internal capabilities.

	Traditional Approach to Supplier Development (Lean)	Toyota's Approach to Supplier Development (TPS)
Goals of Supplier Development	Goal 1: Go fast Goal 2: Go cost reduction	Goal 1 :Go slow Goal 2: Go step by step

FIGURE 13.12
Comparison of supplier development approaches: building interdependence.

systems that aim to stabilize employee relations. The Toyota Way and other business practices are a way to interface with suppliers but not to intervene with suppliers' values or management systems.

The reason why organizations are in a hurry to adopt Toyota's supplier development practices is that they are results oriented and not process oriented. Organizations want cost reductions quickly and view that the best way to obtain results is to overhaul the existing system, similar to business process reengineering (BPR). In this context, most organizations feel that it is quicker to throw out the old system rather than spend time fixing it. Toyota's approach to supplier development is to go slow with suppliers and gradually build up the relationship step by step (Figure 13.12). In this context, the factors that make the system successful can be indentified one by one. From Toyota's point of view, even for a new system, the factors that make it successful must be understood. If the factors cannot be indentified in the existing system, it is unlikely that a new system will be an improvement.

Interestingly, Toyota's approach to building interdependent structures is not to start at the systems level with suppliers (Figure 13.13). Instead, Toyota is largely targeting specific processes rather than overhauling a supplier's complete system. This approach supports the argument that Toyota is keeping consistent with the kaizen philosophy, which means to go small with steady improvements (process design) rather than to start from scratch (system design). From an implementation point of view, this would allow Toyota to start with any supplier (working to improve their existing system step by step) rather than installing a complete system that is new and unknown to the supplier. Simply, the best starting point in building interdependent structures is to start at the organization's current point. In summary, the path to interdependent structures is through process replication and not system replication.

Processes that suppliers are expected to replicate from Toyota are largely focused on daily management practices and the reliance on backup systems (Figure 13.14). For example, Toyota does not expect suppliers to

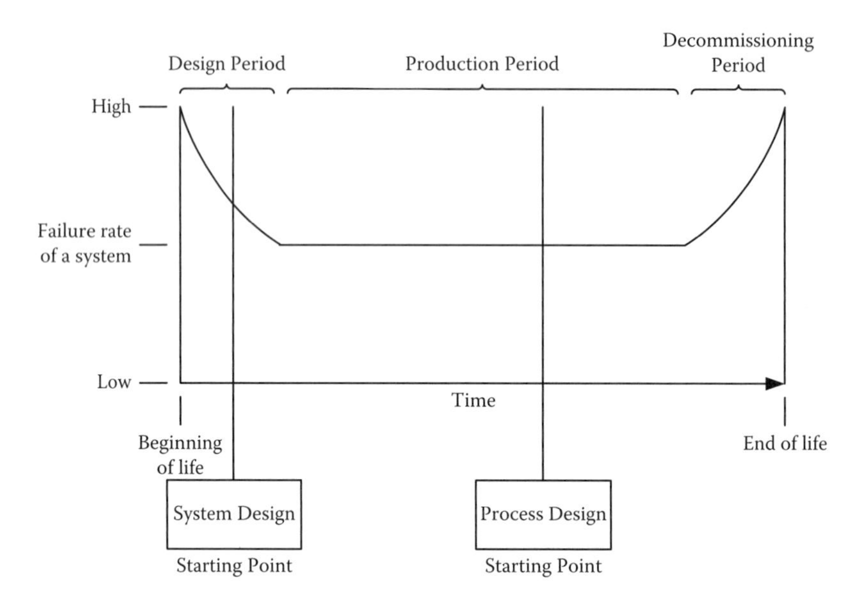

FIGURE 13.13
A comparison of the starting point for system design and process design.

adopt elaborate inventory management systems, ERP systems, or logistical systems. Instead, suppliers are expected to set standards for inventory levels, maintain standards for inventory, and demonstrate stability through tracking. Toyota also does not require suppliers to adopt preventive and predictive maintenance systems or maintenance initiatives such as TPM. Simply, Toyota asks that suppliers track and monitor downtime for equipment and processes.

Interestingly, Toyota does not require suppliers to participate in gemba walks or shop floor management activities. Toyota does ask suppliers to review quality problems daily. One of the most surprising daily management practices not enforced on suppliers is the use of Toyota's eight-step problem-solving process. While the automotive global standard ISO/TS 16949 mandates that suppliers define a specific type of problem technique

System Replication (Lean Approach)	Process Replication (Toyota's Approach)
Inventory management system	Set minimum and maximum levels for inventory
Total productive maintenance	Track and monitor downtime for equipment and processes

FIGURE 13.14
Examples of system and process replication for Lean and TPS.

System Replication (Lean Approach)	Process Replication (Toyota's Approach)
Define a specific type of problem-solving technique that can lead to root cause	Use *one* problem-solving method across departments
Organization shall audit the characteristics of its products	Are QA systems and methods audited by top management?

FIGURE 13.15
Examples of system and process replication for Lean and TPS.

that can lead to root cause, Toyota only checks that suppliers use one system across departments. Toyota believes that using one problem-solving technique consistently across groups is more important than adopting a specific problem-solving methodology such as DEMAIC (Define, Measure, Analyze, Improve, Control), global 8D (Ford Motor Company's global problem-solving methodology), GM's five-phase, Chrysler's seven-step, or Honda's five-step process (Figure 13.15).

Toyota's strong emphasis on standards is possibly one of the most targeted techniques used in supplier development. While most outsiders believe that Toyota expects its suppliers to adopt standardized work, there is more of a focus for suppliers to develop backup procedures and systems that provide a statically defined recovery approach. The ISO 9000 series is largely known for its reaction plans; Toyota would rather focus on standardized recovery plans. The difference is that reaction plans tend to focus on who to call for help, namely, the supervisor. Most suppliers leave it up to the supervisor to decide how to respond. In this context, reaction plans vary considerably. Toyota expects suppliers to define a standardized approach for abnormal activity. This does not mean that every possible alternative should be included in a what–if analysis; Toyota requests suppliers simply to know when a process is operating in nonstandard mode and the planned approaches for getting back to normal (Figure 13.16).

Organizations can learn from Toyota by emphasizing the process characteristics that make supplier development successful. International and automotive standards such as ISO 9000 and TS 16949 (required by American OEMs) do well at raising a supplier's awareness of business functions. Unfortunately, these guidelines do not describe the variables or factors of an effective process. While suppliers have the freedom to decide what makes a process effective, American OEMs such as Ford, GM, and Chrysler tend to place an emphasis on compliance to a supplier system rather than its effectiveness. Toyota does not require suppliers to adhere to

System Replication (Lean Approach)	Process Replication (Toyota's Approach)
Deploy standardized work for *normal* operating systems	Establish backup procedures and systems when there are *abnormalities*
Deploy a system to audit manufacturing process	Audit standardized work

FIGURE 13.16
Examples of system and process replication for Lean and TPS.

ISO 9000 and TS 16949. Toyota places more importance on the effectiveness of the system instead of verifying a supplier is adhering to its own system. Toyota views that adherence to an ineffective system is as bad as not knowing which process variables are important. Instead of Toyota trying to get suppliers to create systems throughout their organization, Toyota simply tries to explain which process variables are important to achieve certain outcomes. This personal exchange of knowledge sharing is extremely important in Toyota's supplier development process because Toyota takes a direct route to help suppliers reach maturity quicker. In some ways, American OEMs have used ISO 9000 and TS 16949 to shield themselves from knowledge sharing with their suppliers. It is much easier for American OEMs to request suppliers to develop a nonconformance system than to explain which factors make their nonconformance system effective. If American OEMs would simply state which process variables are important, suppliers could emulate customer processes and improve customer satisfaction (Figure 13.17).

Another way that Toyota gradually builds interdependent structures with their supply base is through kaizen. The number of kaizen events is a popular metric for OEMs to access a supplier's level of Lean. Kaizen events

System Replication (Lean Approach)	Process Replication (Toyota's Approach)
Adherence to an operating system	Effectiveness of an operating system
Minimize safety issues and risk to employees in organizations	Does the organization perform daily safety checks?
Establish a system to collect and analyze data	Is scrap and PPM tracked for quality improvement
Develop a system to verify setups	Use a process setup sheet

FIGURE 13.17
Examples of system and process replication for Lean and TPS. PPM = parts per million.

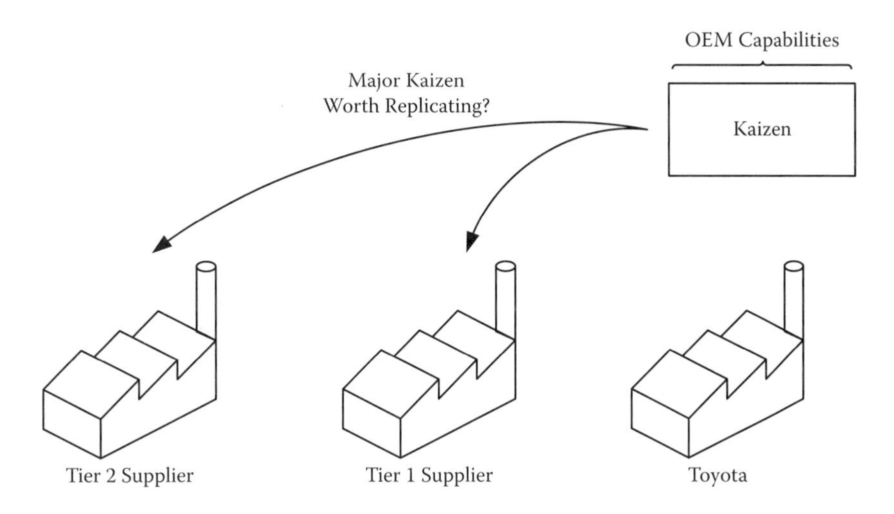

FIGURE 13.18
Outside impression of major kaizen for the supply base from Toyota.

(a form of major kaizen) are focused activities intended to show dramatic improvements within a short amount of time. Most outsiders perceive that since Toyota is efficient at kaizen, its suppliers should be effective at kaizen as well (Figure 13.18). Practitioners wanting to emulate Toyota often visit suppliers when benchmarking best practices. Organizations are curious how kaizen is applied at suppliers and how Toyota encourages kaizen to be applied so frequently and at such a large scale.

Because kaizen means constant improvement, practitioners often have the impression that kaizen events should occur continuously. A major challenge for suppliers is running kaizen events without disrupting normal operations. Suppliers that make changes to manufacturing processes after production has started run the risk of disrupting stable operations and consistent part quality. From this perspective, major kaizen may not be worth replicating, especially if it means disrupting an already-stable and tested process.

Toyota ranks the types of kaizen activities (namely, minor and major) by their approval process (Figure 13.19). As one might expect, the approval process is not equal, and in the case of major kaizen, approval by Toyota is not easy. Most outsiders gain the impression that kaizen occurs frequently and approval is easy.

Toyota describes minor kaizen as activities that reduce cycle time or reduce or allocate manpower or activities that advance standardization (Figure 13.20). Minor kaizen could include improving workability to

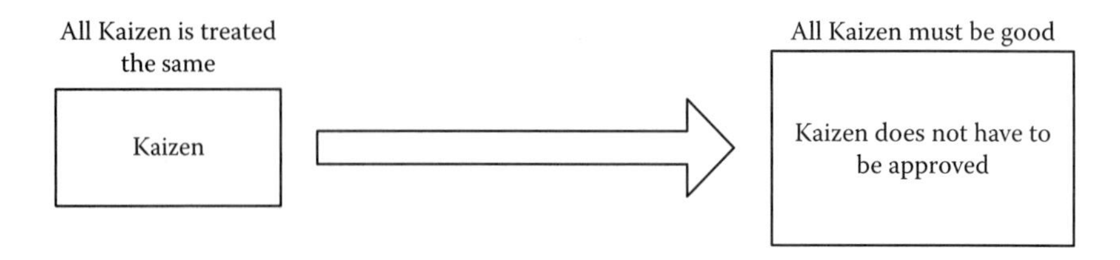

FIGURE 13.19
Traditional view about kaizen and approval.

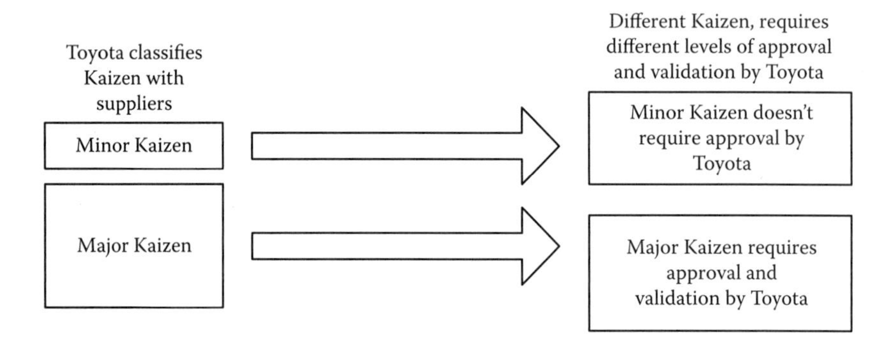

FIGURE 13.20
Toyota's approach to kaizen and approval.

remove a temporary inspection process or conducting equipment activities to extend normal machine wear or tool replacement. Although suppliers are not required to notify Toyota when conducting minor kaizen, suppliers are expected to complete internal quality verification and tracking to ensure that parts meet a standard before, during, and after kaizen.

Toyota's approval process for major kaizen is quite comprehensive compared to minor kaizen. Major kaizen activities include the relocation of equipment and processes or modification or change of a machine, tool, die, or mold. Major kaizen can also include the elimination of a shift or eliminating a permanent inspection process. Toyota also considers using a new machine the same as implementing new technology. In all cases, any change that has an impact on the manufacturing method, such as a material change, part specification change, or any change with an impact on the product's function, performance, durability, or reliability, is considered major kaizen. One of the techniques that Toyota employs in communicating expectations early to suppliers to help differentiate between minor and major kaizen is the use of special symbols on parts and process to designate safety standards, regulation standards, emission standards, Japan export standards, and Federal Motor Vehicle Safety Standards

(FMVSS). Any kaizen that affects these special characteristics are strictly monitored and controlled and require approval by Toyota.

Toyota's approval process for major kaizen includes a variety of detailed steps, reviews, and prenegotiation. First, Toyota purchasing requires pre-approval and approval prior to implementing any change. Major kaizen activities have to be completed during off shifts or prearranged with Toyota purchasing during planned production. Since most companies have limited capacity and labor resources to run trials during off shifts, most suppliers are not willing to engage in these types of activities unless there is a significant return or benefit. Next, purchasing will participate in the suppliers' quality assurance planning activities, including on-site verification to go and see. A process change request is also submitted, which includes a variety of reviews and checks from quality, operations, and production capability. The approval process for major kaizen can take anywhere between three weeks and three months to complete. Toyota utilizes a gate review process for tracking kaizens and, depending on the nature of the change, could include a variety of plant-level departments. The gate review process includes prenegotiation, planning approval, first-off-tool sample approval, midsize trial production approval, and a mass trial production approval. With five levels of approval for suppliers to engage in kaizen, it can be understood how kaizen compares to maintaining part quality. Yes, Toyota would like for suppliers to engage in kaizen, but not at the expense of disrupting a stable and controlled system.

Toyota's approach to kaizen is supportive of process replication because it acts as a way to slowly build, one step at a time, the factors that make a successful supplier relationship. The go slow approach minimizes risk in relationship building and the possibility of developing structures that are competitive in nature or self-protecting, for example, a relationship that serves the need of one side, a relationship that is one directional, and goals that are competing in nature. Toyota's approach to supplier development is an excellent example of relationship building that is based on the idea of interdependent structures that are positive and reciprocal and are achieved through joint decision making.

Last, Toyota views that you cannot give away something that you do not already own. It is not unusual for most companies to want to Lean out their supply chain overnight without implementing Lean internally. An example of this mindset is illustrated by the adoption of ISO. ISO certifications create distance between suppliers and OEMs because they do not encourage information sharing. There is no provision that a successful

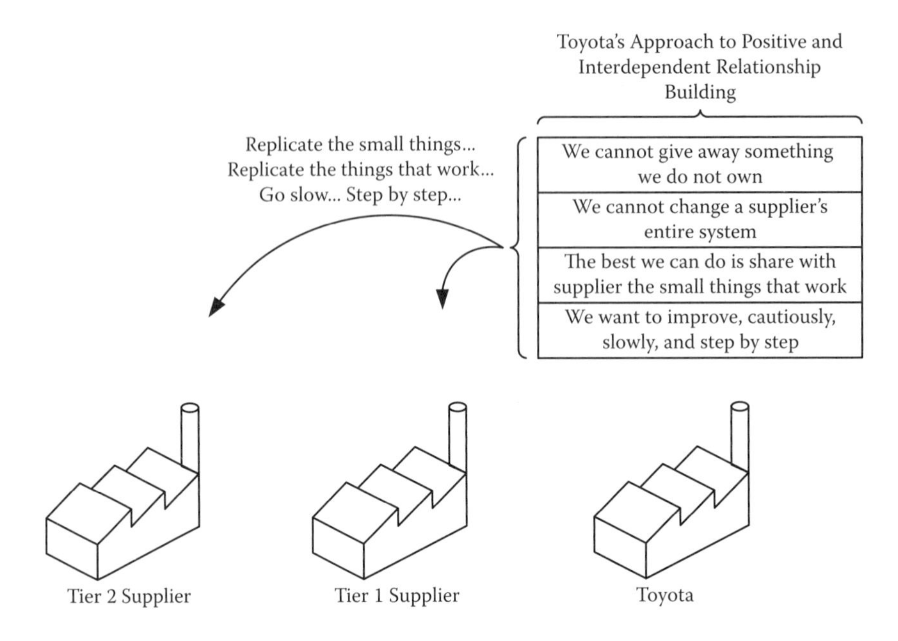

FIGURE 13.21
Toyota's approach to supplier development.

process should be shared with the supplier. What is surprising is that ISO is supposed to strengthen the supplier development relationship through replicating practices to improve the capabilities of suppliers, not add more distance between them. ISO encourages self-protecting structures because it asks the supplier to develop a system on itself without advocating which variables or factors make the system effective. While this hands-off approach may seem ideal for purchasing managers who are already too busy or provide the suppliers freedom to develop their own system, Toyota would rather leverage internal capabilities by sharing specific processes that are known to be successful. This approach requires both groups to be open and willing to share information to work toward common goals. It is obvious that sharing the small things that work is a pretty big commitment to ensure a successful supplier relationship (Figure 13.21).

13.6 SUMMARY

The purpose of interrelationships and interdependency in a system is to create structures that are positive and mutual for achieving purposeful behavior. Otherwise, a system is likely to hinder itself by creating conditions that encourage isolation and competition for resources. An effective system can achieve its identity when its structures are reciprocal in nature and cooperate in meeting systemwide goals.

The best way to explain the common denominator of interdependent structures within TPS was to explore the basic elements of the supplier relationship. The supplier relationship is the last extension of teamwork that makes TPS possible. While it is argued that TPS may contain more exhaustive structures at the company or small team level, the supplier relationship contains the basic building blocks for teamwork to exist.

Toyota's approach for building an interdependent structure is modest, starting at the process level and slowly improving step by step. Process replication was favored over system replication mainly because Toyota would like to share the small things that are known to be successful. Toyota would like suppliers to engage in minor kaizen to improve their system, but not at the risk of disrupting a stable and consistent process. From these observations, it appears that suppliers are successful because Toyota shares those process variables that have been proven to work. By starting small (at the process level), suppliers can improve their own system while minimizing risk.

REFERENCES

Alloo, R. (2010) Executive Lean Leadership Institute, University of Kentucky, College of Engineering, Lean Systems Group—executive in residence, presentation, August.

Bieber, O. (1988) Total involvement requires total commitment. *Journal for Quality and Participation*, Vol. 11, No. 2, A6–A8.

Bluegrass Automotive Manufacturing Association, website, http://www.bama-group.org.

Cagliano, R., Caniato, F., and Spina, G. (2006) The linkage between supply chain integration and manufacturing improvement programmes. *International Journal of Operations and Production Management*, Vol. 26, No. 3, 282–299.

Carr, A., Kaynak, H., Hartley, J., and Ross, A. (2008) Supplier dependence: Impact on supplier's participation and performance. *International Journal of Operations and Production Management*, Vol. 28, No. 9, 899–916.

Chin, K., Yeung, I., and Pun, K. (2006) Development of an assessment system for supplier quality management. *International Journal of Quality and Reliability Management*, Vol. 23, No. 7, 743–765.

Crook, T. R. and Combs, G. C. (2007) Sources and consequences of bargaining power in supply chains. *Journal of Operations Management*, Vol. 25, No. 2, 546–555.

Deutsch, M. (1949) A theory of cooperation and competition. *Human Relations*, Vol. 2, 129–152.

Deutsch, M. (1985) *Distributive Justice: A Social Psychological Perspective*, Yale University Press, New Haven, CT.

Dyer, J. H. (2000) *Collaborative Advantage: Winning through Extended Enterprise Supplier Networks*, Oxford University Press, New York.

Ellram, L. (1991) A managerial guideline for the development and implementation of purchasing partnerships. *International Journal of Purchasing and Materials Management*, Vol. 31, No. 2, 9–16.

Ellram, L. M. (1996) The use of the case study method in logistics research. *Journal of Business Logistics*, Vol. 17, No. 2, 93–138.

Gigch, J. (1978) *Applied General Systems Theory*, Harper & Row, New York.

Govindan, K., Kannan, D., and Haq, A. (2010) Analyzing supplier development criteria for an automobile industry. *Industrial Management and Data Systems*, Vol. 110, No. 1, 43–62.

Hartley, J. L. and Choi, T. Y. (1996) Supplier development: Customers as a catalyst of process change. *Business Horizons*, Vol. 39, No. 4, 37–44.

Henke, J. (2010) In the driver's seat: Why US carmakers have stepped up a gear with supplier relations. *Supply Management*, June 24.

Humphreys, P., Li, W., and Chan, L. (2004) The impact of supplier development on buyer-supplier performance. *Omega*, Vol. 32, No. 2, 131–144.

ISO/TS 16949:2002 (International Standard Organization / Technical Standard) (2002) prepared by the International Automotive Task Force (IATF) and Japan Automobile Manufacturers Association, Inc. (JAMA), with support from ISO/TC 176, Quality management and quality assurance.

Johnson, D. and Johnson, R. (1989) *Cooperation and Competition: Theory and Research*, MNL Interaction, Edina, MN.

Juettner, U. (2005) Supply chain risk management. *International Journal of Logistics Management*, Vol. 16, No. 1, 120–141.

Knemeyer, A. M., Zinn, W., and Eroglu, C. (2008) Proactive planning for catastrophic events in supply chains. *Journal of Operations Management*, corrected proof, Vol. 26, No. 4, 536–554.

Krause, D. R., and Ellram, L. M. (1997) Critical elements of supplier development. *European Journal of Purchasing and Supply Management*, Vol. 3, No. 1, 21–31.

Krause, D. R., Handfield, R. B., and Beverly, B. T. (2007) The relationships between supplier development, commitment, social capital accumulation and performance improvement. *Journal of Operations Management*, Vol. 25, No. 2, 528–545.

Krause, D., Scannell, T., and Calantone, R. (2000) A structural analysis of the effectiveness of buying firms' strategies to improve supplier performance. *Decision Sciences*, Vol. 31, No. 1, 33–55.

Laamanen, T. (2005) Dependency, resource depth, and supplier performance during industry downturn. *Research Policy*, Vol. 34, No. 2, 125–140.

Leenders, M. R. (1966) Supplier development. *Journal of Purchasing*, Vol. 24, 47–62.

Lemke, F., Goffin, K., and Szwejczewski, M. (2003) Investigating the meaning of supplier-manufacturer partnerships: An exploratory study. *International Journal of Physical Distribution and Logistics Management*, Vol. 33, No. 1, 12–35.

Li, W., Humphreys, P., Yeung, A. C. L., and Cheng, T. C. E. (2007) The impact of specific supplier development efforts on buyer competitive advantage: An empirical model. *International Journal of Production Economics*, Vol. 106, No. 1, 230–247.

Lo, V., Sculli, D., and Yeung, A. (2006) Supplier quality management in the Pearl River Delta. *International Journal of Quality and Reliability Management*, Vol. 23, No. 5, 513–530.

MacDuffie, J. P. and Helper, S. (1997) Creating Lean suppliers: Diffusing Lean production through the supply chain. *California Management Review*, Vol. 39, No. 4, 118–151.

Marksberry, P., Badurdeen, F., Gregory, B. and Kreafle, K. (2010) Management directed kaizen: Toyota's Jishuken process for management development. *Journal of Manufacturing Technology Management*, Vol. 21, No. 6.

Marksberry, P. (2011) A new approach in analyzing social-technical roles at Toyota: The team leader. *International Journal of Human Resources Development and Management*, Vol. 10, No.4, 395–412.

Modi, S. and Mabert, V. (2007) Supplier development: Improving supplier performance through knowledge transfer. *Journal of Operations Management*, Vol. 25, No. 1, 42–64.

Mollenkopf, D., Stolze, H., Tate, W., and Ueltschy, M. (2010) Green, Lean and global supply chains. *International Journal of Physical Distribution and Logistics Management*, Vol. 40, No. 1/2, 14–41.

Motwani, J., Youssef, M., Kathawala, Y., and Futch, E. (1999) Supplier selection in developing countries: A model development. *Integrated Manufacturing Systems*, Vol. 10, No. 3, 385–392.

Nauta, A. and Sanders, K. (2000) Interdependent negotiation behavior in manufacturing organizations. *International Journal of Conflict Management*, Vol. 11, No. 2, 135–161.

Newman, R. G. and Rhee, K. A. (1990) A case study of NUMMI and its suppliers. *International Journal of Purchasing and Materials Management*, Vol. 26, No. 4, 15–20.

Peach, R. (1992) *The ISO 9000 Handbook*, CEEM Information Services, Fairfax, VA.

Putt, A. (1978) *General Systems Theory Applied to Nursing*, Little, Brown, Boston, MA.

Sako, M. (2004) Supplier development at Honda, Nissan and Toyota: comparative case studies of organizational capability enhancement. *Industrial and Corporate Change*, Vol. 13, No. 2, 281–308.

Shimokawa, K. and Fujimoto, T. (2009) *The Birth of Lean*, Lean Enterprise Institute, Cambridge, MA.

Srinivasan, M. (2004) *Streamlined: 14 Principles for Building and Managing the Lean Supply Chain*, Thomson, Stamford, CT.

Thompson, J. (1967) *Organizations in Action*, McGraw-Hill, New York.

Toyota Motor Corporation – TMC (2001) The Toyota Way, Internal company document, Toyota Institute, April.

Toyota Motor Corporation – TMC (2006a) Published by Toyota, Toyota Code of Conduct, March.

Toyota Motor Corporation – TMC (2006b) Published by Toyota, Toyota—Green Purchasing Guidelines, Global Purchasing Planning Division.

Toyota Motor Corporation – TMC (2010) Published by Toyota, Toyota Terms and Conditions, Global Purchasing Planning Division.

Toyota Supplier Support Center (TSSC) website, http://www.tssc.com/.

Wagner, S. (2006) Supplier development practices. *European Journal of Marketing*, Vol. 40, No. 56, 554–571.

Wu, Y. (2003) Lean manufacturing: A perspective of Lean suppliers. *International Journal of Operations and Production Management*, Vol. 23, No. 11, 1349–1376.

Zaheer, A., McEvily, B., and Perrone, V. (1998) The strategic value of buyer-supplier relationships. *International Journal of Purchasing and Materials Management*, Vol. 34, No. 3, 20–26.

14

The System Property of Equifinality in TPS

14.1 SYSTEM PROPERTY: EQUIFINALITY AND MULTIFINALITY

One of the last tenets of systems theory is the concept of equifinality and multifinality. This property relates to the ability of a system to attain the same final result from many different initial conditions. Regardless of the starting position or condition of a system, the same end point can be reached through a variety of different paths. The idea is that the system can demonstrate convergent behavior and be equally effective at reaching the same destination regardless of the initial condition. Equifinality is often associated with dynamism and adaptability and is a distinctive characteristic between living and nonliving systems (Figure 14.1). In any closed system, the final state is determined by the initial state or condition of the system. If either the initial conditions or the process is altered, the final state will be changed. In open systems, the final state may be reached from varying initial conditions and in different and equally effective ways. According to the property of equifinality, no matter where one begins, the ending will be the same (Becvar and Becvar, 1982).

Equifinality also demonstrates the idea that a system can be long lasting and maintain a dynamic equilibrium. Living organisms that exhibit the property of equifinality can reach maturity through a stable state–path trajectory. Equifinality and multifinality are used in living systems theory to describe a variety of dynamic processes, such as evolution and entropy. In most physical systems, the final state is determined by the initial condition. When there is a change in the initial condition, the final state changes. Closed systems cannot have equifinality; however, if they have regulatory functions, they can appear as a living system. In open systems,

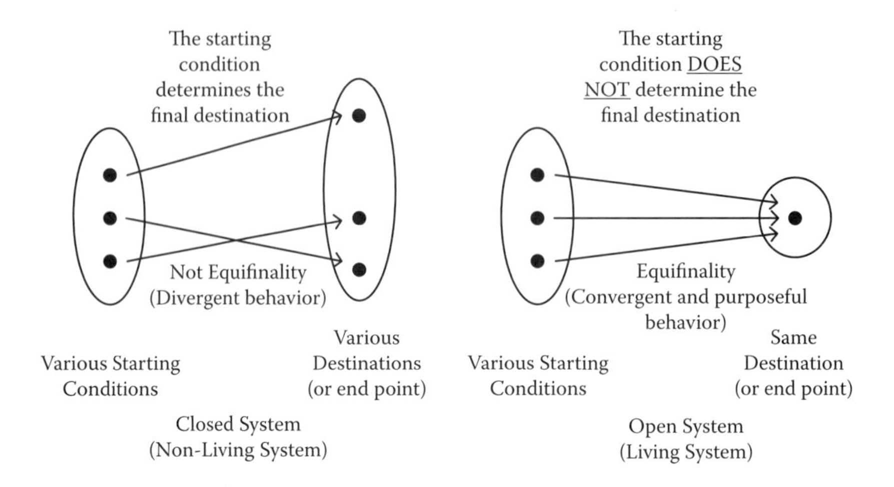

FIGURE 14.1
A comparison between living and nonliving systems.

the behavior of the system can be independent of the initial conditions because they are constantly exchanging materials with the environment. Equifinality is often used to describe living systems because living systems have the tendency to approach equilibrium states through interacting with their environment, exchanging energy and matter, which is more than just the ability to self-regulate.

14.2 THE SYSTEM PROPERTY OF EQUIFINALITY FOR A MANAGEMENT SYSTEM

The aim of a management system is to provide the path for information and action to give shape for organized structure and processes (Barnard, 1938). A management system provides unity of purpose through the organization and represents the fundamental style for decision making. A management system is effective when decision making can occur uniformly through each business function and condition. The performance of an organization is largely determined by how well the management system fits the organization's needs and ability to fulfill its intended purpose. When a management system does not fit an organization's culture, structure, capability, or strategy, counteracting forces affect organizational performance (Figure 14.2). Researchers and scientists show that the best-performing organizations are those with management

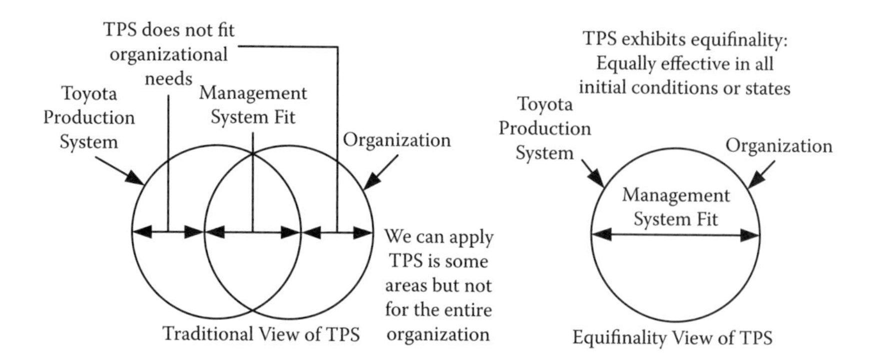

FIGURE 14.2
A comparison of the organizational fit of TPS: a traditional and equifinality view.

systems that are designed to meet internal and external environmental factors affecting the organization (Miller, 1992). Without some minimum fit of an organization's management system, the organization will not survive.

14.3 THE SYSTEM PROPERTY OF EQUIFINALITY FOR NON-VALUE-ADDED WORK ENVIRONMENTS

Much has been written about how the Toyota Production System (TPS) works in a manufacturing environment. TPS is believed to be working when the four basic elements of a production system—transportation, delay, inspection, and inventory—are eliminated. When these four elements are reduced, the ratio of value-added to non-value-added increases. The fifth element of the production system, namely production, is used to describe activities in the construction or service of goods that the customer is willing to pay (Figure 14.3). Eliminating the four wastes in a production system increases efficiency of the system by only expending resources on activities that return an investment. From the end-user's perspective, consumers purchase products and services based on the value they provide, not by the inefficiencies of how they were assembled, put together, or produced.

The four basic elements (i.e., waste) are part of every production system, and no organization is completely free of any of them; however, organizations become more competitive when these elements are eliminated and reduced.

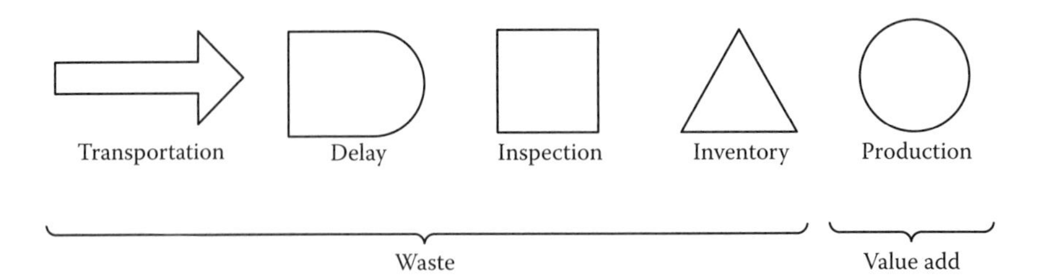

FIGURE 14.3
The five elements of a production system.

Another element in a production system that is often debated regarding whether it is value added or waste is the maintenance function (Figure 14.4). Facilities maintenance is one area for which maintaining equipment does not add any value to the product. No matter how many times equipment is repaired, overhauled, lubricated, or cleaned, the value of the product does not change, although it is a fact that if machinery is not maintained, aging and deterioration of moving parts, motors, or spindles can ultimately affect a product's quality, thus decreasing value. However, the effort applied to keep the equipment at an operational level to meet a product's quality requirements is not readily seen or valued by the customer. In this situation, how can TPS be applied if there is no value to maximize? If a department such as maintenance is considered to be non-value-added, would Toyota not try to eliminate the need for maintenance? Is the thinking behind the Toyota Way head count reduction? How can departments practice TPS if the departments are completely filled with waste?

One of the secrets behind TPS is not how Toyota manufactures cars but how management systems can be applied consistently throughout the organization. As a result, TPS is universal because management systems can be applied anywhere that people perform work. This means that TPS

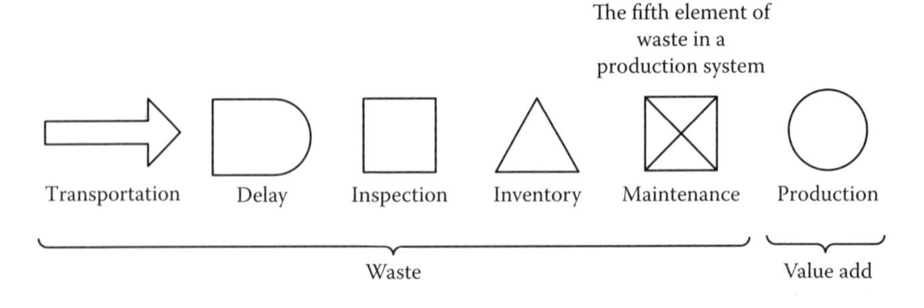

FIGURE 14.4
The five elements of a production system plus maintenance.

is an approach to work that can be shared by all employees in the company rather than a select few. TPS is a way of thinking, a way of communicating, a way of problem solving, and a specific way that employees can give their effort and energy to the company.

According to the principle of equifinality, TPS must be equally effective in non-value-added environments as they are in value-added environments. This chapter presents how TPS can function effectively when there is no value to increase and only waste to eliminate. The maintenance function is also particularly interesting to study because most work in maintenance is not routine, which raises the issues of variability and unpredictability. While there have been numerous works describing total productive maintenance (TPM), which shares some similarities to TPS, there has been little explanation of Toyota's maintenance systems and management approaches. Consequently, there has been little interest in applying Lean to maintenance mainly because most companies fail to implement Lean outside manufacturing or processes that are inherently support functions.

14.4 LITERATURE REVIEW

Global production has forced maintenance functions to achieve uniform quality worldwide and to be responsive to emerging trends in technology (Wireman, 2004; Sakai and Amasaka, 2007; Tsang, 2002). The complexity of equipment and machinery has caused organizations to invest in mechanization, automation, robots, automatic warehousing, and unmanned guided vehicles (Braglia et al., 2006). Consequently, organizations have been under pressure to produce to changing market conditions and adopt just-in-time (JIT) systems to remain competitive. Organizations are forced to move from separate to synchronous systems using interlocking supply chains, which has put enormous stress on equipment availability (Figure 14.5) (Mathaisel, 2005; Ghayebloo and Shahanaghi, 2010). Operations have also undergone extensive changes due to the toughening of societal expectations (i.e., International Organization for Standardization [ISO] 26000, *Guidance on Social Responsibility*) and environmental pressures (i.e., ISO 14000, *Environmental Management*) in an attempt to improve safety for employees as they interact with equipment (Tsang, 2002). In short, maintenance operations play a crucial role in an

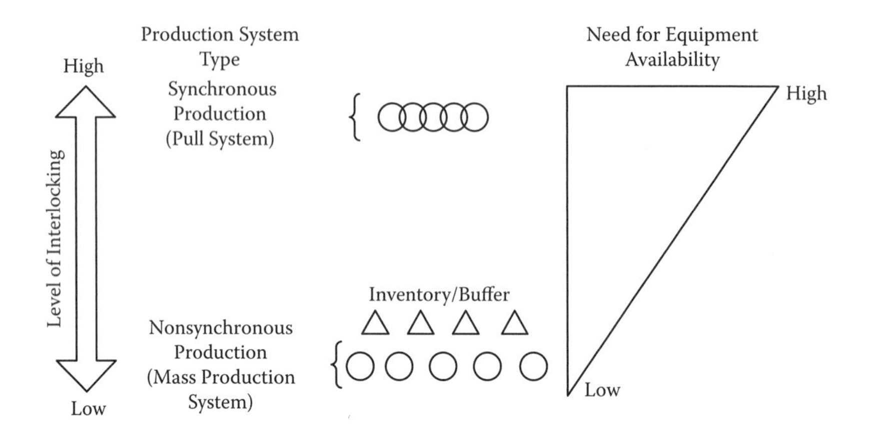

FIGURE 14.5
The need for equipment availability in synchronous production systems.

organization's competitiveness and development of highly reliable production systems (Pollitt, 2010; Karuppuswamy et al., 2007; Ashayeri, 2007).

The primary objective of maintenance is to ensure availability of equipment to meet production demands (Wang and Marvin, 2000; Modarres et al., 1999). Maintenance is a broad and diverse discipline that performs a variety of internal functions. Maintenance ensures compliance to part and equipment specifications through servicing, repairs, and overhauls of equipment and machinery. Maintenance has the role of controlling the maintenance workload by adhering to daily schedules and reducing equipment downtime and increasing availability using predictive and preventive strategies (Smith and Hawkins, 2004a). The type and structure of maintenance systems depend on production line requirements and customer specifications (Moayed and Shell, 2009).

Today, maintenance is seen as a significant factor in achieving an organization's strategic goal; however, in the twentieth century maintenance was considered an isolated and separate function (Garg and Deshmukh, 2006). Earlier maintenance strategies were reactive and added little to the business (Mamber et al., 1999). Breakdown maintenance is one such strategy that encouraged maintenance on machines only after a failure. Breakdown maintenance is not concerned with scheduling inspections or service routines on deteriorating components (Kaiser, 2007). As maintenance systems developed, they became more proactive and efficient. In the 1950s, productive maintenance strategies increased the overall efficiency of the maintenance function by analyzing maintenance processes. In the 1960s, preventive maintenance (PM) encouraged the scheduling of routine inspections and performing necessary upkeep in an effort to prevent and fix problems before failure occurred. In the 1970s, TPM became popular and, like total quality management (TQM), took a more holistic view of departmental functions (Pramod et al., 2007). TPM is based on three interrelated concepts: (1) maximizing equipment effectiveness; (2) autonomous maintenance by operators; and (3) small-group activities (Nakajima, 1988). TPM has the goal of zero defects, zero accidents, and zero breakdowns (Bamber et al., 1999).

In the 1980s, predictive maintenance sought to increase equipment reliability by replacing critical components on machinery before failure. The primary strategy of predictive maintenance systems is to utilize sensors and various monitoring systems to observe characteristics with established standards and specifications to evaluate an equipment's existing condition and projected trend of deterioration (Kaiser, 2007). Unfortunately, predicting the degree of deterioration of equipment month by month is still difficult because most failures are not time based. To this end, many companies have given up on equipment diagnostic technology (Murase, 2007). An improvement over PM is said to be condition-based maintenance. Condition-based maintenance utilizes real-time monitoring of equipment to more accurately assess optimal maintenance intervals (Kaiser, 2007). The primary goal of condition-based maintenance is to more accurately assess an equipment's PM schedule rather than relying on generic build specifications, which can vary greatly depending on equipment use and rate of wear.

Newer trends in predictive maintenance include reliability-centered maintenance (RCM) and dependability management (DM) (Figure 14.6). RCM is a systems engineering approach to maintenance that is used to evaluate potential failure modes in equipment. RCM has replaced selected military standards and specifications in recognition of today's intense and competitive demands for high equipment and machinery reliability (i.e., Society of Automotive Engineers (SAE) JA 10000, 1998). RCM was initially developed for the commercial aviation industry. DM strategies are more general in nature compared to RCM and include the enhancement of maintenance functions in areas of reliability, availability, and maintainability (i.e., AS International Electrotechnical Commission (IEC) 60300 standard).

Despite the ongoing evolution of maintenance strategies, a variety of problems and challenges still exists for the maintenance function. One of the most reoccurring obstacles for maintenance is overcoming the

	1st Generation	2nd Generation	3rd Generation	4th Generation	5th Generation
Strategies	Breakdown Maintenance	Productive Maintenance	Preventive Maintenance	Total Productive Maintenance	Predictive Maintenance
Characteristics	Reactive. Responds to break downs only.	Aimed at increasing productivity and quality of product.	Time-based (routine) maintenance featuring periodic checking, adjusting, and replacing parts to prevent failures.	Systematic plantwide program to achieve zero break downs.	Technique to help determine the condition of equipment in order to predict when maintenance performed.

← 1950s 1960s 1970s 1980s →

	6th Generation	7th Generation	8th Generation
Strategies	Condition Monitoring Computer-Based Maintenance Management Systems (CMMS) Reliability Centered Maintenance (RCM) Dependability Maintenance (DM)	Sustainable Asset Management (SAM)	Cradle-to-grave analysis to assess environmental impacts of an asset throughout its life.
Characteristics	Monitoring and analyzing techniques to diagnose the condition of equipment during operation to identify signs of deterioration.	LCA for assets Value-based approaches to maintenance	Cradle-to-cradle analysis to assess sustainability impacts of an asset through multiple life cycles.

← 1990s 2000s 2010s →

FIGURE 14.6
Review and trend of maintenance strategies.

adversarial role with production (Yamashina, 1995). Traditionally, production views maintenance as the fault for machine failure, while 70% of machine failures are caused by operator error (Figure 14.7). The mentality of "I operate" and "you fix" spurred organizations to adopt TPM to integrate maintenance and production functions gradually (Yamashina, 1995). Unfortunately, numerous organizations have failed to implement TPM successfully, mainly because it requires integration of culture, process, and technology (Liu, 2004). Consequently, TPM programs grow

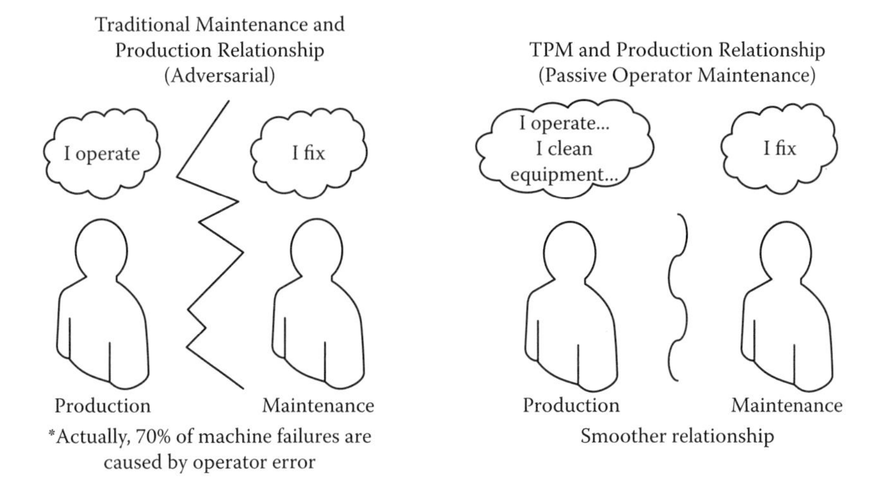

FIGURE 14.7
Traditional and TPM maintenance relationships with production.

weary over time, which causes operators to lose interest (Thomas et al., 2008). Most experts agree that passive operator maintenance like cleaning is not enough for true survival of TPM, which means that operators need to know and understand about repairs, setups, adjustments, and measurements (Nakajima, 1988).

Another problem for maintenance is the national maintenance skill shortage (Pollitt, 2010). Due to the changes in the workforce, the number of skilled workers has declined, which has led to an increase in turnover (Figure 14.8). Sophistication of equipment and improved reliability have also caused fewer opportunities to hand down skills throughout the maintenance trade. Thirty years ago, it was possible for workers to learn the mechanisms of equipment failure by disassembling the equipment and finding the location of the failure to repair (Murase, 2007). With equipment failing much less, there are fewer opportunities for maintenance skills to be developed internally.

Possibly the greatest obstacle for maintenance to overcome is getting beyond the mindset of short-term cost savings (Ghayebloo and Shahanaghi, 2010). Maintenance operations are viewed as an operational expense, which means that cost are regularly monitored and managed (Ahuja and Kumar, 2009). Unfortunately, most strategies to cut cost in maintenance have led to short-term cost savings but have seriously impacted an organization's ability to remain competitive in the long term. Short-term cost savings have led to the outsourcing of specialized maintenance, causing eroding of in-house

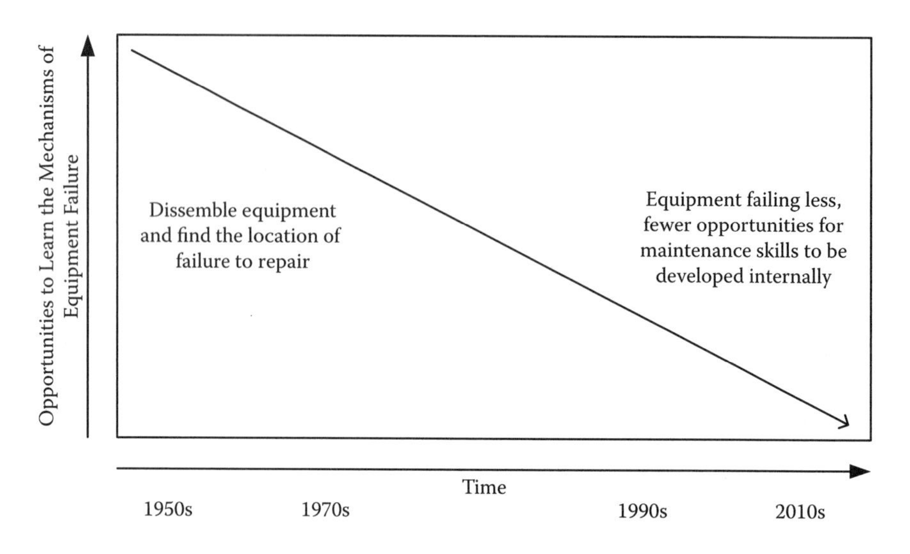

FIGURE 14.8
Maintenance skills: fewer development opportunities due to improvements in equipment.

technical capabilities (Tsang, 2002; Murase, 2007). While maintenance may be an overhead expense that has to be controlled, most organizations still view the maintenance function as a nonessential and non-value-added unit to the business (Waeyenbergh and Pintelon, 2009).

Recently, there has been a variety of different approaches to combat changing trends in the maintenance discipline. The standardization of sensing and control technologies has been developed to prevent failures in equipment (Murase, 2007). The computerization of maintenance systems (i.e., CMMSs, computerized maintenance management systems) continues to develop to support work order management, planning, scheduling, budget cost, spares, and key performance indicators (KPIs) (Smith and Hawkins, 2004b). CMMSs are currently used as a tool to refine PM activities by controlling, analyzing, and tracking work orders.

The most popular efforts in developing the maintenance function continues to be in the area of TPM. Organizations continue to use TPM to reduce unexpected breakdowns, maximize equipment effectiveness, and improve overall efficiency (Ahuja and Khamba, 2008). Some researchers have even proposed Six Sigma frameworks for TPM implementation, which includes the use of DMAIC (define, measure, analyze, improve, control) to reduce variation in maintenance functions (Thomas et al., 2008). One of the most recent developments in TPM is the development of Lean maintenance, which is described by a more holistic application of TPM.

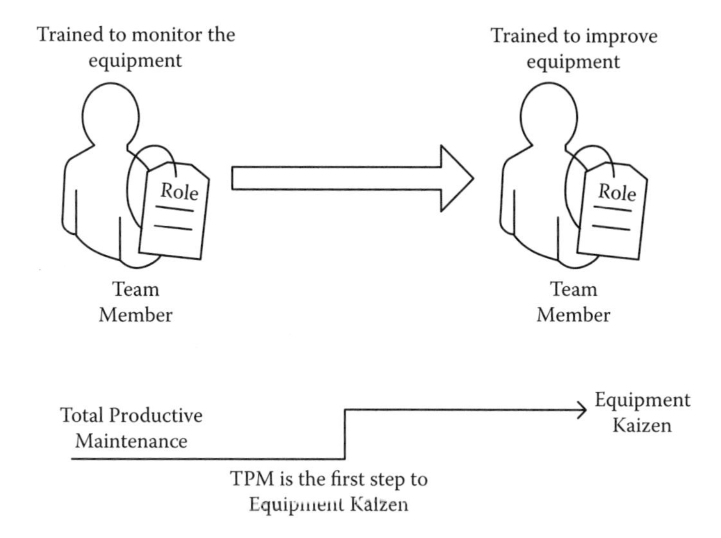

FIGURE 14.9
TPM and the path to equipment kaizen.

Lean maintenance practices are centered on planning and scheduling, documentation, work order systems, predictive maintenance, and root cause failure analysis (Smith and Hawkins, 2004a; Pal and Maiti, 2005). One of the motivating factors for adopting Lean maintenance is the idea of equipment kaizen. There is work that suggested that operators who are better trained to monitor equipment are more likely to improve equipment (Piper and Sumukadas, 1993). Therefore, TPM is a first step to practical equipment kaizen (Figure 14.9) (Sharma et al., 2006). Lean maintenance also offers an effort more centered on maintenance procedures and systems so that repairs are easier to accomplish and achieve through maintenance redesign and modifications (Thomas et al., 2008).

In some ways, Lean maintenance has been described as all the good habits of traditional maintenance. The work of Pal and Maiti stated that Lean maintenance focuses on eliminating downtime from operator error, program error, downtime due to inadequate PM systems, or downtime from chronic wear and stress (Pal and Maiti, 2005). Other articles showed that Lean maintenance is nothing more than improving line availability based on evaluating random failure, initial failure, and wear failure (Sakai and Amasaka, 2007).

Toyota's maintenance system has grown in interest, specifically because of the success in TPS (Arnold, 2010; Harada, 2006). While Toyota is open about its production system, it appears to be closed about other parts of its systems, namely, maintenance (Harada, 2006). To the best of my knowledge, there is little information on Toyota's maintenance systems. There is no theoretical, quantitative, empirical, or qualitative work on their systems or structures or how they interface with production. There is also no scientific investigation of the maintenance function or how TPS is applied within the maintenance discipline.

Toyota's JIT systems are successful, but it is speculated that maintenance plays a critical role. Without machine availability, parts cannot be produced in the right quantity and delivered at the right time. The "Toyota Way" for maintenance is possibly one of the best-kept secrets about Toyota, mainly because it represents TPS for non-value-added operations. Toyota's approach to maintenance may offer some interesting insight into understanding how TPS can be applied to areas that are not repetitive or predictable. Most literature describes Lean maintenance as a set of tools, such as value stream mapping, eliminating the seven wastes, kaizen, jidoka, visual controls, andon, performance boards, and mistake proofing (Hawkins, 2005; Alsyouf, 2006; Bagadia, 2008). Thus, an investigation of Toyota's management systems within maintenance may provide the thinking that is used to apply TPS outside manufacturing.

14.5 THE TPS GOAL IN FACILITIES MAINTENANCE

The goal in facilities maintenance is to eliminate as much cost as possible without compromising availability of equipment (Figure 14.10). The less maintenance that has to be performed on equipment in the assembly and construction of automobiles, the lower the cost will be. Ideally, we would like to design equipment so that it does not require maintenance. The problem becomes a question of management, specifically how to accomplish the work that is required with the resources that are available.

A common approach to improve maintenance functions and systems is the use of initiatives (Figure 14.11). Initiatives (like TPM) are powerful for stimulating improvement, but only for small pockets of the organization. Initiatives are not permanent, long lasting, routine, or mandatory. Initiatives are concerned about getting some groups within the organization to do something part of the time. Initiatives are like part-time voluntary jobs for the company. No one really wants another job, especially if he or she does not get paid and it is not really required.

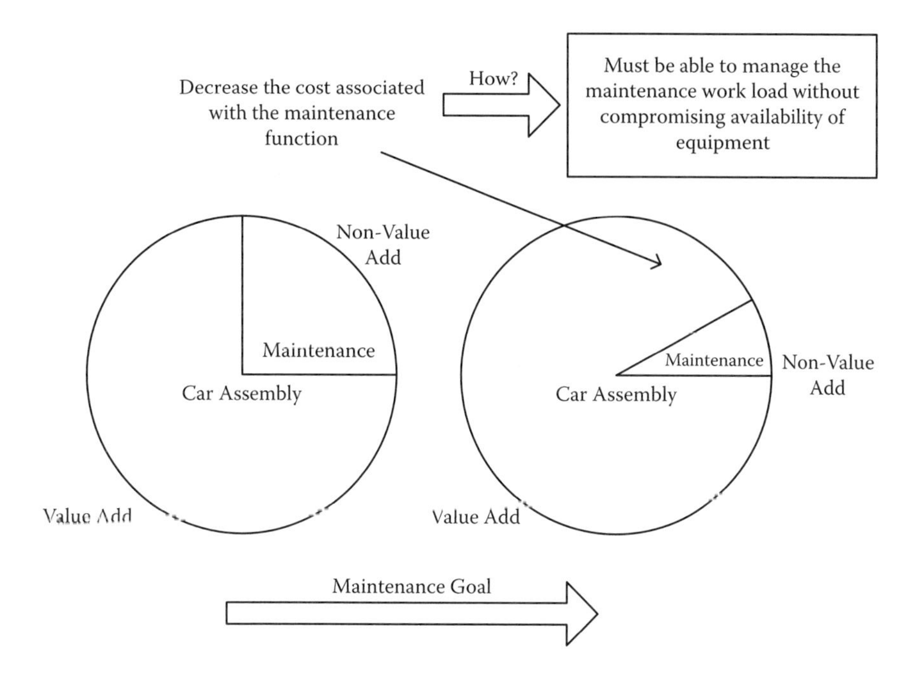

FIGURE 14.10
Maintenance goal in TPS.

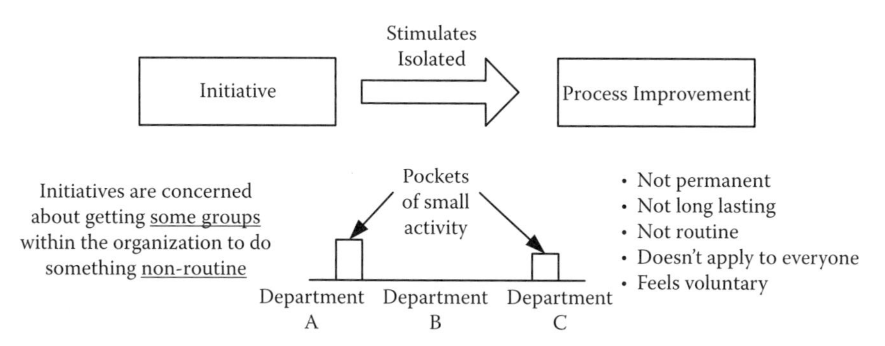

FIGURE 14.11
Initiatives stimulate isolated pockets of improvement.

Toyota's approach to improving the maintenance function begins with building the maintenance management system. For example, Toyota's 4S (i.e., seiri, seiton, seiso, and seiketsu) systems are embedded in production and maintenance roles. Toyota does not use a separate TPM initiative to encourage workplace organization or make it voluntary. Instead, workplace organization is expected to be practiced by everyone so that abnormalities can be identified when they occur. The advantage of a management system

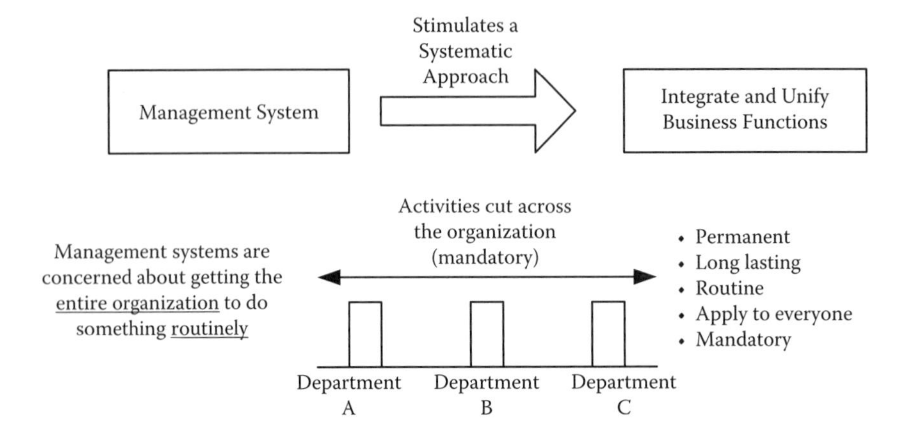

FIGURE 14.12
Management systems build. Integrate and unify business functions.

over an initiative like TPM is that it is more permanent, long lasting, and routine and applies to everyone. Management systems are concerned about getting the entire organization to do something routinely. While most organizations pursue initiatives in implementing Lean, Toyota is more concerned about building a management system that can be used by everyone to perform workplace improvement (Figure 14.12).

One way that Toyota builds the maintenance management system as it relates to managing the workload of the maintenance function is through the use of roles. The maintenance management structure mirrors that of production for the team member (TM), team leader (TL), group leader (GL), assistant manager (AM) and department manager. Both production and maintenance have assigned roles for maintaining equipment. On the production side, the AM has the role to evolve policy relating to production and equipment trouble and to take a leading role for machinery when there is a model change or new model. The AM provides input for inspection necessary for equipment operation and ensures that daily systems are working through spot checks. The AM also has to understand the status of major equipment problems and grasp details to prevent the recurrence of equipment trouble. The AM has the role of personally mastering equipment operation and ensuring operation manuals are standardized. The AM handles administrative issues relating to equipment trouble by attending meetings and advancing requests to maintenance. In all cases, the production AM acts as the final customer to maintenance, which provides the prioritization of work to the maintenance department.

The production GL has the role of ensuring that maintenance training is performed, planned, and adhered for all TLs. The GL is the first responder to notify maintenance when there are problems and to take action within an assigned range. The GL checks the equipment after it is repaired and gives feedback regarding reoccurrence prevention. The production TL has the primary role of performing daily maintenance. The TL is responsible for checking equipment daily and verifying the condition of tools, materials, and parts. The TL also looks at excessive consumption, reports problems to the GL, and is the one who performs the initial investigation of equipment trouble. TLs, like GLs, have an assigned range for which they can troubleshoot and adjust when there is equipment trouble.

On the maintenance side, the TL has the role of performing job instruction training for his or her specific area. TLs teach TMs how to calibrate equipment in their area, how to perform fluid fills in their area, how to perform minor repairs in their area, and so on. While all employees generally favor a higher aptitude for certain maintenance disciplines, the TL is not responsible for completely improving employee skills. For example, a maintenance TM may be a good machinist but not efficient at operating a Programmable Logic Controller (PLC). The TL will teach the TM how to open the PLC program or how to navigate through the program or show the TM how to troubleshoot the PLC program. However, the TL is not expected to teach the TM how to do PLC programming. In this situation, Toyota relies on another aspect of its maintenance development functions.

The GL has the role of worker allocation, bottleneck analysis, and approving standardized work. The maintenance AM has the role of evolving

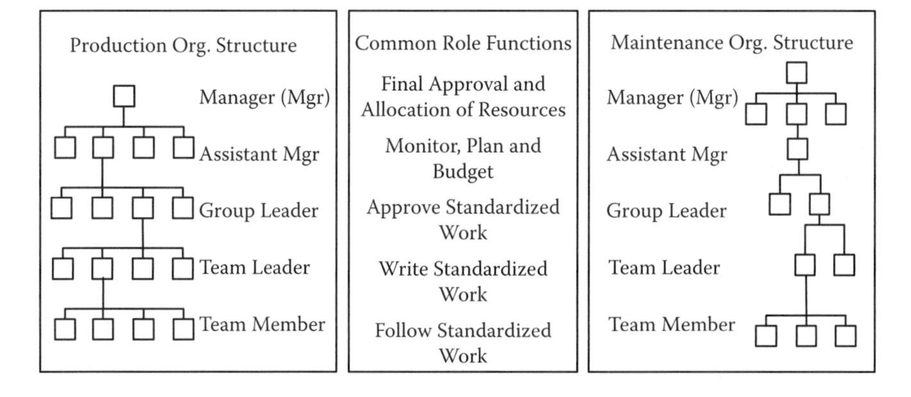

FIGURE 14.13
Comparison of roles between production and maintenance.

policy and establishing assignments based on the production plan. The main difference between production and maintenance is the span of control. Because of the nature of work and the degree of predictability, the span of control for production is typically larger than maintenance (Figure 14.13). It is not uncommon for two TMs to report to a TL and a few TLs to report to a GL. Toyota's span of control for the maintenance function is dependent on the variability and predictability of work rather than the type of job.

14.6 THE TENETS OF WASTE ELIMINATION IN MAINTENANCE

One of the main focal points in TPS as it relates to waste elimination in maintenance is the management of labor. Since the goal is to reduce labor, various strategies are used in the maintenance function to increase flexibility, predictability, and continued learning. This section highlights three of the most widely used tenets of waste elimination in maintenance that describe the behavior of Toyota's management system in a non-value-added work environment.

14.6.1 Establishing Normal Work in Unknown and Nonroutine Work Environments

One of the fundamental aspects of TPS is the industrial engineering mindset that strives at differentiating normal versus abnormal. This type of thinking is throughout Toyota and is the backbone for establishing consistency and predictability in the work environment. Toyota applies this type of thinking not only in work environments that are repetitive, such as production, but also in nonroutine work environments such as maintenance. By differentiating and separating normal and abnormal work, labor can better respond to complexity and variability in the work environment. While no work environment can be completely defined or predicted, this TPS concept is continuously used as a way to bring order to the workplace.

Differentiating or separating normal and abnormal work is only one part of the solution. TPS is successful in maintenance when normal and abnormal work flow has been standardized, written down, and proven to be successful (Figure 14.14). Most organizations think of standardized

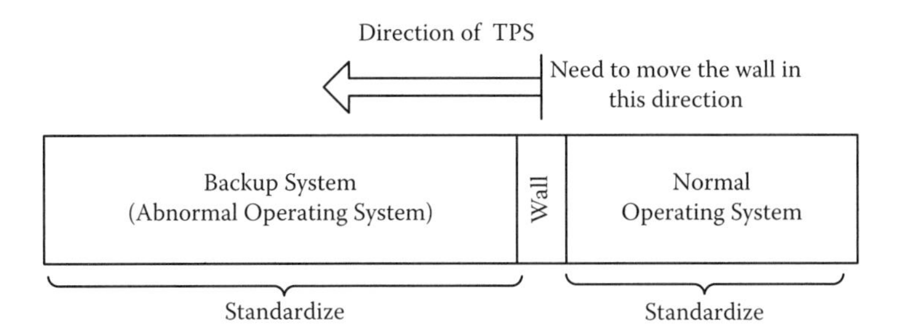

FIGURE 14.14
Standardization of normal and abnormal operating systems.

work as a documented process that relates to workflow that is repetitive and routine. If standardized work is followed, errors are minimized, and workflow can be consistent and predictable. Most companies associate standardization with processes that are normal, not abnormal. When a process experiences trouble, a standardized way to respond to trouble can also be helpful. Unfortunately, most downtime occurs when it is not known how to get back to a normal process. In other words, the system is in a state of abnormality, and there does not exist a recovery plan to return to a routine workflow. According to systems theory, a system is likely to stay in a state of abnormality longer because there exist more scenarios, more things that can go wrong, because the regulator has failed to control the system. One of Toyota's goals is to develop reactionary systems so that operations can quickly return to a safe and normal operating range. One example is how redundant systems are established to keep the main production line operational. If a motor fails, there are redundancies in place that allow maintenance to get the line back up in a short amount of time. Standardizing backup systems is one way maintenance can eliminate waste due to breakdown and be productive in maintaining equipment availability.

Standardization is important for maintenance primarily because most maintenance tasks are not routine or repeated on a daily, weekly, or monthly cycle. Maintenance functions tend to be less predictable, consistent, and repeating. Because there are so many ways to accomplish maintenance tasks, most organizations choose not to standardize or develop ways to make work routine. While it is impossible to make all work routine and predictable, TPS continuously tries to design work so that it can be more routine or follow a set schedule. Routine work is more predictable than nonroutine work and provides a degree of consistency helpful in planning.

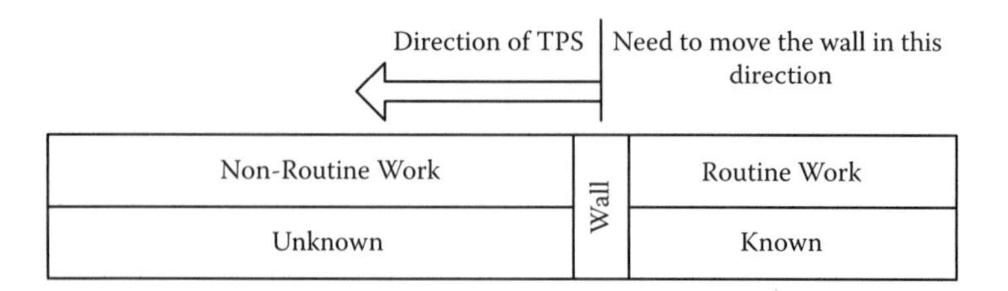

FIGURE 14.15
Toyota's trend to routine and known work.

The end goal is to design work (as much as possible) to be routine so more of the operations exist in a known state rather than an unknown state to plan the use of labor accurately. When work is known and routine, the planning of labor can be done more accurately and efficiently (Figure 14.15).

Two techniques used in maintenance to achieve standardization are standard processes and standard procedures. Standard procedures are written instructions that clearly document how to perform repairs, PM, and other specific maintenance activities. Standard processes are general guidelines that describe maintenance functions that cannot be completely documented. Standard processes include a variety of maintenance activities that relate to troubleshooting and the diagnosing of problems. Standard processes often use decision trees, logic diagrams, and flowcharts to generalize a course of action that could lead in many directions.

Consider a standard process for diagnosing engine failure. A standard process would help users to become acquainted with the problem domain by checking a few of the most common areas that are most likely to fail. If the cause of failure has not been identified by searching those four areas, the user must then rely on experience to determine a solution. At this point, it is often unknown and too difficult to document the course of action. The problem domain is unknown, which makes the solution space unknown. Users cannot follow a standardized troubleshooting path because it cannot be determined if it is relevant. In this example, Toyota would prefer to save time by documenting (i.e., standardizing) what is known by creating a basic guideline. If users cannot distinguish the relevance of information, standard processes are likely to send users down the wrong path. The degree of using standard processes compared to procedures is dependent on an organization's ability to learn. Standard procedures allow users to master the problem domain, while standard processes are likely to acquaint users with the problem domain.

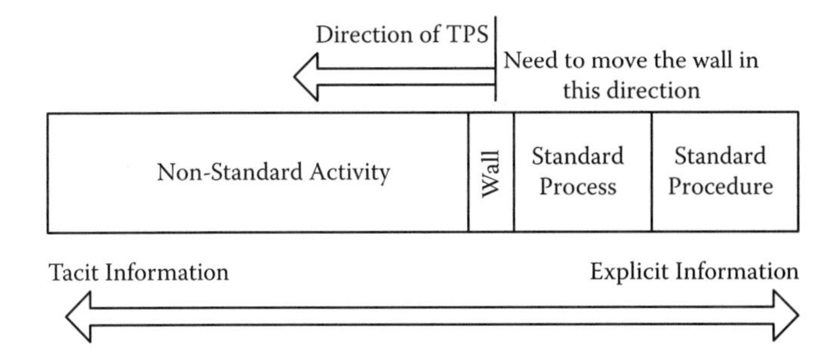

FIGURE 14.16
The use of standard processes and procedures in maintenance.

Toyota utilizes both standard processes and procedures to move work from unknown to known (Figure 14.16). The idea is that the more work is explicit (i.e., decoded and transferred into usable form), the better management can allocate labor. When knowledge is in a tacit form (i.e., not yet decoded or written down), planning and worker allocation cannot occur. Standard procedures and processes allow the management of labor to be controlled, allocated, and utilized efficiently. The trend in TPS is to (as much as possible) convert tacit information into explicit information so that there are more allocation and planning opportunities, specifically more flexibility in how work is performed (Figure 14.17).

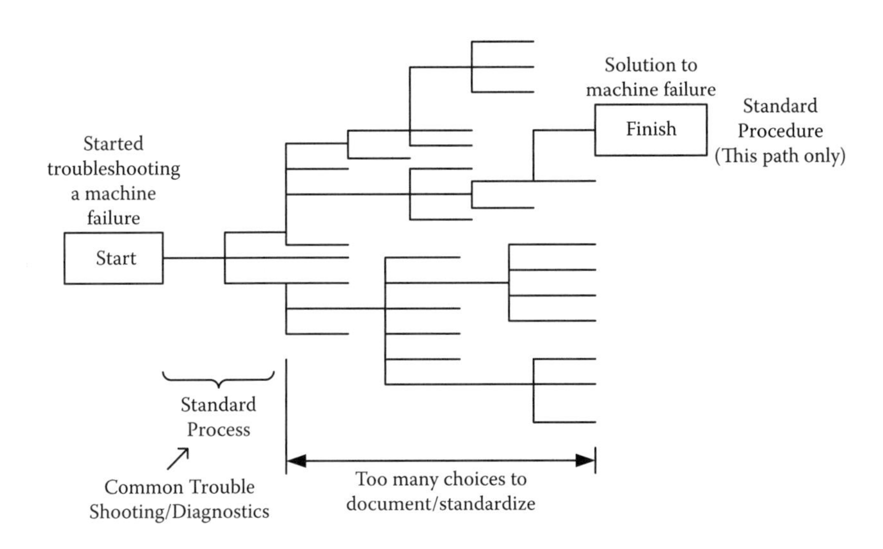

FIGURE 14.17
Comparison of standard process and procedure using a logic tree.

14.6.2 Fewer Higher-Skilled Employees

Union work environments often drive single-skilled workers who can only perform one type of job. In a single-skilled work environment (Figure 14.18), the scheduling of work has few options since only one job can be performed by one worker. This type of system is costly to manage and inflexible since worker allocation cannot meet the changing demands of the work environment. While it can be argued that single-skilled workers may be ideal for routine work environments, where job switching is not required or necessary when there is a high degree of specialization, labor cannot be easily reduced because the system is designed with constraints. The end result is a maintenance schedule that plans for peak resources and skills (Figures 14.19 and 14.20).

One of the main reasons why Toyota encourages a multiskilled maintenance worker is that the organization can be flexible in changing work demands, especially in nonroutine work environments that are variable and not predictable. A multiskilled maintenance TM can work on all aspects of the equipment such as mechanical, electrical, controls, hydraulics, pneumatics, machining, and PLCs. Compared to a single-skilled workforce; the number of jobs that can be completed is limited by the number of employees who have that specific skill. Another advantage of a multiskilled worker in maintenance is that the nature of work is multidisciplinary (Figure 14.21). Since machine failure can be caused by a number of different factors, specialized workers can only troubleshoot in their domain. Often, machine failure is system related, which means problems cannot be properly diagnosed by examining the components of the

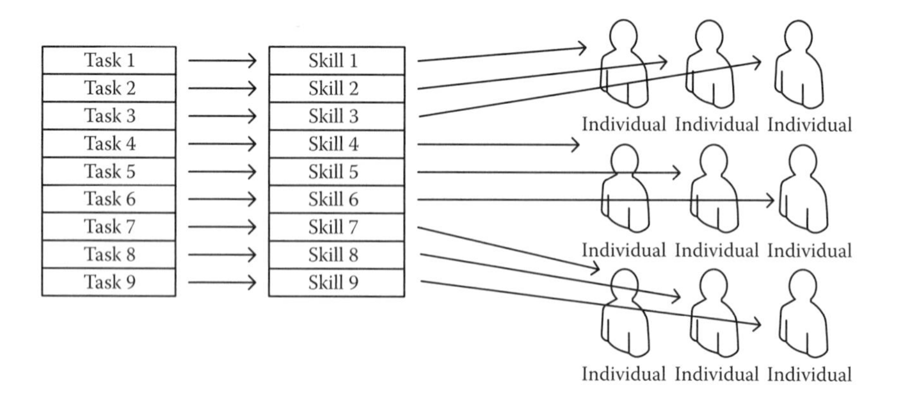

FIGURE 14.18
Traditional single-skilled work environment.

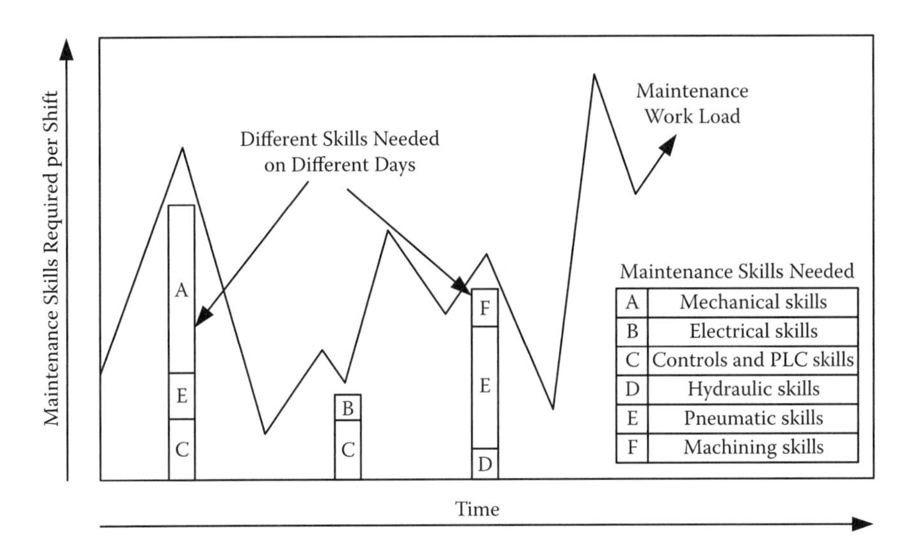

FIGURE 14.19
Traditional maintenance workload.

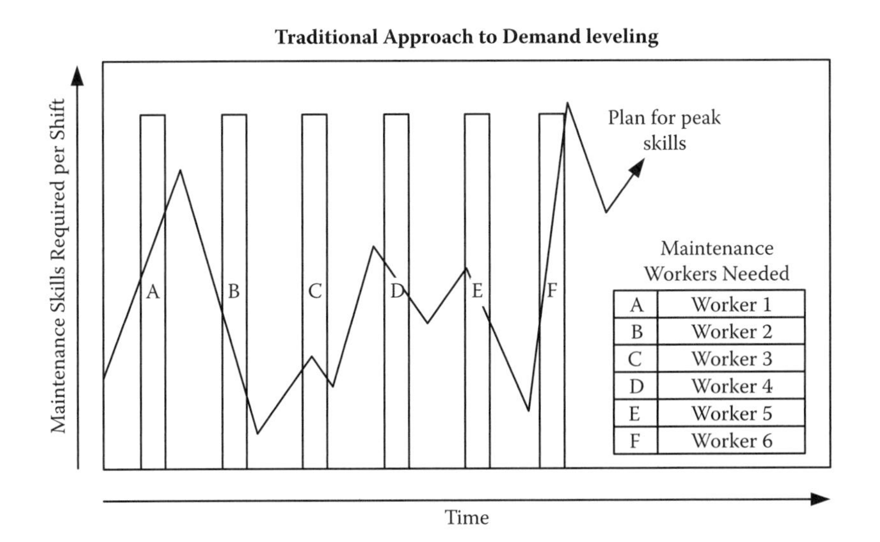

FIGURE 14.20
Traditional approach to maintenance scheduling.

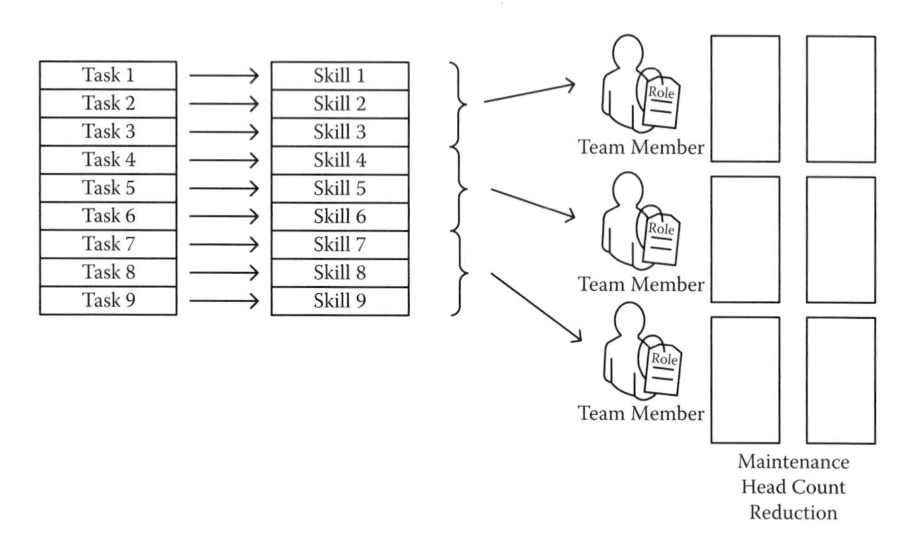

FIGURE 14.21
Multiskilled work environment.

equipment individually. A multiskilled maintenance worker is trained in all areas of the equipment, which allows the worker to be highly skilled in maintenance diagnostics and repair.

While the concept of a multiskilled maintenance workforce is not new, Toyota's approach to achieving a multiskilled workforce is unique. Toyota is committed to the ongoing development and abilities of its TMs. Toyota expects maintenance members to participate in cross training to perform many jobs in their area and to participate in future training opportunities to gain new skills to prepare them for continued career advancement (Figure 14.22). The skilled trades training and development program is a program designed to support the philosophy of "promote from within" and address the long-term needs for multiskilled maintenance employees (Toyota Motor Manufacturing Kentucky, 2006). This adult apprenticeship program allows interested full-time TMs (production or maintenance employees) to participate in extended classroom and on-the-job training over several years. Toyota trainers are certified through the National Vocational Qualification (NVQ) and can certify participants of the program to be awarded qualifications in engineering operations (NVQ level 2) and in engineered maintenance (NVQ 3). Participants are also awarded a level 3 technical certificate that allows them to be trained in electricity, electronics, computer control systems, robots, and pneumatic and hydraulic systems (Toyota Academy, 2010). The skilled trades training and

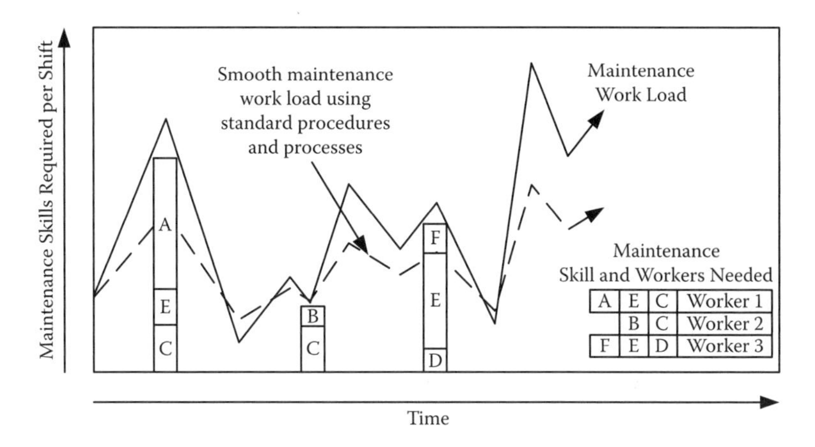

FIGURE 14.22
Toyota's approach to demand leveling for maintenance skills and workers.

development program allows employees to learn new skills and develop their career in a way they never imagined.

At some point, Toyota would expect that the maintenance person can do it all (i.e., work with electricity, mechanics, fluid power, fabrication, etc.). Worker flexibility relates to standardization because when employees are sick or their skills are needed elsewhere, a procedure allows another worker to complete the job. Worker flexibility helps to manage workload during periods of unusual conditions, which largely describes the nonroutine setting of the maintenance environment. In this manner, Toyota is constantly trying to improve TM capability by standardizing work procedures and processes so that the maintenance function can be more flexible to handle nonroutine work.

It is important to note that this does not imply that Toyota expects that fewer workers should work harder. It does imply that the amount of productive work should increase, thus decreasing the amount of waste in the job. Toyota views that all work is safe work, but not all work is considered productive work. Toyota is looking to develop multiskilled employees along with standardized procedures to build the right product, the right skill, for the right breakdown time.

14.6.3 Seibi: Starting Point for Equipment and Maintenance Standards

A specialized work unit that helps in the development of setting and establishing maintenance standards is Seibi. Seibi is a corporate entity inside Toyota that is part of TEMA (Toyota Motor Engineering and Manufacturing North America) to prepare the development of equipment before production. Seibi performs equipment reviews, equipment buy off and equipment sign off. Seibi is a specialized group within Toyota that provides a unique perspective for TPM and provides feedback for equipment and machinery design. Without Seibi, TPM activities would not be possible. Seibi works with equipment manufacturers such as Toyoda Machine Works (internal Toyota equipment manufacturer) to ensure that TPM activities are feasible and effective. For example, the Seibi group would ensure that all gauges, all lubrication points, air pressure, hydraulic pressure, and fluid levels are outside the safety fence so that production employees can perform TPM activities safely.

Seibi employees start off as plant-level employees, often in maintenance, and opt to join Seibi for a period of time for career advancement. Seibi employees travel extensively and become highly experienced and trained in various areas of equipment development. The Seibi group prepares and writes all equipment documentation, including PM, predictive maintenance, repair instructions, and TPM documentation. Seibi employees train maintenance and production engineering (PE) personnel and stay long enough at the plant site to ensure that the equipment and process meet all mass production requirements.

After Seibi finalizes equipment installation, PE has the responsibility for writing the maintenance standard. PE is responsible for controlling and building in quality, which can only be accomplished by establishing and controlling machine conditions. Maintenance is then responsible for controlling the standard and performing work that is within initial limits established by PE. Production has the role of monitoring the maintenance standard through daily production. In this flow, the maintenance standard originates from Seibi, is prepared by PE, is controlled by maintenance, and is monitored by production (Figure 14.23).

14.6.4 Improving Equipment Availability Using Problem Solving

If the maintenance product is to keep the line running, everything that the maintenance department can do to have an impact on line availability is considered a value. Toyota utilizes two metrics as a strategy to measure line availability, mean time between failure (MTBF) and mean time to repair (MTTR). One of the primary activities that the maintenance

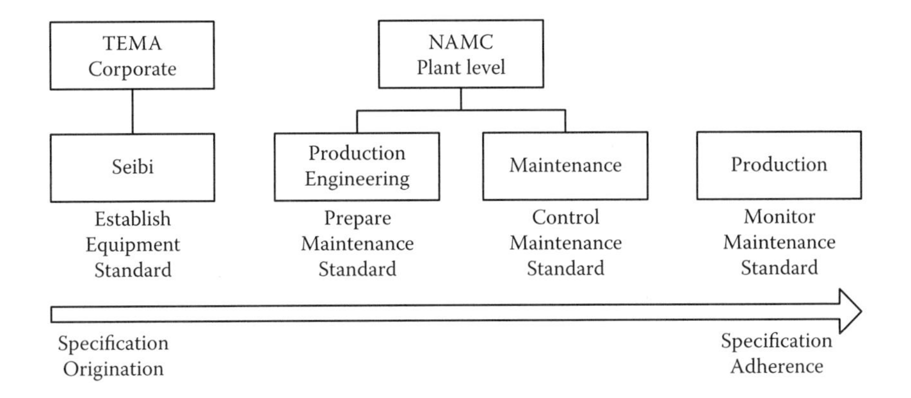

FIGURE 14.23
Origination and adherence of maintenance specification. NAMC = North American Manufacturing Company.

function engages in to improve MTBF and MTTR is team-based problem solving. Each maintenance team is expected to engage in problem solving to identify the causes of machine failure or ways to improve machine repair. Interestingly, most outsiders are not aware that the maintenance function engages in problem-solving activities much more than production or assembly operations. While most practitioners view Toyota's assembly workers as a benchmark for being proficient at performing problem solving, the maintenance department is far more experienced (Figure 14.24).

One of the ways that Toyota improves MTTR is through its troubleshooting techniques and redundant systems. Toyota views troubleshooting as a temporary activity used to get the equipment up and running.

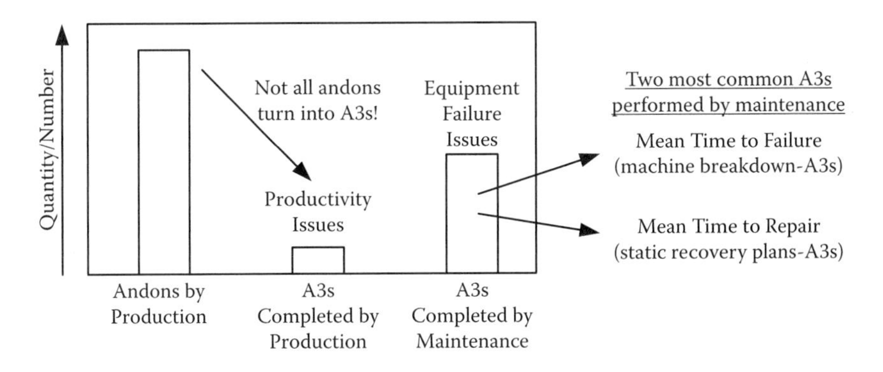

FIGURE 14.24
Illustration of A3s by production and maintenance.

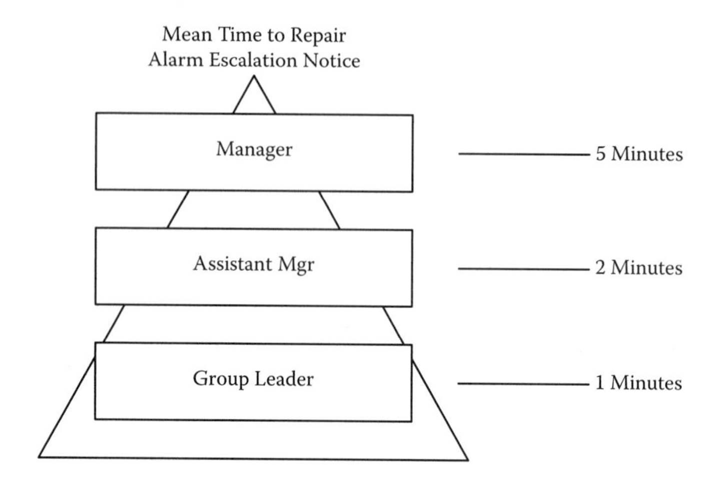

FIGURE 14.25
Example of standard recovery plan: alarm escalation notice.

Troubleshooting should not be confused with problem solving since troubleshooting does not immediately fix the root cause. Toyota utilizes redundant and backup systems to minimize downtime by speeding up troubleshooting so that line availability is less impacted. An interesting measure that is used to evaluate Toyota's MTTR is its alarm escalation notice (Figure 14.25). This recovery standard dictates when to notify the GL, AM, or manager for help. The GL is notified after one minute of downtime; the AM is notified after two minutes of downtime, and the manager is notified after five minutes of downtime. While this notice may seem aggressive, it is Toyota's standard for getting management involved when problems escalate.

14.7 SUMMARY

This chapter proposed how TPS exists in a work setting that does not contain value. According to the property of equifinality, systems exhibit behaviors that allow them to reach the same destination regardless of initial starting conditions. Open systems can react with their environment, exchanging energy and matter to regulate purposeful behaviors in meeting functions and outcomes. This chapter illustrated how TPS continues to be a good fit for Toyota even though there is no value to maximize.

However, as one might expect, waste can always be eliminated. The question that was proposed was the degree that waste should be eliminated. If the maintenance function is considered non-value-added, why should it exist? The answer is simply to ensure line availability. One could argue that Toyota is trying to minimize maintenance resources without comprising equipment availability; however, the techniques discussed here actually raise the value of the maintenance function's greatest resource, its people. Techniques such as standardized processes and procedures, problem solving, and training were used to develop resources to be fully functional regardless of the situation. While Toyota entertained various labor reduction strategies, each approach raised the intellectual capital of the work group. In this setting, in which all activities are overhead, training was not stopped, and the time and effort to convert tacit knowledge to explicit knowledge was never discouraged (i.e., learning). Toyota's approach to maintenance shows that regardless of the work environment, either non-routine or unpredictable, team-based activities are used to strengthen problem solving and consistency in operations.

REFERENCES

Ahuja, I. and Khamba, J. (2008) Assessment of contributions of successful TPM initiatives towards competitive manufacturing. *Journal of Quality in Maintenance Engineering*, Vol. 14, No. 4, 356–374.

Ahuja, I. and Kumar, P. (2009) A case study of total productive maintenance implementation at precision tube mills. *Journal of Quality in Maintenance Engineering*, Vol. 15, No. 3, 241–258.

Alsyouf, I. (2006) Measuring maintenance performance using a balanced scorecard approach. *Journal of Quality in Maintenance Engineering*, Vol. 12, No. 2, 133–149.

Arnold, P. (2010) PM improvements at Toyota lift truck plant. *Machinery Lubrication*, March.

Ashayeri, J. (2007) Development of computer-aided maintenance resources planning (CAMRP): A case of multiple CNC machining centers. *Robotics and Computer-Integrated Manufacturing*, Vol. 23, No. 6, 614–623.

Bagadia, K. (2008) Asset reliability drives Lean maintenance. *Maintenance and Management, Plant Engineering*, May.

Bamber, C., Sharp, J., and Hides, M. (1999) Factors affecting successful implementation of total productive maintenance. *Journal of Quality in Maintenance Engineering*, Vol. 5, No. 3, 1355–2511.

Barnard, C. (1938) *The Functions of the Executive*, Harvard University Press, Cambridge, MA

Becvar, D. and Becvar, R. (2003) *Family therapy: A systemic integration* (5th ed.). Allyn and Bacon, Boston, MA.

Braglia, M., Carmignani, G., Frosolini, M., and Grassi, A. (2006) AHP-based evaluation of CMMS software. *Journal of Manufacturing Technology Management*, Vol. 17, No. 5, 585–602.

Garg, A. and Deshmukh, S. (2006) Maintenance management: literature review and directions, *Journal of Quality in Maintenance Engineering,* Vol. 12, No. 3, 205–238.

Ghayebloo, S. and Shahanaghi, K. (2010) Determining maintenance systems requirements by viewpoint if reliability and Lean thinking: A MODM approach. *Journal of Quality in Maintenance Engineering*, Vol. 16, No. 1, 89–106.

Harada, T. (2006) *Equipment Maintenance and TPS*, Factory Strategies Group, ed. A. Smalley.

Hawkins, B. (2005) The many faces of Lean maintenance, maintenance and management. *Plant Engineering*, Vol. 59, No. 9, 63–65.

ISO 26000 (2010) *Guidance on social responsibility*, Geneva, Switzerland, web source: http://www.iso.org/iso/home/standards.htm.

ISO 14000, *Environmental Management*, Geneva, Switzerland, web source: http://www.iso.org/iso/home/standards.htm.

Kaiser, K. (2007) A simulation study of predictive maintenance policies and how they impact manufacturing systems, thesis, University of Iowa, Iowa City, July.

Karuppuswamy, P., Sundararaj, G., and Elangovan, D. (2007) Application of computerized maintenance management system coupled with risk management techniques for performance improvement of manufacturing systems. *International Journal of Business Performance Management*, Vol. 9, No. 1, 7–21.

Liu, E. (2004) A methodology for implementing total productive maintenance in small manufacturing enterprises: Using the transform enterprise, Dissertation, The University of Texas at Arlington.

Mathaisel, D. (2005) A Lean architecture for transforming the aerospace maintenance, repair and overhaul (MRO) enterprise, International Journal of Productivity and Performance Management, Vol. 54, No. 8, 623–644.

Miller, K. (1992). A framework for integrated risk management in international business. *Journal of International Business Studies,* Vol. 23, No. 2, 311–333 I.

Moayed, F. and Shell, R. (2009) Comparison and evaluation of maintenance operations in Lean versus non-Lean production systems, *Journal of Quality in Maintenance Engineering*, Vol. 15, No. 3, 285–296.

Modarres, M., Mark, K., and Vasiliy, K. (1999) *Reliability Engineering and Risk Analysis*, Dekker, New York.

Murase, Y. (2007) Issues and aims (visions) for JIPM (Japanese Institute for Plant Maintenance), study activities, web source: http://www.jipm.or.jp/en/.

Nakajima, S. (1988) *An Introduction to TPM*, Productivity Press, Cambridge, MA.

Pal, B. and Maiti, J. (2005) Lean maintenance—Concept, procedure and usefulness. *Journal of Mines, Metals and Fuels*, Vol. 52, No. 11, 318–321.

Piper, J. and Sumukadas, N. (1993) The continuous improvement of machines, their operators and their maintenance technicians. *Proceedings of the Administrative Sciences Association of Canada (ASAC),* Vol. 14, No. 7, 45–54.

Pollitt, D. (2010) Today's production workers are the craftsmen of tomorrow at Toyota. *Human Resource Management International Digest*, Vol. 18, No. 2, 18–20.

Pramod, V. R., Devadasan, S. R., and Jagathy Raj, V. P. (2007) Receptivity analysis of TPM among internal customers. *International Journal of Technology, Policy and Management*, Vol. 7, No. 1, 75–88.

SAE JA 10000 (1998) Reliability Program Standard: 45 Reliability, Maintainability, Supportability and Logistics, SAE Standard, Published by the Society of Automotive Engineers, Warrendale, PA.

Sakai, H., and Amasaka, K. (2007) The robot reliability design and improvement method and the advanced Toyota production system. *Industrial Robot: An International Journal*, Vol. 34, No. 4, 310–316.

Sharma, R., Kumar, D., and Kumar, P. (2006) Manufacturing excellence through TPM implementation: a practical analysis. *Industrial Management and Data Systems*, Vol. 106, No. 2, 256–280.

Smith, R. and Hawkins, B. (2004a) *Lean Maintenance*, Elsevier Butterworth-Heinemann, New York.

Smith, R. and Hawkins, B. (2004b) Pre-planning for Lean maintenance. *Lean Maintenance*, 105–124.

Thomas, A., Barton, R., and Byard, P. (2008) Developing a Six Sigma maintenance model. *Journal of Quality in Maintenance Engineering*, Vol. 14, No. 3, 262–271.

Toyota Academy (2010) Toyota Motor Corporation, web source: http://www.toyota.co.za/.

Toyota Motor Manufacturing Kentucky (TMMK) (2006) *Team Member Handbook*, Published by Toyota, Georgetown, Kentucky.

Tsang, A. (2002) Strategic dimensions of maintenance management. *Journal of Quality in Maintenance Engineering*, Vol. 8, No. 1, 7–39.

Waeyenbergh, G. and Pintelon, L. (2009) CIBOCOF: A framework for industrial maintenance concept development. *International Journal of Production Economics*, Vol. 121, No. 2, 633–640.

Wang, J. X., and Marvin, L. R. (2000) *What Every Engineer Should Know about Risk Engineering and Management*, Dekker, New York.

Wireman, T. (2004), *Benchmarking Best Practices in Maintenance Management*. Industrial Press, New York.

Yamashina, H. (1995) Japanese manufacturing strategy and the role of total productive maintenance. *Journal of Quality in Maintenance Engineering*, Vol. 1, No. 1, 1355–2511.

15

Summary

15.1 SYSTEMS THEORY AND THE MODERN THEORY

The modern theory of the Toyota Production System (TPS) is a way to explore and understand Toyota's management system using general systems theory (GST) and living systems theory (LST). System theories, like GST and LST, provide a perspective that gives managers a chance to understand the factors and variables that make a system successful or ineffective. All systems are unique, but they contain common properties that give insight in how they function and work. System properties show how the system can achieve its goal and are instrumental for designing, controlling, and implementing systems.

Various models and frameworks were presented to help explain how Toyota's management system compares. In most cases, many of Toyota's practices can be explained by extending or combining various management theories in use today. The system properties of Toyota's management system were illustrated in this book to answer several questions about the Toyota Production System.

- What is the identity and purposeful behavior of TPS?
  The identity and goal-seeking property of TPS is the industrial engineering and problem-solving mindset. Toyota uses problem solving as the main mechanism to improve and correct organizational systems. Toyota has remained shielded from the influences of contemporary industrial engineering to improve workplace productivity so that more employees of the company can use and apply simple techniques. The strength in numbers for applying simple industrial engineering techniques is more important than a few elite industrial engineers.

- What regulates TPS?
  Toyota's approach for regulating TPS is through quality circles and jishukens. Both of these activities, led by team members or by managers, are to reinforce the identity of TPS. Regardless of the organization state, quality circles and jishukens are used to regulate TPS thinking among the workforce. Toyota uses these activities to provide information about the degree TPS is functioning and is being maintained and the means TPS is effective in reaching its goals in the system.
- Which aspects of TPS have to be differentiated or cannot be substituted?
  Differentiation is when certain parts in a system have to perform specialized functions to better adapt to the open and dynamic nature of its environment and surroundings. Leadership is essential because it provides the destination, the path, and the justification for the existence of TPS. Consequently, only leadership can promote cooperation and collaboration among units in achieving organizational goals that cut across sections, departments, plants, divisions, and regions.
- What gives TPS order and disorder?
  Entropy or disorder can be generated in TPS when employees no longer feel open or engaged in workplace improvement. Progressive human resource functions are used within Toyota to prevent TPS from deteriorating. The human resource function must stabilize the basic employment terms and conditions to place employees in a free state, absent of fear and full of trust.

  TPS cannot carry out its most important functions (problem solving and industrial engineering) if employees are not interested in pursuing kaizen. Toyota employs various organizational learning techniques to position employees to be creative in their work environment. Employees are more willing and able to learn when they have ownership of the entire job and there is not the fear of failure for trying something new.
- Are there aspects of TPS that have to be as complicated as the company's surroundings?
  Toyota's approach to production leveling is one of the best illustrations of variety handling in TPS. TPS manages the variety of the environment by taking a proactive stance against variety, which in the long run requires less variety handling. Technical and

social structures within production leveling manage uncertainty when there are natural tendencies to go against departmental cooperation and teamwork.

- What are the basic transformation processes in TPS?

The transformation process of TPS can best be described as the way ideas for kaizen are converted into action. TPS encourages change at the group level by utilizing various conversion processes, such as incremental, transitional, and developmental change. The immediate supervisor is one of the primary ways that ideas are collected in the workplace. Combined with problem solving, supervisors can sustain change using a systematic approach for organizing information that is ready for implementation.

- How much interdependence is needed for TPS to function?

One of the objectives of TPS is to ensure that all subsystems work together and contribute toward the system's objectives. A way to understand the amount of interdependence of Toyota's management system is to examine how TPS is replicated at suppliers. TPS encourages suppliers to engage kaizen, but not at the risk of disrupting a stable and consistent process. TPS encourages steady improvements over system replication, mainly because system kaizen requires suppliers to start from scratch. Toyota does not expect suppliers to adopt their managerial values, principles, or problem-solving technique, but they are expected to apply some of the same basic operational parameters that have proven to be successful at Toyota.

- Are there equally effective ways in applying TPS?

Toyota's management system is equally effective in both value-added and non-value-added work environments. The maintenance function illustrates how TPS has the ability to attain the same final result regardless of the initial condition. While facilities maintenance does not create any value, TPS encourages standardized processes, problem solving, and team member development. The primary goal of facilities maintenance is to maintain line availability; TPS promotes labor reduction strategies in non-value-added work environments by investing in employees skills.

- What does the modern theory of TPS assume?

Systems theory assumes that each acting element in a system has a relation to the whole. A part of Toyota's management system cannot be applied without implementing or considering another part. The modern theory of TPS assumes that there is coordination and integration, that the design and implementation of Toyota's management system occurs simultaneously. The modern theory assumes that there is a degree of connectivity and closeness between two or more groups or elements within Toyota and requires a series of exchanges from one part to another to be successful. The modern theory of TPS assumes that there is a tendency to grant and prioritize the goals of the larger system to a smaller system.

15.2 MANAGEMENT SYSTEM PROPERTIES CAN BE SATISFIED IN MANY WAYS

It is not the intent of this book for practitioners to emulate the properties of Toyota's management system without understanding how the properties of systems work. When companies copy the practices listed here, there is often a strong desire to fix the structures in the organization regardless of the business climate. Systems theory is based on the interplay and dynamic nature of variables, which require systems to be adaptable and flexible. The success of a management system is largely dependent on how the properties of the system are organized when the conditions of the company change. Understanding how properties of systems work is more beneficial than emulating practices that work in only one unique business climate and organizational state.

15.3 ORGANIZING THE PROPERTIES OF TPS FOR IMPLEMENTATION

System properties generally do not follow an order or precedence since all systems are presumed to obey the same principles. Consequently, no single property can determine or influence the properties of another. Making arguments about the order of system properties can be simplistic and misleading; however, there is a general concern that more critical system properties should get higher priorities than less-critical ones. There is also the belief that the distinction of system properties should be prioritized due to the common good or interest in the part they play in achieving the system's goal. While arranging system properties according to the variety of their complexity can make implementation easier to manage, all system properties should be given consideration.

	Phase 1	Phase 2	Phase 3
Property No. 1 (Roles)	Start → Finish		
Property No. 2 (Industrial Engineering)		Start → Finish	
Property No. 3 (Hoshin Kanri)			Start → Finish

Implementing the properties of a system does not follow a sequence.

The challenge is timing. Management systems have to fit the needs of the organization, which makes their implementation completely dependent on the business climate and organizational circumstance. Implementing TPS in a sequence of steps is not nearly as important as knowing the variables that have an impact on the system. Spending too much time on any single property, variable, or tool of TPS without knowing how it can impact other areas is not fruitful. Since the approach is based on learning, the trick becomes how quickly the variables can be understood. In the beginning, companies wanting to install a management system simply need to experiment with a little of everything. Regardless of the organizational state, if leadership does not understand how the properties function, it is unlikely that they will be implemented correctly even if timing is right.

	Phase 1	Phase 2	Phase 3
Property No. 1 (Roles)	Start →		
Property No. 2 (Industrial Engineering)	Start →		
Property No. 3 (Hoshin Kanri)	Start → } Understand the variables		
Property No. 4 (Jishuken)	Start →		
Property No. 5 (Human Resources)	Start →		

Implementing the properties of a system in parallel to encourage learning.

Development of systems has to be driven by the needs of the organization. Companies have to work on the right things at the right time. Management systems have the goal to focus the company's resources to solve problems that are meaningful and time sensitive. The better management can coordinate the company's resources, the better the company can meet its internal needs. Developing a system based on need is not sequential or necessarily something that can be implemented in parallel. System development does not follow a straight path or is not something that can be completely predicted at the time of planning. Implementing the properties of systems requires leaders to be flexible and adaptable.

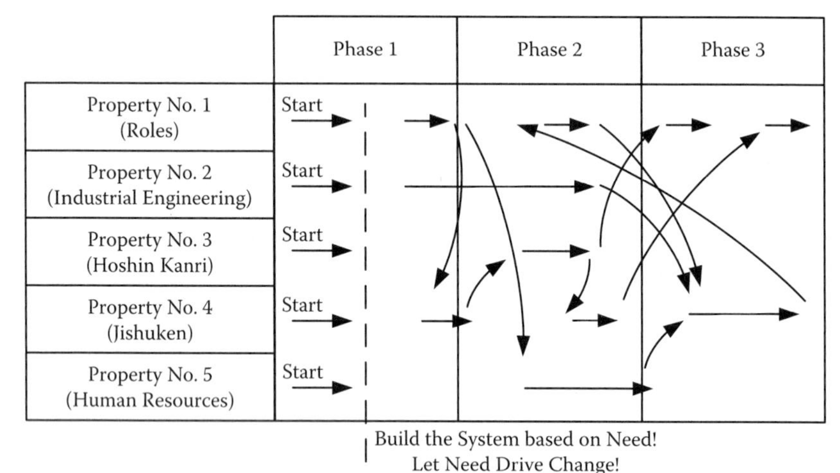

Implementing the properties of a system based on need.

15.4 TIMING AND PACE

The time it takes for systems to develop is dependent on how directive or participative leadership chooses to drive change. When leadership articulates how to improve the system, the pace of development can be quick. While not all members may agree or understand, directions and assignments are carried out on a time scale. When change is completely participative, members have the opportunity to learn on their own and apply their previous experiences to improve the system. Timing and pace are problematic when members do not know how to improve the system themselves. If leadership cannot help member's break through by showing or demonstrating the skill firsthand, direction will be meaningless. Members must have a way to interpret and decode the organizational priorities.

15.5 HOW DO YOU KNOW WHEN THE MODERN THEORY OF TPS IS WORKING?

A management system is said to be fully implemented when all of its members agree to use it or they believe it is a better way to go than to work to another system. By this definition, the management system is installed when members begin to use it. The good news is that implementing or installing a management system is much quicker than a never-ending journey of continuous improvement. A standardized management system will drive improvement that is coordinated from the entire organization. The modern theory of TPS is a tool for leadership to fit the organization's management system to the needs of the business. The modern theory initiates inquiry to help managers and leaders to deepen their understanding so that they can be more effective at driving purposeful behavior. Leaders and managers are invested with the authority to intervene and to correct the problems of the company. The modern theory helps managers to manage the gray, which requires the ability to drill down, ask questions, get facts, and act on information.

Index

F

G

H

I

M

N

O

P

Q

R

U